# Boreal Forests Under Global Warming

This groundbreaking book presents an innovative predictive concept in geographical ecology that reveals local mechanisms of global environmental changes. Through an empirically established display of background bioclimatic trends using catenary biogeocoenotic systems, the work explores regional structures of localized natural zonality. The landscape–ecological forecast methodology employs ecological niches of studied objects, with spatial coordinates transformed into time coordinates according to the fundamental properties of environmental ergodicity, creating a powerful framework for predicting geo-ecosystem restructuring.

The research offers significant advancements in ecological modeling through its application of discrete mathematics for processing extensive empirical data from field and laboratory landscape studies. By incorporating specific examples and empirical simulation techniques, this work provides environmental scientists with formalized tools for analysis and prediction that can be applied to real-world ecological challenges. The methodology presented transforms geographical ecology into a more precise and quantifiable discipline capable of addressing complex environmental dynamics.

This monograph is primarily intended for specialists in environmental conservation, ecological modeling and forecasting, and ecological planning in forestry and agriculture. Additionally, as it draws from the Russian school of landscape science – an approach not widely familiar to many Western European and American researchers – this book offers valuable theoretical and methodological insights for both landscape scientists and ecologists across various specializations. Researchers seeking innovative approaches to understanding environmental change will find this work particularly valuable for its unique perspective and practical applications.

# Boreal Forests Under Global Warming

## Empirical Simulation Predictive Modeling

Erland G. Kolomyts

CRC Press
Taylor & Francis Group
Boca Raton London New York

CRC Press is an imprint of the
Taylor & Francis Group, an **informa** business

Designed cover image: Erland G. Kolomyts

First edition published 2027
by CRC Press
4 Park Square, Milton Park, Abingdon, Oxon, OX14 4RN

and by CRC Press
2385 NW Executive Center Drive, Suite 320, Boca Raton FL 33431



*CRC Press is an imprint of informa UK Limited*



*British Library Cataloguing-in-Publication Data*
A catalogue record for this book is available from the British Library

ISBN: 978-1-041-22595-9 (hbk)
ISBN: 978-1-041-22599-7 (pbk)
ISBN: 978-1-003-73311-9 (ebk)

DOI: 10.1201/9781003733119

Typeset in Times
by codeMantra

# Contents

*Acknowledgements* vii
*Foreword* viii
*Introduction* xi
*About the Author* xv

**1 Empirical Simulation Base and Predictive Modeling in Geographical Ecology: Framework Concept 1**

**2 Landscape–Ecological Strategy for Ground Geo-Ecological Monitoring of the Forests 17**

**3 Boreal Ecotone and Geography Zonality 31**

**4 Methods of Local and Regional Landscape–Ecological Modeling 49**

**5 Regional and Local Geosystems as the Objects of Predictive Research 72**

**6 Landscape–Geophysical Foundations of Empirical-Simulation Predictive Modeling 100**

**7 Empirical Simulation of Climate-Genic Changes in Biological Cycle 119**

**8 Peculiarity of Landscape–Ecological Forecasting 146**

**9 Empirical-Simulation Predictive Modeling of Forest Ecosystems 165**

**10 Regional Landscape–Ecological Prognoses 179**

**11 Cartographic Models of Regional Landscape–Ecological Forecasting 202**

**12 Local Landscape–Ecological Prognoses 216**

**13 Phytocoenotic and Soil Signs of Current Global Warming 233**

**14 Mechanisms of Stability of Forest Ecosystems** **244**

**15 Role of Boreal Forests in Regulation of the Carbon Cycle and Mitigation of Present Global Warming** **264**

**16 Empirical Simulation of Changes in Productivity and Carbon Stock in Forest Ecosystems under Climate Variations** **299**

**17 Brief Account of Predictive Landscape–Ecological Conception Using Empirical Simulation Models** **312**

*Conclusion* 334
*References* 336
*Index* 355

# Acknowledgements

Special thanks to the Institute of Basic Biological Problems, Pushchino Scientific Center of the Russian Academy of Sciences, for their support in the project.

# Foreword

The questions of forecast analysis of the state of natural ecosystems concerning their possible response to global climatic changes have been quite insufficiently covered in the world literature. Cognition of the mechanisms of continental biosphere sustainability in a changing climate is connected in many respects with studying the biotic regulation of the carbon cycle in the soil–vegetation–atmosphere system. Boreal forests play an important role in the global carbon cycle. There are indications in the literature that the evaluation of sources and sinks of carbon in boreal forests of the world interferes with the uncertainty in the data available from Russian forests.

A principal new landscape–ecological approach to the decision of the complex prognostic problems has been realized in this monograph, which expounds the regional prognostic conception in geographical ecology. The author describes the empirical simulation models created over the past 40 years and the landscape–ecological concepts developed on their basis: geo-ecotonic, landscape-zonal polymorphism, paleo-forecast (in light of global climate changes), forecast-topological, and high-mountain geo-ecological.

This book presents a strategy of empirical-statistical simulation modeling of forest ecosystems as objects of terrestrial ecological monitoring. Spatial functional monitoring of forests is based on empirically established local and regional landscape–ecological connections, which are considered mechanisms of metabolic reactions of forest geo(eco-)systems to certain climatic trends. This is the novelty of the prognostic empirical simulation concept of local (and regional) geo-ecological forecasting developed by the author.

Prognostic landscape-ecological scenarios of the nearest future of biosphere and their paleo-geographical analogs as a single system of global changes in the environment have been considered for the first time. The mechanisms of local and regional response to the anticipated global warming in the 21st century have been assessed by the example of forest ecoregions of the Volga River basin headwater, which are attracted to southern boundaries of the boreal belt on the Russian Plain and form an economic and demographic "core" of European Russia.

These concepts reveal the mechanisms (including processes) of the formation of landscape-ecological systems at both regional and local levels, their natural and anthropogenic dynamics, as well as evolutionary trends. The forecast-ecological analysis itself is based on the method of numerical landscape-ecological forecasting for the first time developed by the author. The technique has been brought up to the prescription level. It organically combines, on the one hand, a fairly strict formalized approach to solving forecasting problems, and on the other, the availability of procedures for collecting, processing and analyzing empirical material for a wide range of researchers.

The main peculiarity of this approach is that the phenomenon of ecological consequences of global warming is considered in landscape terms, thereby allowing for diverse types of this consequence determined by the spatial distribution of natural landscapes and their simple units (biogeocoenoses – eco-elements, or land-type phases) under different zonal-regional conditions. This made it possible to find the origins of the mechanisms of transformation of regional and local geo(eco-)systems and to relieve the phyto-biota effect on the carbon exchange between the earth surface and the atmosphere and, finally, to identify those mechanisms that ensure the resistance of forest ecosystems to climatic changes.

For the first time the author has developed the numeral methods of complex regional and local landscape-ecological forecast on the basis of empirical-statistical models of natural ecosystems with the using of discrete mathematic methods. The estimates of the past and future structure-functional characteristics for forest ecosystems are based on the empirical regularities of current landscape geophysical relations in the region, including the relations with vegetation cover, soil, and primary bio-productivity. This is the new empirical-simulation method of landscape-ecological forecasting proposed and developed by the author. The regional and local predictive landscape-ecological scenarios and paleogeographical reconstructions developed by the author is distinguished by a qualitatively new level of spatial resolution and much more profound ecological analysis as compared with the known analogous investigations for the Russian Plain.

The conception of regional and local manifestations of global environmental changes is undoubtedly of fundamental importance as a whole for global ecology because global biosphere processes and events are perceived most deeply through the level of ecological regions, which is aided by the high diversity of their nature-territory structures, as well as the complexity of functioning and evolutionary transformations of regional and local ecosystems.

To solve these problems, E.G. Kolomyts used a unique and rather extensive material from large-scale landscape–ecological surveys carried out at experimental sites according to a method specially developed by him. The author also created a procedure for local landscape–ecological forecasting, described in detail in the book, based on discrete empirical modeling of geo(eco-)systems, for which the methods of information theory and the theory of "fuzzy" sets that have proven themselves in geography and ecology are widely used. The development of predictive landscape–ecological scenarios is based on predictive climate models, which give a different degree of global warming and thus allow for obtaining appropriate variants of ecological forecasts.

The author of this book, E.G. Kolomyts, is known as an expert with encyclopedic knowledge in physical geography and landscape science. He has pursued research in Siberia, the Far East, and the high mountains of the Great Caucasus. In recent decades, he has been engaged in exploring complex ecological and geographical problems of the Russian Plain, including the investigation of ecotones as the most dynamic structural units of the biosphere. He is the executor of several pilot research projects of the Russian Foundation for Basic Research. His original thinking and approaches to solving relevant ecological problems evoke enhanced interest in his works. This monograph is a result of many years of studies by the author in the field of geographical ecology and landscape–ecological prediction.

The monograph is intended for specialists in environmental conservation, ecological modeling and forecasting, and ecological planning in forest economy and agriculture. We would like to note that the Russian school of landscape science, which forms the ideological basis of the present work, is not sufficiently well known to many West-European and American researchers. For this reason, the given book may be of theoretical and methodological interest to both landscape scientists and ecologists of broad profiles.

*Christian Opp*
*Professor of Physical Geography*
*University of Marburg, Germany*

# Introduction

This monograph presents the author's paradigm of experimental geographical ecology and its core – landscape ecology. The results of more than 40 years of scientific and methodological research by the author are presented. The created empirical-statistical models and the ecological–geographical concepts developed on their basis reveal the mechanisms of formation of landscape–ecological systems at regional and local levels, their natural and anthropogenic dynamics, as well as evolutionary trends. Landscape–ecological analysis is carried out on specific examples, using methods of empirical imitation of calculated predicted situations. In the created models, geographical ecology acquires an effective formalized tool for analysis and forecasting, which employs methods of discrete mathematics for processing and generalizing the mass empirical material obtained in field and office landscape studies.

The monograph expounds the main statements of the original topo-ecological concept of prediction: "Global Changes on the Regional and Local Levels," as a basis for regional–local bioecological and geosystem monitoring under global anthropogenic climatic changes. The objects of research are forest and forest–steppe landscape-zonal systems of the headwaters of the Volga River basin. Forecast analysis was based on the materials of large-scale landscape–ecological surveys conducted by the author and his colleagues at seven experimental test sites in the Middle Volga Region. These test sites encompass a wide range of zonal transitions from mixed forests to northern steppe on the Russian Plain. The author terms this landscape-zonal spectrum as the *boreal ecotone*.

The ecological safety of the continental biosphere largely depends on the state of zonal–regional types of natural ecosystems and, first of all, forest cover, which plays a vital role on land in preserving water resources and mitigating climate fluctuations. This problem is becoming urgent in the context of global warming, with its very diverse and sometimes difficult-to-predict ecological consequences. The forest cover of Russia makes up about 20% of the planet's forests, with boreal forests occupying more than 90% of this area. Thus, the problem of preserving forest ecosystems and reproducing forest resources on the southern border of the temperate forest zone, where forest communities are in conditions close to critical, is one of the fundamental ecological problems.

The solution to the above problems has revealed new approaches and methodological decisions for the quantitative substantiation of the above theoretical statements of the conception. First of all, it is necessary to mention the starting ideological prerequisite, which was the basis of the methods of regional paleogeographical reconstructions and complex landscape–ecological prognoses. This prerequisite may be formulated as follows: the aggravation of lateral (inter-complex) natural interactions in the periods of disturbance of previously achieved (to this or that extent) stability of natural ecosystems under the influence of new climatic signals is the origin and the driving force of landscape–ecological reconstructions, which occurred in the past and are anticipated in the future.

This book presents a strategy of empirical-statistical simulation modeling of forest ecosystems as objects of terrestrial ecological monitoring. Spatial functional monitoring of forests is based on empirically established local and regional landscape–ecological connections, which are considered mechanisms of metabolic reactions of forest geo(eco-)systems to certain climatic trends. This is the novelty of the prognostic empirical simulation concept of local (and regional) geo-ecological forecasting developed by the author.

For the first time in forest ecology, a full cycle of climate-genic monitoring of forests – "observation (assessment of the state) – control (forecast) – management (feedback, regulation)" – was carried out using a specific ecoregion (the Volga River basin) as an example. The development of scientific and methodological bases for spatial functional monitoring of forests is presented. The predictive analysis is based on empirical-statistical modeling of ecosystems and on multifaceted descriptions of their spatial organization. The property of polyzonality of local geo(eco-)systems as a way of their reaction to global climate changes is revealed. An ordination analysis of landscape connections is carried out based on the main discrete parameters of the biological cycle. A local empirical imitation of a regional bioclimatic trend by a hydro-thermo-edaphic system of biogeocoenoses is conducted, and patterns of transitions of forest communities to critical states are revealed. Catenary landscape–ecological structures that form regional systems of localized natural zonality capable of simulating the main directions and scales of geosystemic reorganizations are described.

The landscape–ecological forecast itself is of an experimental nature. It is presented as a system of operations with ecological niches. In the calculation models, the course of the predicted processes is reproduced using their empirical imitation by spatially distributed parameters of the basic ecological niches. Empirical simulation predictive models describe future scenarios of forest ecosystems according to the laws of their basic spatial organization, with the spatial coordinates replaced by time coordinates in the forecast procedure, in accordance with the fundamental properties of ergodicity of the natural environment.

The first step toward understanding the local mechanisms of global changes included a methodical construction with the working title: "Empirical Simulation of the Regional Bioclimatic Trend by Local Ecosystems." Within each physico-geographical region (regional ecosystem), forest communities are clearly differentiated by their varied zonal affiliations, so that the functional and structural transitions of local ecosystems must adequately reflect the background shifts of the zonal–regional bioclimatic system. The changes in zonal biogeocoenosis (landscape facie) groups may be the basis for assessing the extent to which local metamorphoses are able to transform the whole zonal–regional background of the given territory under the effect of this or that climatic trend.

Spatial and functional monitoring of forests is based on empirically established local and regional landscape–ecological relationships, which are considered as mechanisms of metabolic reactions of forest ecosystems to certain climatic trends. This is the novelty and prospects of the basic and predictive empirical simulation concept of regional and local geo-ecological monitoring developed by the author.

The entire analysis is accompanied by the construction of the appropriate empirical simulation models. Local mechanisms of global changes are revealed through the empirically established reflection of the background bioclimatic trend with a cadenauric system of biogeocoenoses. Regional structures of localized natural zonality are considered capable of simulating the main directions and scale of geo(eco)systemic restructuring. The orderly analysis of inter-component and inter-complex landscape relations is aimed at identifying the transitions of forest communities to critical conditions, which cause structural restructuring up to a zonal–regional scale.

The landscape–ecological forecast, as a central monitoring link, is experimental. It is described in the form of a system of operations with ecological (hydrothermal) niches of the studied objects. The empirical imitation of landscape–ecological forecasts is realized as the reproduction of the produced scenarios of biogeocoenotic systems according to the laws of their basic spatial organization, with the replacement of spatial coordinates with temporal coordinates, in accordance with the fundamental properties of the ergodic environment. The mechanisms for the adaptation of forest ecosystems to global climatic signals are considered through the prism of their functional resistance to these signals. The methodology for the quantitative assessment of the stability of forest ecosystems has been developed by the author.

As is known, natural processes and events at the regional hierarchical level are characterized by the greatest diversity and high discreteness; therefore, the regional response to global climatic changes inevitably takes the form of multiple reactions of vegetation, soils, and landscapes as a whole to background climatic signals. So far, there is no distinct notion of this multiplicity because the measure of sensitivity of soil-biotic components to climatic changes in different zonal-climatic and geomorphological conditions has not yet been estimated. This is of particular concern for the ecosystems at the local level. The regional hydrothermal trends are generally described by two initial parameters: temperature and precipitation, the connection of which with elementary landscape units (biogeocoenoses) is weak and often statistically unreliable. It was necessary to establish the causal mechanisms of local response to global and regional signals by revealing transfer functions in landscape connections.

Thus, the problem of regional prognostic conception is quite up to date. The author has developed a working version of this conception based on the example of the main catchment area of the Volga River basin, which is occupied mainly by forest and forest–steppe landscapes. This version considers prognostic landscape–ecological scenarios for the near future of the region as a single system of global changes in the environment.

This book presents the experience of analytical and cartographic prognostic scenario modeling of the near and distant future for a given ecoregion. The mechanisms of local and regional responses of forest ecosystems to the expected global warming in the 21st and 22nd centuries are estimated. Two global climate models, GISS-1993 and HadCM3-2000 (version A2), are used for forecasts. The quantitative methods of landscape–ecological forecasting developed by the author are described in detail.

This book presents prognostic landscape–ecological scenarios of climategenic changes in the biological cycle and carbon balance in forest ecosystems as leading factors of biotic regulation of the environment. This made it possible to identify those zonal–regional types of forest formation that will promote the stability of terrestrial ecosystems in a changing climate (according to Le Chatelier's principle). It is predicted that under conditions of global warming, various temperate forest associations and formations will carry out both positive and negative biotic regulation of the carbon cycle in the soil–vegetation–atmosphere system.

We have established empirically that the main channel of connections between regional and local ecosystems and the background climatic system passes through the summer soil moisture content, which, on the one hand, serves as a rather reliable geophysical indicator of the state of ecosystems and, on the other hand, is the most powerful local ecological factor. It has also been determined that the object of priority in landscape–ecological prediction over the next century must not be the structural evolution of ecosystems but rather the directed change in their functioning, i.e., the shifts in the small biological cycle (phytomass production and degradation).

The transition from the local to the regional level of prediction, with the construction of maps on the scale of 1:2,500,000, was carried out using our method of induction-hierarchic extrapolation. This method is based on the empirically established phenomenon of polyzonality of local geo(eco)systems as a form of their response to global climatic changes.

Understanding the mechanisms of continental biosphere sustainability in a changing climate is closely connected with studying the biotic regulation of the carbon cycle in the soil–vegetation–atmosphere system. These issues have already been highlighted in numerous papers both in Russia and abroad. However, significant emphasis is placed on the procedure for determining carbon pools in different components of phytobiota and in soil, as well as the "work" of the carbon cycle and the current carbon content in forests, marshes, and other ecosystems. The problems of quantitative assessment of the responses of different biotic components of the carbon cycle to global climatic changes are illustrated much less. Accordingly, there are virtually no predictions of the climate-genic dynamics of factors affecting the biotic regulation of carbon dioxide content in the atmosphere.

Further development of the concept of biotic regulation is viewed as a study of local–regional mechanisms of the biological cycle, which provides stability for the natural environment in accordance with Le Chatelier's principle. This problem occupies an important place in the final part of the given monograph. Analytical and cartographic empirical-statistical predictive models have made it possible to reveal local mechanisms of biotic regulation and identify the zonal–regional types of forest formations capable of stabilizing the continental biosphere in a changing climate.

# About the Author

**Erland G. Kolomyts** was a scientist at the Institute of Geography of Siberia and the Far East, Russian Academy of Sciences, Irkutsk, from 1960 to 1972; a scientist at the Pacific Institute of Geography, Russian Academy of Sciences, Vladivostok, from 1972 to 1977; the Head of the Laboratory of Geo-ecology at the High Mountain Geophysical Institute, Nalychik, from 1977 to 1986; the Head of the (sub-)faculty of Physical Geography at Nizhny Novgorod Pedagogical University, from 1986 to 1990; and the Head of the Laboratory of Landscape Ecology at the Institute of Ecology of the Volga River Basin, Russian Academy of Sciences, Togliatti, in 1990. Since December 2022, he has been working as a leading researcher at the Institute of Basic Biological Problems of the Pushchino Scientific Center for Biological Research of the Russian Academy of Sciences. His honors and awards include Doctor of Geographical Sciences (1976), Professor (1986), Meritorious Science Worker of the Russian Federation (2003), and the Grigoryev Prize Laureate from the Russian Academy of Sciences (2015), Honorary Doctorate of Letters by the International Biographical Centre, Cambridge, England (2016), Honorary Doctorate of Philosophy by the International Biographical Centre, Cambridge, England (2018).

# Empirical Simulation Base and Predictive Modeling in Geographical Ecology 1

## *Framework Concept*

## 1.1 INTRODUCTORY PROVISIONS

According to the decisions of the United Nations Conference on Environment and Development, held in Rio de Janeiro in 1992, the strategy for the sustainable development of open-source education is the preservation of the natural resource potential of natural ecosystems, primarily their biotic components, in order to meet the vital needs of present and future generations. To solve the problems of sustainable development, it is necessary, first of all, to have indicators of the sustainable development of environmental elements and systems. This constitutes the subject and objectives of geo-ecological monitoring.

Ecological security of the continental biosphere depends substantially on the state of zonal–regional types of natural ecosystems, and primarily, the forest cover, which plays a major role on land in preserving water resources and in mitigating climatic fluctuations. This problem has become a current concern under global warming, with its highly diverse and sometimes difficult-to-predict ecological consequences. The forest cover in Russia is about 20% of the planet's forests, and boreal forests occupy more than

DOI: 10.1201/9781003733119-1

90% of this area. Thus, the problem of maintaining forest ecosystems and reproducing forest resources on the southern boundary of the temperate forest zone, where forest communities are present in states close to critical, is among the fundamental ecological problems.

The book presents a strategy for studying the mechanisms of functional and structural organization of forest ecosystems as objects of terrestrial geosystem monitoring. The strategic goal of geo-ecological monitoring is to reveal the ecological potential for a transition to adaptive forestry based on revealing the mechanisms of sustainability of forest ecosystems in the context of modern global warming. The author initiated the development of scientific and methodological bases of spatial functional monitoring of forests based on empirically established local and regional landscape–ecological connections, which are considered the mechanisms of metabolic responses of forest ecosystems to particular climatic trends.

Catenary landscape–ecological structures that form regional systems of localized natural zonality are capable of imitating the main directions and scales of geosystemic restructurings. Landscape–ecological forecasting is experimental in nature. It is presented as a system of operations with ecological (hydrothermal) niches of the objects under study. In calculation models, the course of predicted processes is reproduced using their empirical imitation by spatially distributed parameters of basic ecological niches.

The description of future states of biogeocoenotic systems is carried out according to their current spatial organization in accordance with the fundamental properties of ergodicity of the natural environment. The ambiguous nature of the transformation of forest topogeosystems is revealed at a fixed value of the regional geophysical trend, when their new state can have features of not one, but several basic states. A working algorithm for predictive landscape–ecological calculations is presented. Based on geo-ecological forecasts, the problem of regulating the quality of the natural environment under global climate change through carbon balances and the functional stability of forest ecosystems is addressed.

## 1.2 STRATEGY OF SCIENTIFIC AND METHODOLOGICAL SEARCH

Analytical and cartographic landscape–ecological prognostic models for the territory of the Volga River basin and its surroundings made it possible to establish the regularities of functional and structural responses of zonal–regional geo(eco)systems to global climatic signals, leading to the development of landscape–geophysical connections that provide wider and more efficient use of the method of actualism in retrospective and numerical geo-ecological prognosis. Priority was given to revealing, typifying, and analyzing the stability of climatic niches of boreal and sub-boreal forest ecosystems, as well as to disclosing the mechanisms of their transformations under given theoretical (calculated) scenarios of changes in the global climatic system. This pressing problem is associated with the preservation and reproduction of forest resources in the south of the boreal belt in light of envisaged (and probably already starting) global warming.

The empirical basis of prognostic constructions is formed by three groups of regional models:

a. Landscape–geophysical – a system of chain reactions in component interrelations in response to external impacts, which opens the main channels for the transmission of hydrothermal signals and is closed at vegetation, soils, and primary biological productivity;
b. Bioclimatic – an ordered ensemble of climatic niches of soil–plant cover in the space of landscape–geophysical factors, which serve as starting objects of climatic prognoses and paleo-reconstructions;
c. Functional – a group of parameters describing small biological turnover, which experiences the primary impact of the climatic system and therefore must be a top-priority object of landscape–ecological prognosis.

It needs to be emphasized that this research covered only the structural–functional organization of natural (indigenous) plain landscapes in different zonal conditions of the temperate belt but did not address the involvement of anthropogenic geo-components in this organization. The authors posed the problem of re-interpretation of one of the core aspects of traditional complex physical geography: the theory of geographical zonality based on empirical modeling and system analysis. The introduction of various attributes of anthropogenic transformation of geo(eco-)systems into the models under study would complicate the data matrix so much that the empirical models based on them would lose their main property – sufficient observability and representativeness for solving the posed problem.

The goal of this research was to consider the mechanisms of territorial organization of zonal geo-spaces composed of transit, i.e., functional-dynamic, geo-components consisting of structural units (landscapes) distinguished by the fixed components: conservative (lithogenic) and soil-biotic. The matter in question is the study of mechanisms for the transformation of transit, or micro-substrate, geo-components (first of all, heat and moisture fluxes) under the influence of the lithogenic (orographic + edaphic) frame into territorial macro-substrate structures (Solntsev 1981). In other words, it is necessary to make a "transition from a territorial complex to a system" (Armand 1988a, p. 123). This must imply the shift of the emphasis from the genetic interdependence of geo-components to their functional ordering (Isachenko 1991). Thus, the main goal of the research is the elucidation of the cause-and-effect relations between structural elements of the regional level and factors of their formation, i.e., connections integrating these elements into a comprehensive whole: landscape–zonal geo-space.

This goal is achieved by solving two rather large-scale problems corresponding to two different trends of scientific search. The first trend includes the ascertainment of the character and degree of influence of hydro-climatic factors determining the appearance of zonal geo-spaces on the territorial ordering of landscape classification units distinguished as elements of these geo-spaces. It is necessary to reveal the system-forming significance of zonality factors for landscapes as the lowest regional unit, transitional from the regional level of biosphere organization to its local (topological) level. Generalization of research experience shows that there is a qualitative leap in the structural–functional characteristics between the above hierarchic levels of the biosphere (Armand et al. 1969; Sochava 1978; etc.). When solving the first problem, one should answer the following question (Solntsev 1981): what physical interactions underlie the formation of natural complexes distinguished by "non-physical" criteria? This problem concerns one of the promising trends of modern landscape science: the theory of the poly-structure of landscape systems (see Chapter 3, Section 3.2), which, as is shown by N.A. Solntsev (1973), does not contradict the known law of zonality (Dokuchaev 1949) but even enriches it with new content.

In essence, the matter in question is the regional and local diversity of zonal ways of biosphere organization and the respective diversity in the formation of discrete natural-territorial units (regional landscapes) in the system of continuous hydro-climatic fields of the zonal geo-space. These fields of forces make up a matter-and-energy backbone, which interacts with the lithogenic frame, thereby creating a whole series of partial landscape structures: landscape "projections" according to Armand (1988a).

On the other hand, when considering the landscape organization of natural zones and sub-zones, we contemplate revealing to what extent these rather large elements of the biosphere are complex formations and, accordingly, to what extent these bioclimatic structures can be objects of landscape–ecological research.

The second trend addresses the no less important problem concerning the further development of fundamental points of the monosystem landscape paradigm: "the concept of natural complexes as stable, regular combinations of components" (Armand et al. 1969, p. 6). It is necessary to determine to what extent the "unity", "internal consistency" (or "complete community"), and "closest interaction" of natural components and processes in regional landscapes, as well as the respective "aggregate homogeneity" of the latter, manifest themselves in reality (Dokuchaev 1949, p. 357). The structural–functional integrity of natural complexes has long been considered their basic property (Berg 1947; Grigoryev 1965; Isachenko 1965; Mil'kov 1967; Solntsev 2001; etc.). The a priori conceptions of the "harmonicity" of regional landscapes (as a combination of geo-components), its genetic uniformity, and relative homogeneity persist to this day.

However, it is still unclear to what extent the "complete accommodation of landscape elements to each other" manifests itself in a real landscape and what mechanisms form "a landscape, where all parts are integrated into an ordered whole" (Berg 1947, p. 6, 8). This problem has not lost its relevance because the matter in question essentially concerns the forms and extent of manifestation of causality in the biosphere and its separate parts, the "degrees of freedom", and the critical values of geo-component connections, beyond which the geo(eco)system loses its stability and qualitative definiteness. It is considered that the landscape is more stable when geo-components are better fitted to each other and to the external matter–energy parameters of the environment and more tightly interconnected through the biological cycle of materials (Perel'man 1975; N.A. Solntsev 2001). This thesis is too general and can hardly be interpreted unambiguously because the relationships of isomorphism between geo(eco)system structure and stability are most likely absent (Svirezhev 1977).

## 1.3 MAINTENANCE OF THE WORK AND ITS PURPOSE

The book outlines the new concept of geo-ecological forest monitoring at the present stage of global warming. For the first time, the full triad of monitoring in its classical definition is considered: «observation (state assessment) – control (prediction) – management (adaptation, feedback, regulation)". The framework for this concept was proposed 50 years ago but has yet to be developed.

Geosystem monitoring is an integral part of geo-ecological monitoring. Its main content consists of (1) a complex analysis of the state of geo(eco)systems as integral natural formations and as differential structural units of the biosphere; (2) an estimate of their resistance to external impacts, as well as the forecasting of their anthropogenic changes. The extraction of the anthropogenic component in the global changes of the environment is included in the strategy for the transition to sustainability.

The ultimate scientific–practical goal of monitoring the mechanisms of sustainable forest ecosystems in a changing climate, set out in this monograph, is to create theoretical, methodological, and factual prerequisites for unlocking the ecological potential of forests, which can ensure the transition to adaptive forestry. The problem is solved by implementing the full triad of geosystem forest monitoring under current global warming in the form of base and predictive ecological experiments, with the construction of analytical and cartographic empirical–statistical models.

The research strategy is based on the ideological and scientific–methodological provisions of the concept of geosystem monitoring as a landscape–ecological search system "Global changes at the local level" put forward by the author. Spatial functional monitoring of the forests was conducted on the basis of empirically established local and regional landscape relationships, which are considered mechanisms of metabolic response of forest biogeosystems to climate trends, including current global warming. This is the experimental nature of the monitoring study itself.

In this aspect, the problem of geosystem monitoring remained, until recently, poorly developed, since, firstly, there was no factual basis required, and secondly, no sufficiently rigorous methodology for local and regional landscape–ecological forecasting, a key link in monitoring itself, was created. In our scientific and methodological searches, which are presented in this book, the ways of solving these problems are covered in sufficient detail. Using the boreal ecotone of the Volga River basin as an example, we have substantiated the methods for determining those parameters of structural and functional organization of forest ecosystems of the southern boreal belt and the northern sub-boreal belt that are most sensitive to global climatic changes and which, therefore, can be used for local and regional geosystem monitoring of forests.

The methods of local landscape–ecological forecasting developed by the author on the basis of empirical–statistical modeling are stated. The mechanisms of local and regional response to the global warming anticipated during the 21st century have been assessed by the example of forest ecosystems of the Volga River basin. Here, we present the empirical substantiation of the known conceptual propositions regarding the ecological resources of forest cover as its ability to additionally absorb greenhouse gases by the mechanisms of carbon cycle regulation under the conditions of climate change. The adsorption potential of primary and derivative boreal and nemoral forests was determined, and their ability to mitigate climate changes, i.e., to reduce anthropogenic warming, was assessed.

The mechanisms of adaptation of forest ecosystems to global climate signals are examined through the prism of their functional stability to the effects of these signals. For the first time in domestic and world literature, it was possible to develop quantitative methods for calculating and mapping the functional stability of forest ecosystems as integral chorological units, rather than according to their individual structural or functional characteristics, as was practiced earlier. Two stability levels – labile phytocoenotic and inertial soil-biotic – are specified. The resistance and resilience indices of forest biogeosystems are calculated for each level.

This approach is fully landscape–ecological. Its effectiveness lies in its reliance on the discrete parameters of the small biological cycle. Using the Euclidean distance metric, we calculated the indices of mobile (phytocoenotic) and inertial (soil-biotic) stability of two types of biogeocoenoses: resistant and elastic–plastic. This allowed us to reveal the metabolic diversity of these types of resistance.

Developed on the example of forest ecosystems of the Volga River basin, discrete empirical–statistical models describe the overall future picture of the biotic regulation of the carbon cycle in the forest–atmosphere system according to well-known climate scenarios, taking into account the functional and structural changes in biogeocoenoses. Based on predictive models, certain local series and zonal–regional types of regulation of atmospheric $CO_2$ content are revealed during the development of climatic thermo-aridization in each ecoregion. In accordance with the areal distribution of these types, the total regional contribution of forest phytobiots to the stability of this part of the continental biosphere is estimated.

Quantitative predictive estimates of the regulation of the carbon cycle by forest ecosystems are provided for scenarios of modern global warming. The further development of the known concept of biotic regulation of the carbon cycle in the biosphere is outlined through the study of local mechanisms in the small biological cycle, which provides stability to the natural environment in accordance with Le Chatelier's principle. It was revealed that the functional states of forest ecosystems influence the biotic regulation of the carbon cycle. Local and regional maps of carbon balance are presented as well.

Experiments were conducted to numerically solve the dual-purpose task set by the Paris Agreement (2015): to assess $CO_2$ absorption from the atmosphere by forest communities under current global warming and the adaptation of forests to climate change. Forest stability, as a direct environmental factor in greenhouse gas absorption, was estimated in statistical terms. A clear understanding of the significant increase in the ecological resources of boreal and nemoral forests was obtained: their adsorption capacity increased with an increase in adaptation potential. Thus, one of the possible ways to achieve the environmental goals set forth by the Paris Agreement on Climate Change has been implemented.

Biogeocoenoses and forest formations were established as priority communities for selecting the most optimal assortment of tree species in various zonal, geomorphological, and edaphic conditions. This will be a decisive step in achieving the final goal – unlocking the ecological potential of forests and ensuring the transition to adaptive forestry. Creating forests with high carbon-absorbing potential is one of the strategic goals of adaptive forestry. This is especially important for afforestation in currently deforested areas in light of expected climate changes, as well as for reforestation after fires and afforestation after logging.

## 1.4 ORIGINALITY OF THE WORK

Any major monograph dealing with the most pressing scientific problems is not written from scratch but is the result of a theoretical and methodological synthesis of many years of research by the author, during which inevitable publications were produced on certain issues related to this problem. This monograph is a summary result of more than 40 years of scientific research by the author in the field of experimental geographic ecology and landscape–ecological forecasting, which is confirmed by the extensive list of the author's works published in international journals.

For the first time in forest ecology, a complete cycle of climate-genic monitoring of forests – "observation (assessment of the state) – control (forecast) – management (feedback, regulation)" – was carried out using a specific ecoregion (the Volga River basin) as an example. A unified system for modeling forest ecosystems in the past, present, and future was developed, identifying the corresponding direct and feedback links between forests and climate. On this basis, the ecological potential of boreal and sub-boreal forests was identified, ensuring the transition of forestry to an adaptive strategy.

The spatial and functional monitoring of forests is based on empirically established local and regional landscape–ecological connections, which are considered mechanisms of metabolic reactions of forest ecosystems to certain climatic trends. This represents the novelty and prospects of the basic and predictive empirical simulation concept of regional and local geo-ecological monitoring developed by the author.

The entire analysis is accompanied by the construction of corresponding empirical simulation models. Local mechanisms of global changes are revealed through the empirically established display of the background bioclimatic trend by the catenary system of biogeocenoses. Regional structures of localized natural zonality are considered capable of simulating the main directions and scales of geo(eco-)systemic reorganizations. Ordination analysis of intercomponent and intercomplex landscape connections aims to identify the transitions of forest communities to critical states that cause structural reorganizations up to the zonal–regional scale.

Landscape–ecological forecasting, as the central link of monitoring, has an experimental nature. It was developed by the author as a system of operations with ecological (hydrothermal) niches. Empirical imitation of landscape–ecological forecasts describes future scenarios of biogeocoenotic systems according to the laws of their basic spatial organization, with the replacement of spatial coordinates in the forecasting procedure with temporal ones, in accordance with the fundamental properties of the ergodicity of the natural environment.

The following were used in the forecasting procedures (as well as paleoreconstructions): (1) polymorphism of landscape–zonal systems, ensuring their functional adaptation to regional climatic changes; (2) regional structures of localized natural zonality, capable of modeling the main directions and scales of geo(eco-)systemic restructurings; (3) local empirical imitation of a regional bioclimatic trend based on models of binary hydro- and thermo-edaphic ordination of forest topogeosystems; (4) properties of poly-zonality of local ecosystems as a way of their response to global climate changes; (5) empirical imitation of climatically determined changes in the biological cycle based on hydrothermal ordination analysis of its parameters; (6) empirical imitation of changes in productivity and carbon sink in forest ecosystems during climate fluctuations.

The boreal bioclimatic ecotone of the Volga River basin is used as an example to carry out the experience of analytical and cartographic prognostic scenario modeling of the near and distant future for the given ecoregion and its paleogeographical analogs. The mechanisms of local and regional responses of forest ecosystems to the anticipated global warming in the 21st and 22nd centuries have been assessed. Two global climate prediction models, GISS-1993 and HadCM3–2000 (version A2), were used for prognoses. The quantitative methods of landscape–ecological forecasts developed by the author are described in detail. Prognostic landscape–ecological scenarios of climate-genic changes in the biological cycle and carbon balance in forest ecosystems, as leading factors of the biotic regulation of the environment, are presented. This made it possible to reveal zonal–regional types of forest formation that will promote the sustainability of terrestrial ecosystems in a changing climate (according to Le Chatelier's principle). It is predicted that, under global warming, different temperate forest associations and formations will carry out both positive and negative biotic regulation of the carbon cycle in the system of soil–vegetation–atmosphere.

Matrix maps of the stability of forest facies and formations are compiled using the example of the Oka–Volga River basin. They characterize the sensitivity of the forest units to climate fluctuations and their functional regeneration potential. The overwhelming majority of forest formations are characterized by high sensitivity to external disturbances but, at the same time, possess sufficiently developed mechanics that personify their elastic–plastic stability, i.e., sufficiently high adaptive capacity.

## 1.5 NOVELTY OF THIS SCIENTIFIC SEARCH

We will describe the main strategic directions of the theory and methods of geo-ecological monitoring, which constitute the general scientific novelty of the given monograph.

I. In recent decades, possible functional and structural transformations of natural ecosystems and their components under the influence of anthropogenic climatic changes, including current global warming as an urgent environmental problem of mankind, have been actively studied. Much progress has been made in developing integrated global monitoring frameworks to detect the advance of environmental changes. However, the extrapolation of methods for global monitoring of the "ocean–atmosphere–continent" system to the regional and, especially, local levels encounters significant factual and methodological difficulties. Here, it is necessary to develop independent monitoring methods.

Global remote data on vegetation distribution and dynamics, as well as the environmental maps based on them, have been considered a potential means for monitoring and modeling the impacts of global climate change on plant communities (Franklin et al. 2017). According to the BERMS Program, measurements of $CO_2$ fluxes over the boreal forests of Saskatchewan province (Canada) during 1993–2003 revealed a negative role of drought in carbon uptake by deciduous stands (Kljun et al. 2007).

The range of similar studies in Russia is very significant. Numerous experiments have been conducted to study the influence of modern meteorological and climatic conditions on the state, phenology, and trends of plant communities in different regions of the Russian Plain and the Middle and Southern Ural Mountains (Budnik 2017; Ivanova and Skok 2019; Bogdanovich and Lipka 2020; Kukhta and Popova 2020; Maksimova et al. 2021; Solovieva and Shikhova 2011; etc.). The study of the long-term dynamics of biosphere processes in the Central Chernozem Biosphere Reserve has made it possible to formulate some unsettled issues regarding their predictive interpretation (Vlasov et al. 2021).

Thus, each of the three triads of geo-ecological monitoring has already received independent scientific and methodological development, but *the integration of all these results into a single operational monitoring system remains very problematic*. The framework concept of a complete triad of geo-ecological monitoring, proclaimed more than 50 years ago, has not yet been implemented. The task remains to implement a full cycle of climate-genic monitoring of forests – "observation–control–management" – using the example of a specific ecoregion, with the development of a unified system for modeling the state of forest ecosystems in the past, present, and future, as well as the corresponding direct and feedback relationships between forests and climate. The experience of solving this scientific and methodological problem in the field of ecological ecology is presented in this monograph.

II. There are fundamental differences between our modeling methods in monitoring and most of the methods in the studies mentioned above. These studies are usually based on *simulation (continual, dynamic) modeling* of the natural environment, which essentially does not go beyond the limits of autecology and synecology, leaving out of sight the geographical (spatial) ecology. These are models with concentrated parameters; they do not provide a territorial sweep of supra-parcelular landscape units (biogeocoenoses, stows, localities) in their organization and in response to external disturbances, and this is a certain limitation of simulation models.

> In our investigations, geosystem monitoring was carried out on the basis of *discrete empirical and statistical models*, according to Rozenberg (1984). These models can be classified as self-organizing models. They are substantially different from the known simulation models (Printice et al. 1992; Smith et al. 1992; Holten et al. 1993; Kudeyarov 2007; Zavarzin 2007; et al.), which are more or less a priori. Empirical-statistical models are a posteriori, i.e., the results of field studies are used not for verification of some calculation data but as a factual basis for modeling per se. This yields results of much higher spatial resolution and empirically substantiated, rather extensive geographic generalizations. In contrast to simulation models, the proposed and implemented *method of local discrete modeling of natural ecosystems* can be much more reasonably considered a biogeocoenotic method – in the spirit of V.N. Sukachev's theory (1972). It is important to emphasize that such models allow predictive analysis with wide use of the results of traditional field landscape–ecological surveys, as well as various cartographic materials. The spatial and functional monitoring of forest ecosystems described in this monograph is based on empirically established local and regional landscape–ecological connections, which are considered mechanisms of metabolic reactions of forest ecosystems to certain climatic trends.

Thus, we are talking about a fundamentally new empirical simulation basic and forecast modeling of natural ecosystems in their climate-genic monitoring, which is reflected in the subtitle of the given monograph. This modeling served as the main tool for pre-forecast landscape–ecological analysis and was then included in the algorithm of the predictive calculations themselves.

At the same time, the landscape–ecological forecast, as the central link of monitoring, has an experimental nature. It was developed by the author as a system of operations with ecological (hydrothermal) niches of the objects under study. Empirical imitation of landscape–ecological forecasts is implemented as the reproduction of predicted scenarios of biogeocenotic systems according to the laws of their basic spatial organization, with the replacement of spatial coordinates in the forecast procedure with temporal ones, in accordance with the fundamental properties of the ergodicity of the natural environment.

In the procedures of forecasting, the following were used: (1) local empirical imitation of a regional bioclimatic trend using models of binary hydro- and thermo-edaphic ordination of forest topo-ecosystems; (2) use of the poly-zonality property of local ecosystems as a way of their reaction to global climate changes; (3) empirical imitation of climate-induced changes in the biological cycle based on hydrothermal ordination analysis of its parameters; (4) empirical imitation of changes in productivity and the carbon sink in forest ecosystems during climate fluctuations.

The simulation and empirical–statistical models cannot be considered interchangeable. Each of these classes of models has its pros and cons, similar to any type of modeling. As shown by comparative analyses of the models of plant responses to climatic changes (Lischke et al. 1998), simulation (dynamic) modeling provides a detailed "time scan" of the studied structural or functional ecosystem parameters but has no opportunity for reasonable spatial extrapolation of the results. In the meantime, empirical–statistical models, being static, can describe the territorial diversity of predicted situations but suffer from uncertainty over time. In this context, we accept that the priority object in landscape–ecological prediction within a century must not be the structural evolution of ecosystems but the *directed change of their function*, i.e., the shifts in the small biological cycle (phytomass production and decomposition, as well as humidification), which take the first few years in the taiga zone and occur during the year in the sub-zone of broadleaf forests (Kolomyts 2003). These characteristic times of functional relaxation are proportionate to exogenous changes in the carbon content of forest phytomasses (living and dead) and mobile humus of the soil (Osipov 2004). This approach is in complete agreement with the subject of scientific research.

## 1.6 NUMERICAL METHODS OF LANDSCAPE–ECOLOGICAL FORECASTS

I. The author has developed and implemented *a numerical method* of regional and local landscape–ecological forecasting. The landscape–ecological approach to modeling the functional parameters of local ecosystems is based on N.V. Timofeev-Ressovsky's bio-chorological concept (1970) concerning the spatial organization of the matter-and-energy cycles by living matter in biogeocoenoses (facies) as discrete, elementary structural units of the biosphere. According to theoretical developments in this field, the biosphere is regarded as a statistical ensemble of biogeocoenoses as bio-chorological units that weakly interact with one another but have a highly ordered internal organization (due to stabilizing selection). Each ecosystem at the level of biogeocoenoses or their groups is assigned specific functional characteristics related to the structure of local matter-and–energy fluxes. This approach permits the prediction of the behavioral patterns of forest communities under different geomorphological and edaphic conditions based on comprehensive empirical data collected in the course of field studies while retaining the statistical methods of their analysis.

Procedures were carried out with the ecological niches of landscape–zonal and local ecosystems based on their inclusion relations as descriptive vectors, giving the probabilities of stability (stabilization) of each object and its transitions into other objects according to the given climatic trend. Original methods were used to construct the matrices and oriented graphs of probabilities of landscape and facies transitions, describing the general picture of ecosystems' response to climatic changes.

The technique of calculations has been brought to the prescribed level and can be included in the scientific and methodological arsenal of geographical ecology. It organically combines a fairly rigorous formalized approach to solving forecasting problems with procedures for collecting, processing, and analyzing empirical material accessible to a wide range of researchers. In contrast to the well-known Russian and non-Russian approaches, this method accounts for the multiple nature of the transformation of natural complexes for a fixed climate trend.

A significant feature of the methodology of ecological and geographical forecasting is its *experimental character.* Computational models reproduce the variation in predicted processes by *empirically imitating* the spatially distributed parameters of the basic ecological niches of the studied objects. The forecast itself is essentially *a controlled experiment*, as defined by Helbert (2008a,b). The researcher sets the input variables and, at the output, obtains the pattern of predicted structural and functional states of the studied objects in this statistical sample, with the identification of new objects outside the sample. The calculations use the methods of the theory of information, descriptive ("fuzzy") sets, and Markovian chains. The ways and efficiency of using these principles in the assessment of the past and future states of natural ecosystems are demonstrated as well.

Predictions of ecosystem transformations include two stages of analysis: (1) evaluation of the probabilities of changes in the functional states of ecosystems and (2) calculations of the rates of ecosystem transformations. Quantitative predictive analysis is carried out using operations with hydrothermal niches of regional or local ecosystems. The algorithms of predictive calculations are detailed for both stages of analysis. In conclusion, the methods of calculation and construction of average weighted (by territory) matrices and oriented graphs of landscape–ecological transitions are discussed.

The procedure for predicting the state of forest ecosystems, based on empirical–statistical simulation, has been carried out. The prediction strategy is based on the actualism method. It consists of the original identification of plant units with specific parameters of basic climatic conditions and subsequent quantitative estimation of the most probable transformations of these objects corresponding to the expected climatic changes within a specified time. This approach assumes multifold, probabilistic transformations of ecosystems at the same value of the hydrothermal trend, whereby a new state may have characteristics of several states existing at the initial moment, rather than a single state. Note that functional relaxation, as the primary response of ecosystems to an external factor, is a priority object of landscape–ecological prediction. The main prediction, based on ecological estimation of the forthcoming global climate warming, is expected to be functional rather than structural–morphological.

The strategy of landscape–ecological prognosis may be organized into the following algorithm: first, to carry out an identification of selected ecosystem objects (zonal and sub-zonal phytocoenological systems) in relation to current climatic conditions, and then to estimate the most probable transformation of the identified ecological (hydrothermal) niches of these objects according to expected climatic changes for a given prognostic date. In this regard, it is suggested that there are multiple potential transformations of ecosystems at a fixed value of the hydrothermal trend, where a new state may exhibit characteristics of not one but several states recognized at present moments in time.

The general principle of the prognosis is expressed as follows: the value of the climatic-caused transformation of one ecosystem into another is greater when the degree of intersection of their climatic niches in the initial states is lesser; that is, the greater the contemporary contrasts of their functional states, the larger the range of overlap of niches after the convergence of ecosystems due to the given climatic factor. The maximum values of transformation allow for estimating the most probable directions and the degree of conversion of the ecosystem. At the same time, an unambiguous character of transformation of regional ecosystems is proposed at a fixed value of the geophysical trend, where the new state may exhibit features of not one but several states existing at a given moment.

The results of the local prognostic analysis are presented in both graphic–analytical models and large-scale maps. The transition from the local to the regional level of prognostic modeling is achieved using inductive hierarchical extrapolation, a method developed by the author based on the empirically established phenomenon of the poly-zonal nature of local ecosystems in response to global climate changes.

The proposed regional prognostic conception uses the Volga River basin as an example to consider the prognostic landscape–ecological scenarios for the near future of this region as a single system of global environmental changes. A progressively increasing hermos-arid bioclimatic trend is predicted, with a general shift of zonal boundaries to the north and corresponding changes in the water regime and plant cover structure of the territory. As a general predicted result, it is assumed that significant shifts in the heat energy level of natural ecosystems across all natural zones of the Volga River basin will occur by the middle of the 21st century, and the ecological effects of such shifts will be more drastic in the south. The process of soil aridization will inevitably exacerbate the problems of preserving and reproducing forest resources, as well as crop production in the territory of the modern forest–steppe zone, including the sub-zone of nemoral forests.

II. One of the main provisions of the Paris (2015) Agreement on Climate Change is to address the problem of ecosystem adaptation – strengthening their adaptive capacity, increasing resistance, and reducing vulnerability to climate change. We have considered the mechanisms of adaptation of forest ecosystems to global climate signals through the prism of their resistance to the impact of these signals. The resistance of forest ecosystems is a prerequisite for their effective implementation of stabilizing and protective functions in the landscape.

The mechanisms of adaptation of forest ecosystems to global climate signals are examined through the prism of their *functional stability* to the effects of these signals. In Russian geo-ecology, stability is still evaluated by qualitative gradations such as slight, medium, and high (Buks 1977; Rozhkov and Kozak 1989; GanzeI 2024; etc.), and rigorous classification criteria are sometimes absent.

In non-Russian works on forest ecology, ecosystem stability is usually qualitatively described using quantitative indicators that characterize stability indirectly. The growth rate, productivity, compactness of the tree stand, species and functional diversity, plant damage, and other parameters are used for stability evaluation (Holling 1973; Chapin et al. 1997; Diaz and Cabido 2001; Thompson et al. 2009). A number of works by non-Russian researchers are devoted to the ratio between various types of forest stability: initial resistance and subsequent resilience.

It is also assumed that there is a basic ratio between biological diversity, productivity, and stability of forests and that this ratio is of practical importance for the adaptive monitoring of forests in response to climate changes (Begon et al. 1986; Thompson et al. 2009). In general, the more complicated the community (the greater the species diversity), the higher its stability (Ricklefs 1976). This theoretical statement was not only adopted in general ecology (Odum 1983), but it also found its place in landscape ecology (Sochava 1978). Nevertheless, it is often rejected (Begon et al. 1986) and is still discussed (Whittaker 1975; Loreau et al. 2002).

There are comprehensively elaborated mathematical models of ecosystem stability in general ecology. The models describe a narrow range of relatively simple ecological systems and are usually deterministic. They analyze particular populations or their combination (a biotic community) (Smith 1974; Logofet and Svirezhev 1983; Svetlosanov 1990; etc.). The stability of these ecosystems is calculated as the ratio between the deviation of the disturbing effect and the response of the chosen parameters to this effect or as the ratio between the homeostasis value and the parameter of distribution non-uniformity of a dependent variable (Fedorov 1974; Fedorov 1975). Stability is calculated by linear differential equations. Great attention is paid to the stability of the hierarchy of food chains (of the predator–prey type in particular).

Unfortunately, approaches involving mathematical modeling of the stability of ideal ecosystems are not widely applied in landscape ecology, where the investigation subjects are considerably more complicated, poly-component, and discrete, composed of heterogeneous elements with different characteristic times. The ecosystems at the landscape level are distinguished by the probabilistic nature of internal and external interactions (Armand et al. 1969) and non-linear behavior related to a regulated network of positive and negative feedbacks (Zeidis et al. 2001; Svetlosanov 2009).

The stability and functioning of these systems are primarily determined by geochemical cycles (annual and long-term turnover of organic matter), not by the number and dynamics of plant and animal species (Svirezhev 1983). We found such a method for quantitatively assessing stability when studying the patterns of structural and functional organization of topoecosystems of the Middle Volga Region (Kolomyts et al. 1993; Kolomyts 2005). It was subsequently significantly developed, with the construction of not only analytical but also cartographic models (Kolomyts et al. 2012).

The ideology and methods of quantitative evaluations of forest ecosystem potential sustainability are described in the given book. Two levels of sustainability are distinguished: labile phytocoenological and passive soil-biotic. Statistically significant relations in equations have been obtained that describe the dependence of these indices on the metabolic features of forest biogeocoenoses and on environmental factors. The presented large-scale and small-scale matrix maps of resistance and resilience stability of forest facies (by the examples of Samarskaya Luka and By-Oka-Terrace Reserve) and forest formations (for the Oka–Volga River basin) characterize their sensitivity to climatic changes and subsequent succession/restoration potential, respectively. Local and regional maps of these indexes have been created. A comparative analysis of resistance and reduction potential (elastic stability) of forest ecosystems is presented for various climatic zones.

The methods of calculating and large-scale mapping of the two main types (resilience and resistance) of potential forest ecosystem stability were elaborated. They permit the evaluation of the stability of biogeocoenoses as integral elementary chorologic units rather than by some structural or functional characteristics. The elaborated approaches may be inherently assigned to landscape–ecological methods. They are effective because they are based on discrete parameters of biological turnover. These are: (1) productivity to phytomass ratio; (2) the ratio of forest litter mass to the mass of green substance produced annually; (3) humus mass. These parameters are responsible for the main mechanisms of natural ecosystem stability and provide a great amount of empirical data for stability calculation.

With the use of Euclidean distance, we have calculated the mobile and inertial stability of forest biogeocoenoses. The obtained data reveal the metabolic diversity of the two main stability types: resistance and resilience. They are expressed as corresponding indices that represent the complex parameters of the functional status of biogeocoenoses. The relative role of each metabolic factor in a particular index was quantitatively evaluated, and the calculated models were verified. Based on the obtained data, matrix maps of the biogeocoenosis stability indices were calculated using new geomorphometric approaches. The stability indices were initially calculated for statistical samples based on matrices of environmental factors, for which the statistical correlations revealed by multiple regression were the closest. The calculated stability values were then interpolated and extrapolated for the compilation of maps of natural ecosystem stability.

The analytic and cartographic modeling of forest biogeocoenosis stability in the southern part of the forest belt in the Middle Volga Region allows for the following conclusions. At the initial stage of an unfavorable external effect, the reaction of those forest biogeocoenoses that tend to survive consists of a decrease in the intensity of the annual turnover of the above-ground phytomass and/or the enhancement of mineral plant nutrition by substances from the humus soil horizon. After that, the regeneration potential is realized by the activation of destruction processes of dead organic matter in the forest litter and then by an increase in the annual production of green mass, i.e., by a general enhancement of biological turnover.

One of the most important ways to achieve the goals stipulated by the Paris Agreement on Climate Change (2015) is to solve a two-fold task: the adsorption of $CO_2$ by forest communities from the atmosphere during global warming and their adaptation to these climate changes, which should ensure the effectiveness of the adsorption itself. As an indicator of adaptation, the labile resilient (elastic–plastic) stability of forest ecosystems was considered. For the territory of the Oka River basin, using multiple regression methods, a numerical experiment was conducted to assess the effect of the resilient stability of forest formations and the predicted climatic conditions on the carbon balance. A total of eight linear equations were obtained (with a significance level of $P < 10^{-6}$), which are proposed to be included directly in the "Guidelines for the quantitative determination of the adsorption volume of greenhouse gases".

In the upcoming 100-year forecast period, the overall elastic–plastic stability of forest formations should increase, and to the greatest extent with extreme warming. Accordingly, one should expect a significant increase in the ability of boreal forests to adsorb greenhouse gases. A comparison of the carbon balance values of forest formations obtained with initial (base) and final (final) stability indices gives an unambiguous picture of a significant increase in the adsorption capacity of boreal forests with an increase in their regenerative potential. A decisive contribution to increased adsorption is made by the growth of reforestation adaptation.

## 1.7 MINE INFERENCES

A strategy for predictive empirical–statistical modeling of natural ecosystems as objects of terrestrial geosystem monitoring is presented. Local mechanisms of global changes are revealed through an empirically established display of the background bioclimatic trend by a catenary system of biogeocoenoses. Regional structures of localized natural zonality capable of simulating the main directions and scales of geo(eco-)system restructurings are considered. The landscape–ecological forecast is described as a system of operations with ecological (hydrothermal) niches of the studied objects. Empirical simulation of forecasts is implemented as a reproduction of predicted scenarios of biogeocoenotic systems according to the laws of their basic spatial organization, with the spatial coordinates replaced by time coordinates in the forecasting procedure, in accordance with the fundamental properties of ergodicity of the natural environment.

# Landscape–Ecological Strategy for Ground Geo-Ecological Monitoring of the Forests

# 2

## 2.1 REGIONAL ECOLOGY AND THE PROBLEM OF GLOBAL CHANGES

Fundamental problems of ecology and geography include, as is known, the problem of global changes, which is the core of the "International Geosphere-Biosphere Program" (Kotlyakov 1992). The Program is designed for a long-term outlook and envisages the development of scenarios for the near future of the biosphere in terms of physical models describing basic processes and events. One of the most dynamic natural processes on a planetary scale, which has efficiently influenced biosphere evolution in the past and determines its condition in the future, is changes in the global climate caused by the altered chemical composition of the atmosphere, with the corresponding demonstration of the greenhouse effect. The coming global climatic changes will be associated, first of all, with the technogenic growth of the content of $CO_2$ and other greenhouse gases in the atmosphere, which may disturb the natural carbon cycle in the biosphere and lead to large-scale ecological consequences, including the reorganization of the landscape–zonal structure of entire continents (Budyko 1977; Leemans 1992; Houghton et al. 1996).

Thus, the problem of global changes has a sufficiently large practical aspect. Ecological safety of large territorial subunits of the continental biosphere significantly

DOI: 10.1201/9781003733119-2

depends on the state of the zonal–regional types of natural ecosystems, first of all, forest cover. Therefore, the problem of maintaining forest ecosystems and the reproduction of forest resources on the southern boundary of the temperate forest zone, where forest communities are present in states close to critical, is among the fundamental ecological problems. It has always been of high priority for the East-European countries, where the wide transitional zone from forest to steppe, i.e., the zonal forest–steppe ecotone, is the industrial and demographic core of this large region. At the moment, the questions of the stability of natural ecosystems, as a natural–historical basis of stable development of the region, are gaining increasing significance in the solution of this problem.

As is known, global geosystem monitoring is most up to date and actually realizable on the scale of individual ecological regions (Gerasimov 1985). However, natural processes and events on the regional hierarchical level are characterized by the greatest diversity and high discreteness (Sochava 1978); therefore, the regional response to global climatic changes inevitably takes the form of multiple reactions of vegetation, soils, and landscapes as a whole to background climatic signals. So far, there is no distinct notion of this multiplicity because the measure of sensitivity of soil-biotic components to climatic changes in different zonal-climatic and geomorphological conditions has not yet been estimated.

The known prognostic ecological developments are, with few exceptions, of a very schematic character and aimed mainly at the assessment of the general future condition of the biosphere as a planetary system by a quite limited set of geophysical parameters (for the most part, air temperature and atmospheric precipitation). Accordingly, prognostic and paleogeographical maps are too small scale, and cartographic prognosis confines itself to schematic demonstrations of changes in the boundaries of natural zones and areas of some forest species (Emanuel et al. 1985; Kobak and Kondrasheva 1992; Smith et al. 1992; Sirotenko and Abashina 1998; Menzhulin and Savvateev 2002; Mokhov et al. 2005). The regional and especially local levels of geo-ecological prognoses have still not been developed enough due to the insufficiency of factual material and the methodological difficulties of transferring hydro-climatic prognosis from global to regional and local levels.

The solution to all the above problems is somehow or other associated with an integrated regional *paleo-prognostic conception* presented in this monograph as a working version. Prognostic landscape–ecological scenarios for the near future of the biosphere and their paleogeographical analogs, as a single system of global environmental changes, have been developed and assessed using the example of a large region – the Volga River basin, which is the economic and demographic "core" of European Russia. The main attention was paid to the mechanisms of shifts in the patterns of vegetation, soils, and landscapes on the model territory under given scenarios of disturbing impacts of the climatic system, which are envisaged in the foreseeable future (to the middle of the 21st century) and whose analogs could occur in the geological past – the optima of the Mikulino (Eemian) interglacial period (around 125,000 years ago) and the optima of the Holocene (5–7,000 y.a.). For this purpose, regional analytical and cartographic (on a much vaster scale than before) prognostic models of landscape–ecological conditions have been developed, as well as models of the above two paleogeographical transects, in light of future and past scenarios of global climatic changes.

The known prognostic climate models GISS and HadCM-3, version 2 (see Section 3.8), transformed for the regional level, with appropriate spatial resolution, were used as the starting conditions for the development of corresponding ecological models.

The solution to the multipurpose tasks of ecological forecasting may depend to a great extent on the results of system analysis of natural-territorial structures, which are most sensitive to external effects, including anthropogenic ones. The scientific search in this direction focuses more and more attention on the natural boundaries – both individual and complex – where the most significant natural or anthropogenic shifts in the structure and function of geo(eco)systems are observed. Considering one or another natural boundary as a vector (connection, cascade, para-dynamical, etc.) landscape system with a clearly defined spatial polarization of its different properties, we obtain a "fast-flowing" model of state response and resistance of geo(eco)systems to the action of certain ecological factors.

The conception of regional and local manifestations of global environmental changes is undoubtedly of fundamental importance not only for regional but also for global ecology, because global biosphere processes and events are perceived most deeply through the level of ecological regions, which is aided by the high diversity of their natural-territory structures, as well as the complexity of functioning and evolutionary transformations of regional ecosystems. The ecological region of the main drainage area of the Volga River basin is quite a favorable natural model for paleo-prognostic analysis because it is entirely included in the East-European boreal ecotone – a system of close zonal boundaries that is very sensitive to global climatic changes. Here, zonal boundaries are characterized by high dynamism, as indicated by the entire Holocene history of aboriginal flora in the Middle Belt of the Russian Plain.

At present, researchers from different countries have conducted certain studies on the problem of global climatic and the most common ecological scenarios of expected changes in the environment, which are powered by the accumulation of anthropogenic $CO_2$ and other greenhouse gases in the atmosphere. To use these results in further scientific and practical research in the field of environmental protection, it is necessary to evaluate the common prognostic landscape–ecological situation for each continental region of the Earth. However, the direct use of global prognostic parameters at regional levels is impractical and can lead to errors. Global prognoses sketch the broad outlines of possible changes in the environment and are often accompanied by schematic maps. Unfortunately, the scales of such maps are too small (no more than 1: 30–40,000,000), and only shifts of boundaries of nature zones are shown on them (Leemans 1992; Houghton et al. 1996; Kobak et al. 2002). Ecological predictions themselves are based, for the most part, on palaeogeographical reconstruction (palaeoanalogies) by a very sparse network of points with palynological dates (Velichko 2010), so they do not describe the regional and local diversity of ecological situations.

The greater impact appears to be in the adoption of the theoretical (calculated) estimation of global climatic changes derived from the models of general atmospheric circulation – AOGCMs (Budyko et al. 1992; Houghton et al. 1996). However, special systematic difficulties emerge here. The main problem is the transformation of global climatic signals into regional and local ecological ones on a rather large scale of natural landscape types, including sub-regional and local plant societies. Up to now, regional tests have not been carried out adequately to determine the most probable directions and relative rates of spreading of external signals both in the system of connection of the nature component and along the lines of landscape chains in space. Therefore, the response of ecological (climatic) niches of these phytocoenoses to global climatic changes is not clear.

## 2.2 CURRENT STATE OF THE CONCEPT OF GEO-ECOLOGICAL MONITORING IN A CHANGING CLIMATE

According to its classical definition, monitoring includes "observations of the factors of impact and condition of the environment, forecasting of its future state and assessment of the actually predicted state of the natural habitat" (Izrael 1984, p. 11). Meanwhile, "the problem of controlling (and managing) the quality of the environment relies on ecological forecasting", requiring the construction of ecological-economic models (in place cited 16).

General monitoring consists of two parts – bio-ecological and geosystemic (Gerasimov 1985). *Bio-ecological monitoring* includes the monitoring and forecasting of anthropogenic changes in individual (selective) natural phenomena and processes that are directly or indirectly related to public health. These changes can be both catastrophic and longer-lasting. For example, toxic indicators of air and water pollution can be related to an increase in the incidence of cardiovascular problems, infections, colds, etc.

Geosystem (geo-ecological) monitoring, including the biosphere, is the second stage (block) of general monitoring, which serves as a necessary addition to bio-ecological monitoring. The main content of geosystem monitoring, as defined by I. P. Gerasimov (1975, 1985), consists of (1) a complex analysis of the state of geo(eco)systems as integral natural formations and as differential structural units of the biosphere; (2) an estimate of their resistance to external impacts, as well as forecasting of their anthropogenic changes. The extraction of the anthropogenic component in the global changes of the environment is included in the strategy of transition to sustainability (Kotlyakov 2001). Our research deals mainly with *geosystem monitoring* as a subject of ecological–geographical studies.

Traditional geo-ecological monitoring considers the ecological consequences of mainly focal anthropogenic impacts on the environment: agroforestry, industrial, and transport. Up to date, the most developed system is the integrated background monitoring of anthropogenic pollution in the environments – atmospheric air, soil, surface water, and vegetation (Izrael 1984; Izrael et al. 1986; Buyvolov et al. 2021). In recent decades, possible functional and structural transformations of ecosystems and their components under the influence of anthropogenic climatic changes, including current global warming as an urgent environmental problem of mankind, have been actively studied (Kurz et al. 2002; Kondrat'ev et al. 2003; Mokhov et al. 2005; Kljun et al. 2007;

Brown et al. 2011; Beets et al. 2011; Izrael and Ryaboshapko 2011; Zamolodchikov et al. 2011; Khamedov and Mazurov 2015; Franklin et al. 2017; Romanovskaya 2019; Leskinen et al. 2020).

Much progress has been made in developing integrated global monitoring frameworks to detect the advance of environmental changes (Cairns and van der Schalie 1980; Hansen et al. 1984; Israel and Munn 1985; Leemans 1999; Mokhov 2006; Fatichi et al. 2019). However, the extrapolation of methods for global monitoring of the "ocean–atmosphere–continent" system to the regional and, especially, local levels encounters significant factual and methodological difficulties. Here, it is necessary to develop independent monitoring methods.

According to the Gerasimov–Izrael conception, the whole system of environmental control can be presented by the following formula: "observation (state assessment) – control (forecasting) – management (adaptation, feedback, regulation)". From the very beginning, Yu. A. Israel emphasized that monitoring in its full scope should include not only "tracking" (repeated observations) but "also an assessment and prediction of the state of the environment … and control of the environment quality" (1984, p. 11, 12), i.e., the implementation of its entire operational triad (see above). Unfortunately, this most important methodological provision of the doctrine of geo-ecological monitoring is rarely carried out, especially in regional and local environmental studies, although the term "monitoring" is mentioned every time. In both Russian and foreign literature on monitoring, the vast majority of works are limited to analyzing the initial (basic) state of natural and anthropogenic ecosystems and, at best, identifying cause-and-effect relationships between the dynamics of soil-biotic components and climate change as the basis for environmental forecasts. This simplification of the original concept of ecological monitoring can, to a certain extent, explain the fact that, as noted by A.A. Tishkov (2021), the idea of developing background monitoring stations based on Russian biosphere reserves has not yet been implemented.

Let us give some examples that outline specific ecological studies and characterize, in one way or another, only the first stage of geo-ecological monitoring. Planetary remote data on vegetation distribution and dynamics, as well as the environmental maps based on them, have been considered a potential for monitoring and modeling the impacts of global climate change on plant communities (Franklin et al. 2017). According to the BERMS Program, measurements of $CO_2$ fluxes over the boreal forests of Saskatchewan province (Canada) during 1993–2003 revealed a negative role of drought in carbon uptake by deciduous stands (Kljun et al. 2007). Statistical models of species distribution, as well as ecological factors of their occurrence, were considered (Dorman 2007). New Zealand's afforestation inventory complex (improperly named a "monitoring system") has been described, which, in combination with the models, is used to facilitate the prediction of carbon stocks in forests (Beets et al. 2011).

The range of similar studies in Russia is very significant. New approaches have been developed to estimate climate-controlled net products of forest ecosystems as an objective index of their response to climatic changes (Shvidenko et al. 2008). Seven-year year-round observations of $CO_2$ fluxes from derno-podzolic soils under forest and grassland vegetation were carried out in the Oka River basin (Kurganova et al. 2007). This experiment has been named "multi-year monitoring" in its simplified interpretation. As a result of multi-year field measurements and the use of multispectral satellite information, data were obtained on daily and seasonal dynamics of greenhouse gases in forest, swamp, and fallow lands in the southern taiga of North-West European Russia. The models built enable us to predict the values of $CO_2$ fluxes under climatic changes (Alferov et al. 2017). Numerous experiments have been conducted to study the influence of modern meteorological and climatic conditions on the state, phenology, and trends of plant communities in different regions of the Russian Plain and the Middle and Southern Urals (Budnik 2017; Ivanova and Skok 2019; Bogdanovich and Lipka 2020; Maksimova et al. 2021; Solovieva and Shikhova 2011; etc.). Using the forests of the Arkhangelsk region as an example, the experience of identifying signs of deterioration in the physiological condition of the tree stands based on multi-spectrum satellite imagery and terrestrial forest survey data is discussed (Bogdanov and Aleshko 2011). The study of the long-term dynamics of biosphere processes in the Central Black Earth Biosphere Reserve has made it possible to formulate some unsettled issues regarding their predictive interpretation (Vlasov et al. 2021). The methodological issues of creating a database of satellite and cartographic data on the state of forest resources under the influence of the oil and gas complex of Western Siberia are also outlined. These materials are presented as a "satellite monitoring system" (Khamedov and Mazurov 2015).

Separate from the subjects of the above and many other similar studies are very informative works of the second stage of monitoring – on regional and planetary ecological and geographic forecasts, as well as on the prediction of the carbon forest balance. Here we can note the predictive scenarios of climate-genic changes in zonal and regional ecosystems of the Russian Plain, Western and Middle Siberia (Emanuel et al. 1985; Kobak et al. 2002; Kolomyts 2005; Tchebakova et al. 2009). A picture of general regularities of the global carbon cycle and the greenhouse effect of the atmosphere in the Holocene and in the recent period due to temperature changes has been described (Kondrat'ev and Demirchan 2000). Based on climatic scenarios of the predictive models ECHAM4/OPYC3 and HadCM3, possible changes in the productivity of vegetation in Russia in the 21st century were calculated and mapped on the basis of its modern geophysical ordination (Golubyatnikov et al. 2005). Predictive analytical and mapping models of the carbon balance of Russian forests based on integrated ground and satellite information have been developed (Zamolodchikov et al. 2011; Zamolodchikov 2014; Korzukhin and Korotkov 2018; etc.). A combination of the DGVM (Dynamic Global Vegetation Model) and the model of general atmospheric circulation made it possible to calculate and map carbon stocks and balance in forests and soils of Central Siberia based on satellite data (Quegan et al. 2011). Based on the model developed by the Forest Service of Canada, the carbon sequestration potential of the controlled forests in Russia up to 2050 was predicted, with an estimate of a 3–4-times reduction in carbon sequestration for the next 30 years (Zamolodchikov et al. 2013).

Finally, let us turn to the third and final stage of monitoring – management (regulation). This stage is still mostly in the state of proposed definitions, conceptual hypotheses and proposed scientific and methodological programs. First of all, it should be mentioned the proposed concept of "climate geo-engineering" as a targeted change in the parameters of climatic system in order to prevent catastrophic ecological consequences of the global warming (Izrael and Ryaboshko 2011). Such changes can, and have throughout the history of the biosphere, been caused by forest cover through regulation of the carbon cycle (Gorshkov 1994; Kondrat'ev et al. 2003). Global predictions of the role of forest cover in regulation of the atmospheric greenhouse effect are presented in small-scale scenarios of the carbon budget of circumpolar boreal forests of Eurasia and North America based on correlation relations of their biomes with temperature and precipitation ranges (Gamon et al. 2004; Gauthier et al. 2015; Shvidenko et al. 2017).

One of the most important ways to achieve the goals stipulated by the Paris (2015) Agreement on Climate Change is to solve a twin challenge problem: the absorption of $CO_2$ from the atmosphere by forest communities during global warming and their adaptation to these climate changes, which should ensure the effectiveness of absorption itself through feedback interactive communication. In this regard, the proposed concept of "adaptation monitoring" is very constructive, as it expands the existing climate monitoring system and represents its final stage as "monitoring of climate activity" – mitigation and adaptation (Romanovskaya 2018; Romanovskaya 2019). Adaptation of ecosystems is understood as a set of measures to ensure their sustainability and normal functioning in a changing climate. The global goals and objectives of sustainable development in Russia are analyzed from the perspective of the presence of a climate component in the relationship system (Bogdanovich and Lipka 2020). The expected impacts of climate change on the forests of Russia and the mechanisms of forest adaptation are considered for the use of the latter as a means of mitigating climate signals. This is part of the concept of transition to sustainable forest management in Russia, developed in the works (Shvidenko et al. 2014; Shvidenko et al. 2017). This concept assumes management and the use of forests in such a way as to maintain their biodiversity, sustainability (elasticity), and productivity, as well as their ability to perform ecological functions – in our case, mitigation. The concept considers various models of vegetation–climate feedback and describes a system of sequential actions for sustainable forest management: substantiation of regional indices of forest sustainability, creation of a new unified system of forest inventory and operational monitoring, biophysical and economic assessment of ecosystem functions and services of forest communities on a zonal–typological basis, etc. In order to further develop forest science as a basis for sustainable forest management, the program "Ecological and socio-economic threats of forest degradation in Russia and ways to prevent them" has been proposed (Lukina et al. 2015).

In general, a concrete solution to environmental problems to assess the role of forests in mitigating global warming has not yet been widely used. Consequently, there have been few studies on the potential for this mitigation through forestry (Shvidenko et al. 2017). In the summary of the European Forest Institute "Forests of Russia and climate change. What Science Has to Say" only questions are raised about sustainable and climatically optimized forest management and reforestation to mitigate the effects of climate change and develop a "closed-cycle forest bio-economy" (Leskinen et al. 2020). A rare exception is the detailed development of strategies for managing the forests of US National Parks in order to increase their ability to bind atmospheric carbon and adapt to the expected climatic changes (Ontl et al. 2020). We will discuss this work in more detail below.

A fluent (and very incomplete) analysis of Russian and foreign literature shows that each of the three triads of geo-ecological monitoring has already received independent scientific and methodological development, but the integration of all these results into a single operational monitoring system remains very problematic. The framework concept of a complete triad of geo-ecological monitoring, proclaimed more than 50 years ago, has not yet been implemented. The task remains to implement a full cycle of climate-genic monitoring of forests – "observation–control–management" – using the example of a specific ecoregion, with the development of a unified system for modeling the state of forest ecosystems in the past, present, and future, as well as the corresponding direct and feedback relationships between forests and climate.

A significant scientific and methodological breakthrough in this area should be noted: the search carried out in 2014–2020 under the Federal target program "Research and Development in priority areas of development of the scientific and technological complex of Russia" on the subject "Decision support for ecosystem services of forests in Europe – value determination, synergetic effects and compromises", within the international FP7 ERA – Net Sumforest-POLYFORES project. Applied and experimental research was carried out by a multidisciplinary creative team from various institutes of the Russian Academy of Sciences, focusing on biogeocenoses selected in the forests of Karelia, as well as the Nizhny Novgorod and Moscow regions (Lukina 2020; Shanin and Chertov 2020). Scientific and methodological approaches to assess and predict the compromises associated with climate change mitigation and biodiversity conservation at the national level have been developed.

The described studies of ecological services of forests are undoubtedly fundamental, but they do not constitute any competition to the project proposed by us. They are based on simulation (continual, dynamic) modeling of the natural environment (Sokolov and Bazilevich 1983; Schimal et al. 2001; Zavarzin 2007), which essentially does not go beyond the limits of autecology and synecology, leaving out of sight the geographical (spatial) ecology. These are models with concentrated parameters; they do not provide a territorial sweep of supra-paracelular landscape units (biogeocoenoses, stows, localities) in their organization and in response to external disturbances, and this is a certain limitation of simulation models. In our investigations, geosystem monitoring was carried out on the basis of discrete empirical and statistical models, according to Rozenberg (1984). These models allow for operating with a relatively small number of the most informative signs and obtaining results with a sufficiently high spatial resolution and, most importantly, predicting the behavior of forest biogeocoenoses in the form of integral formations, according to Sukachev (1972). Favorably differing from simulation models by the ability to more fully describe the properties of geographical space, empirical–statistical models have their own significant disadvantages, which are especially noticeable in predictive ecological analysis. The main issue is their "uncertainty in time", but they allow for a quick assessment of the state of ecosystems and their changes over large areas (Lischke et al. 1998). In particular, they can adequately reflect the processes of stabilizing selection as a reaction of biota to climatic disturbances exceeding the threshold of its adaptation.

Therefore, it is proposed to carry out geosystem monitoring in the volume of ecological and geographical research, as defined by I.P. Gerasimov (see above). It is necessary to carry out a complete cycle of climate-genic monitoring: "observation–control–management" on the example of a specific ecoregion, with the development of a unified system of empirical and statistical modeling of the state of forest natural complexes in the past, present, and future, and the identification of appropriate direct and inverse forest–

climate relationships. On this basis, it is necessary to reveal the ecological potential of the forests, ensuring the transition of forestry to an adaptive strategy. Such an ecological–geographical experiment of a local order was carried out in our study.

## 2.3 BIOCLIMATIC ZONALITY AS AN OBJECT OF ECOLOGICAL–GEOGRAPHICAL RESEARCH

The concept of geographic space developed along with the theoretical base and terminological apparatus of geography itself. Defined initially as some (Semenov-Tyan-Shansky 1928, p. 9), this concept was further deepened significantly. In modern interpretation, geographical space is a lot of Earth surface objects, consisting of individual elements, which have certain substrate properties and multichannel territorial connections – both internal and external (Gettner 1930; Gokhman et al. 1968; Rodoman 1999). Moreover, any medium transmitting a signal can serve as a communication channel for actions from factor to phenomenon (Puzachenko and Sankovsky 2005).

In the study of geospatial space, the central place is V.V. Dokuchaev (1948), who developed and founded by A.A. Grigoriev (1966), the concept of the integrity and inseparability of the geographic environment, which received the most complete knowledge in the theory of geographical zoning. This theory, describing a general planetary bioclimatic phenomenon (Köppen 1938), is closely related to landscape studies and ecology, reflecting the general trend of convergence between geography and ecology (Forman and Gordon 1986). The ecological approach allows for an expansion of the scope of an already traditional object of physical geography, such as bioclimatic zonality (Blüthgen 1973; D.L. Armand 1975; Gerasimov 1985; Bazilevich et al. 1986; Isakov et al. 1986; Glazovskaya 1992; Bailey 1995). The present report is aimed at rethinking the fundamental principles of the theory of geographical zonality based on empirical modeling and a systematic approach. The analysis is carried out using the example of the Russian (East-European) Plain, which has a very solid history of studying the zonal principles of the organization of geographical spaces in the temperate belt.

The scientific conceptions of natural zonality were developed by more than one generation of foreign and domestic naturalists, primarily in the field of plant geography, with the most important contributions made by A. Humboldt (1807), V.V. Alekhin (1951), and H. Walter (1968, 1976). However, the theory of complex (landscape) zonality became a complete concept in the well-known works of V.V. Dokuchaev only about 100 years ago. He considered geographical zonality as the highest form of interdependent functioning of the structural parts of the biosphere (Dokuchaev 1948). This conceptual proposition was further developed in the fundamental works of N. Savitsky (1927), V. Köppen (1938), L.S. Berg (1947), A.A. Grigor'ev (1966), V.R. Volobuev (1963), J. Blüthgen (1973), M.I. Budyko (1977) and F.N. Mil'kov (1986), concerning zonal landscape complexes, material and energy factors, and processes determining geographical zonality.

A general characteristic of natural zonality as an object of ecological and geographical research is given in our work as well (Kolomyts 2022). One should particularly note the close link between the subsequent elaboration of the theory of zonality as a global bioclimatic phenomenon and the development of landscape science and then geoecology, representing the general tendency toward convergence between geography and ecology. The structure and function of zonal types of landscapes and natural ecosystems, particularly the complex structure of phyto-biota and its productivity, seem to be manifestations of the higher organizational form of the biosphere.

Thus, the most optimal form of biota for the stable, interdependent functioning of the components of the biosphere is created (Gerasimov 1985). In the phenomenon of zonality, we see "an integral picture of as many different, but related, features of the natural environment as possible" (Hagget 1975, p. 151). We consider the phenomenon of zonality as an integral pattern of the greatest number of different but related characteristics of the natural environment. This phenomenal peculiarity of the zonal organizational form of the biosphere was emphasized and very figuratively stated by V.V. Dokuchaev (1949, p. 320): "You must not even expect the greater analogy, the greater parallelism, the greater connection and genetic affinity … and – even, so to say, worldwide co-assistance and love between separate elements and separate natural kingdoms".

According to the Dokuchaev paradigm, the zonal features of the organization are characteristic to one degree or another of all levels of the hierarchy of natural complexes (Solntsev 1973); however, they are manifested to the greatest extent in the types of geographical environment (Grigor'ev 1970). It was the zonal, first of all, type of landscape that at one time meant L.S. Berg, defining the landscape as "a community of a higher order,... in which all parts are connected to each other into an ordered whole" (1947 6, 8). Much later, this definition was filled with environmental content. Thus, with the characteristics of the main directions of geosystem monitoring, I.P. Gerasimov (1985) singled out the zonal categories of landscape complexes as the main types of natural ecosystems of the land, paying particular attention to the degree of isolation and balance of substances, as well as the speed of their internal circulation as the most important indicators of the general perfection of the organization and sustainability of ecosystems.

The modern theory of the zonal structure of the biosphere and its parts allows the coexistence of various forms of particular (component) and landscape zonality: flat (horizontal) and mountain (high), latitudinal and meridional, paragenetic and paradinical (Mil'kov 1986), and also expositional (including circulation, solar, and wind) and the so-called dislocation zonality – the form of the most complex combination of zone-forming factors (Gartsman 1977). We will consider those properties of zonal geospatial plains that are expressed by the territorial ordering of regional landscapes as in the "core of typicality" of the natural zone, according to D.L. Armand (1975), and on its periphery, including the zonal border itself.

The main goal of our regional research is to develop (by the example of the Russian Plain) a series of empirical mono- and polysystem models of landscapes and zonal geo-spaces and, on their basis, to reveal the mechanisms of spatial organization and stability of zonal–regional geosystems (Kolomyts 2022). This large territory, due to its natural peculiarities, has been a traditional object for the study of zonal forms of organization of the geographical environment and has become a basis for the fundamentals of the theory of worldwide land zonality.

The fundamental concepts of global ecology are based on the close relationship between natural zonality and the heat-to-moisture ratio within the same geographical belt (Grigor'ev 1966; Budyko 1977). In the Russian geographical literature, the most widely used is the well-known Vysotsky–Ivanov annual atmospheric humidity factor $F_{hum}(1)$. This coefficient is the ratio of the annual sum of precipitation $r_{year}$ to the annual evaporativity $E_0$ (potential evapotranspiration):

$$F_{hum}(1) = r_{year}/E_0. \quad (2.1)$$

Parameters $E_0$ and $F_{hum}(1)$ are formed mainly by the average July temperature ($t_{July}$). For the territory of the Russian (East-European) Plain, we obtained the following relationships, with high correlation ($R$) and determination ($R^2$) factors:

$$E_0 = 1384 - 161.6\, t_{July} + 6.245\, t_{July}^2;\ R = 0.93;\ R^2 = 0.87; \quad (2.2)$$

$$F_{hum}(1) = 12.09 - 0.9095V\, t_{July} + 0.01744\, t_{July}^2;\ R = 0.94;\ R^2 = 0.88. \quad (2.3)$$

The connection of the annual coefficient of humidification with the average January temperature ($t_{Jan}$) is non-significant. That means that during global warming, the main factor of future ecological reconstructions is the change in vegetation period temperature. Statistical analysis revealed highly significant correlations between the positions of zonal (sub-zonal) boundaries and the $F_{hum}(1)$ parameter (Kolomyts et al. 2012). For the western and eastern sectors of the Volga River basin, the boundary values of the annual atmospheric humidity factor are equal:

| | | | | | |
|---|---|---|---|---|---|
| **West** | 1.88 | 1.63 | 1.22 | 1.09 | 0.90 |
| Middle taiga | South taiga | Mixed forests | Broadleaf forests | Typical forest–steppe I | Northern steppe |
| **East** | 1.62 | 1.35 | 1.00 | 0.97 | 0.75 |

Covariance is ≤4%–6%, and only for the boundaries of the forest–steppe zone is it up to 10%–11%.

It is important to emphasize that this research covered only the structural–functional organization of natural (indigenous) plain landscapes in different zonal conditions of the temperate belt but did not touch upon the involvement of anthropogenic geo-components in this organization. The authors posed the problem of re-interpretation of one of the core aspects of traditional complex physical geography: the theory of geographical zonality based on empirical modeling and system analysis. The introduction of various attributes of anthropogenic transformation of geo(eco)systems into the models under study would complicate the data matrix to such an extent that the empirical models based on them would lose their main property: sufficient observability and representativeness for the solution of the posed problem.

## 2.4 SONAL STRUCTURE OF THE EAST-EUROPEAN SUB-CONTINENT

The system of natural zones of any plain territory is formed under the influence of two initial climatic factors: latitudinal zonality and longitudinal sectorality. The former factor is based on the radiation–thermal regime of the territory; the latter factor is formed by the planetary and regional mechanisms of atmospheric circulation. At the same time, the sectorial refraction of the zonal background throughout the continent can be so strong that "within a zone we have not the identity of landscape but only the similarity of some of their properties" (Polynov 1956, p. 50).

As is known (Alisov 1956; Isachenko 1985), the Russian (East-European) Plain is classified as a sub-oceanic sector – one of the transition sectors of Eurasia – dividing it into two major climatic sectors: oceanic (West European) and continental (East Siberian). The territory of the Russian Plain includes two geographical belts of the highest taxonomic rank: sub-Arctic and temperate (Figure 2.1). The temperate zone, in turn, is divided into two bioclimatic belts of a lower level: boreal and sub-boreal. The first of these includes the natural zones of taiga and mixed forests (sub-taiga), and the second includes zones of forest–steppe (with a sub-zone of broadleaf forests), steppe, and semi-desert (northern desert). The wide border strip between the boreal and sub-boreal zones, covering mainly the sub-taiga and broadleaf forests, is called the boreal ecotone (Kolomyts 2005).

This transitional sector is characterized by the typical narrowing and subsequent pinching out of two axial natural zones (sub-zones) of the boreal ecotone: mixed and broadleaf forests (see Figure 2.1), indicating a quite rapid increase in climate continentality from west to east. This increase is so strong that the very structure of geographical zonality changes from the Western Dvina and Dnepr river basins to the Pre-Urals. In the early 40s, A.A. Grigor'ev showed (1970) that the range of zonality in the western and, to a lesser extent, axial meridional strips of the Russian Plain is formed under conditions of strong refraction of solar radiation factors predetermining latitudinal zonality by circulation factors, which considerably "blur" this zonality by making the climate of all zones similar to a maritime climate.

The eastern meridional strip of the Russian Plain, beginning from approximately $48^0$ E, is comparatively weakly influenced by the Atlantic Ocean; therefore, here we have "the most complete exposure of the most typical features of the physical–geographical process of the temperate belt and the sub-Arctic in its most typical zonal modifications" (Grigor'ev 1970, p. 241). This fact is confirmed by a very close correlation between the boundaries of geobotanic zones and the energy barriers in this meridional strip, which are not so distinct in the western strip of the Russian Plain.

Thus, A.A. Grigor'ev brought to light the causal mechanisms of the superposition of latitudinal zonality and longitudinal sectorality on the Russian Plain, which underlies a rather complex system of geographical zonality of the temperate and sub-Arctic belts of this climate transition sector of the Eurasian continent.

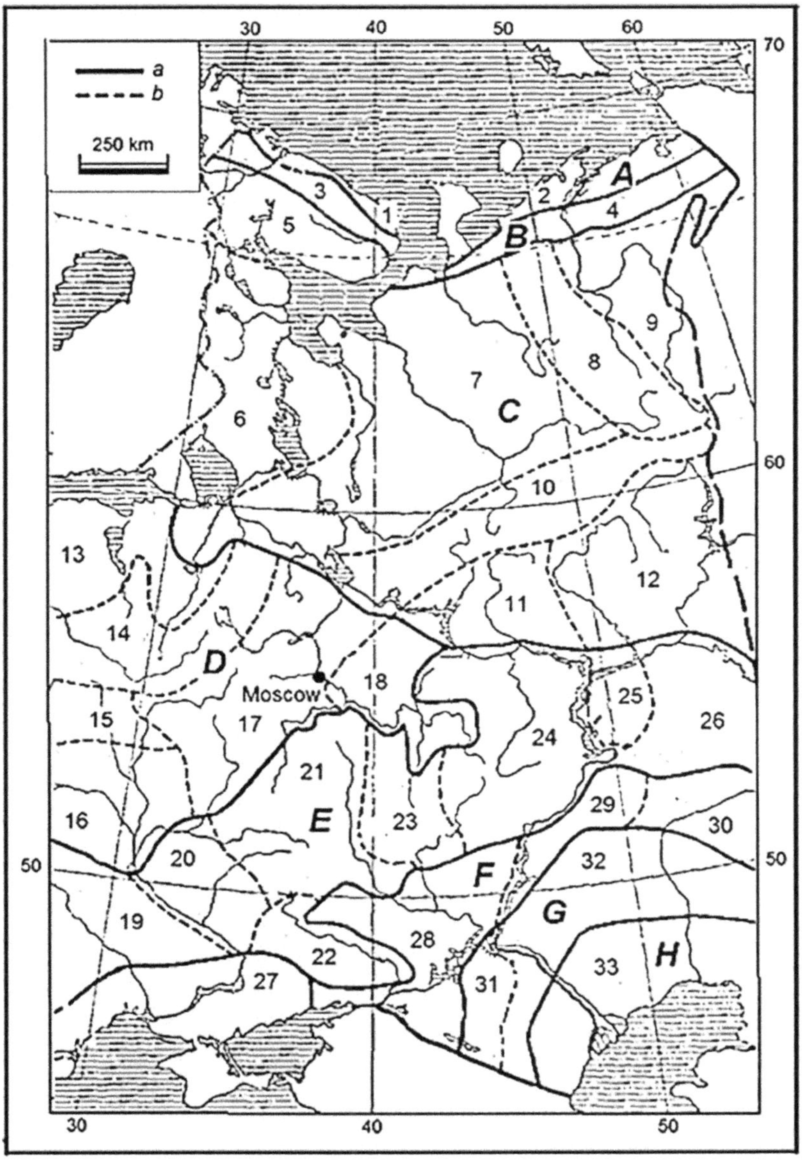


**FIGURE 2.1** Bioclimatic (plant) zones on the Russian (East-European) Plain, by Mil'kov and Gvozdetsky (1986). Zones: *A* – tundra; *B* – forest–tundra; *C* – taiga; D – mixed forests; *E* – forest–steppe; *F* – steppe; *G* – semi-desert; *H* – desert. 1÷33 – numbers of physical–geographical provinces. Boundaries: *a* – zones; *b* – provinces.

At the same time, judging by the outlines of zonal boundaries, the increase in continentality from west to east varies significantly between different natural zones of the Russian Plain. This contrast is especially distinct if we schematize the lines of zonal boundaries (see Figure 2.1), eliminating their sub-regional deviations caused by local lithogenic factors – e.g., according to the known regularities of the forward and reverse vertical differentiation of plain landscapes (Mil'kov 1981). It can be clearly seen that in the boreal belt, beginning from the middle taiga, zonal boundaries are actually sub-latitudinal, with the meridional vector of the main bioclimatic changes. In contrast, the zones and sub-zones of the sub-boreal belt are delineated by oval–concentric boundaries with a diagonal position vector directed from the southeast (from the regional core of continentality localized in central Kazakhstan and Central Asia) to the northwest. The *a* vector characterizes the direction and magnitude of the dominant gradients of landscape–geophysical fields causing their zonal differentiation.

The resulting scheme model of the natural zonality of the Russian Plain fits well into the general scheme of the temperate and sub-Arctic geographical belts and natural zones in the western half of the hypothetical (ideal) continent in the North Hemisphere (Figure 2.2). This diagram also clearly shows (1) the wedging out of sub-oceanic zonal types of landscapes (15, 22, 24) in the transitional zone of the temperate belts from the western margin into the interior of the continent; (2) a concentrically oval outline of the sub-boreal natural zones in the inland continental space of the temperate belt.

The boreal biogeographic ecotone of the Volga River basin is used as an example for consideration of theoretical and scientific–methodical problems of geographical zonality – a fundamental ecologo-geographical concept at the present stage of biosphere evolution. The crucial differences in the mechanisms of formation of zonal geospace have been revealed in the "core of typicalness" of the boreal belt of the Russian Plain and on its southern and northern periphery (boreal and sub-Arctic ecotones).

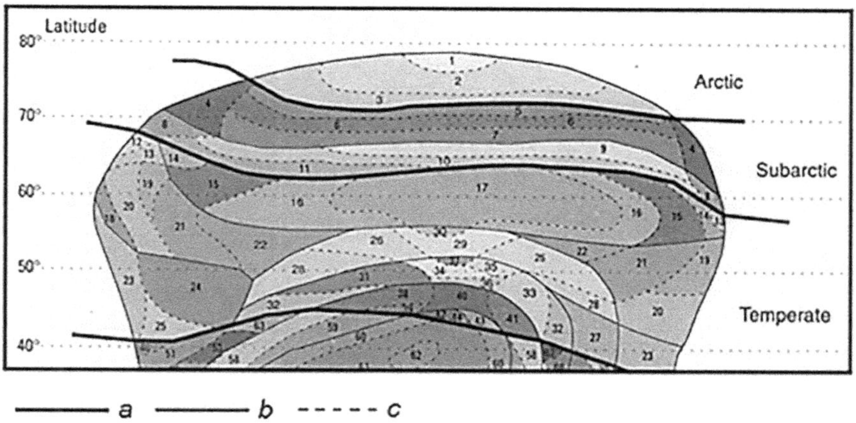


**FIGURE 2.2** Geographical belts, zones, and zonal types of landscapes on the hypothetical (ideal) continent in the North Hemisphere, fragment by Bailey (1995). Boundaries of: *a* – geographical belts; *b* – geographical zones; *c* – landscape–zonal types. 1, 2, 3, ... – numbers of landscape–zonal types. The zonal landscape types 12–17, 21, and 22 belong to the boreal belt, and types 24–41 belong to the sub-boreal belt.

# Boreal Ecotone and Geography Zonality 3

## 3.1 THE MAIN PROVISIONS FOR THE THEORY OF GEOGRAPHICAL ECOTONES

The solution of multipurpose tasks of ecological forecasting may depend to a great extent on the results of system analysis of natural-territorial structures, which are most sensitive to external effects, including anthropogenic ones. The scientific search in this direction focuses more and more attention on the *natural boundaries* – both individual and complex – where the most significant natural or anthropogenic shifts in the structure and function of geo(eco)systems are observed. Considering one or another natural boundary as a vector (connection, cascade, paradynamic, etc.) landscape system (Semenov-Tyan-Shansky 1928; Glazovskaya 1964; Chorly and Kennedy 1971; Armand and Kaydanova 1976; Sochava 1979; Mil'kov 1981) with a clearly defined spatial polarization of its different properties, we obtain a "fast-flowing" model of state response and resistance of geo(eco)systems to the action of certain ecological factors.

The systematic approach to the study of natural environment organization would be expediently realized by the example of marked and large enough biogeographical and landscape borders – *ecotones*, which are the zones of transition between contrasting natural environments. These are the most dynamic structural–functional subdivisions of the biosphere, with enhanced "tension" of matter–energy interactions and with the most pronounced manifestation of new evolutionary tendencies in the environment (Odum 1971; Sochava 1979; Zaletaev 1984; Kolomyts 2005). Landscape contents are a critically important element of the wide range of modern notions of ecotones. As a result, a concept of landscape ecotone, or geo-ecotone, has been developed (Preobrazhensky 1986).

The lateral fluxes and the respective inter-complex connections along geo-ecotones create the spatiotemporal order of natural complexes at all hierarchical levels: from continental or oceanic sectors to landscape facies (Mil'kov 1981). Longitudinal climatic sectors or morphotectonic belts, on the one hand, and soil–geochemical catenas with the respective phytocoenotic spectra, on the other hand, comprise the two extreme levels of the hierarchy of geographical ecotones.

At the same time, a simple identification of a geo-ecotone with any natural boundary is unlikely to have meaningful significance. Based on the initial meaning of the

DOI: 10.1201/9781003733119-3

concept of ecotone (Greek *oikos* – house, *tonos* – tension), the geographical ecotone should be understood only as a border zone, strip, or line through which the interaction between neighboring territories is carried out. Geo-ecotones are characterized by increased horizontal landscape connections that unite these territories into vector landscape structures. A geo-ecotone can be thought of as a spatial alternative to a geo-ecotope – a type of location of any taxonomic rank, the properties of which serve as guiding features in identifying genetically or typologically homogeneous geosystems of a given hierarchical level. D.L. Armand (1975) called geo-ecotopes "nuclei of typical" landscapes. In them, the geo-field gradients are minimal, and therefore the density of horizontal bonds is the smallest.

The higher the functional integrity of the paragenetic (connection, cascade, etc.) geosystem, the more closely its individual links are interconnected; the faster and more anthropogenic impact propagates in the direction of connecting geo-processes. On the other hand, the more intense the flow, the more pronounced the spatial structure of such a lateral system. Consequently, the study of geo-ecotonic structures of different taxonomic levels creates the necessary prerequisites for geosystem monitoring of vast territories. The velocities of geo-points are generally determined by the degree of heterogeneity of the interacting natural environments and the corresponding gradients of the energy fields. For this reason, the principle of contrast at all hierarchical levels, in contrast to the principle of homogeneity, which corresponds to the traditional approach to the study of natural interactions and regionalization of territories, should become the methodological basis for the identification and analysis of such geosystems.

One more important circumstance of a methodological nature should be noted. In contrast to homogeneous geosystems, vector or "synergistic" natural complexes are closest to technical systems in their structural features and functioning methods; therefore, their understanding, in the first approximation, can be reduced to the study of relatively simple processes of material–energy metabolism, which are considered from an engineering (in the broad sense) point of view (Preobrazhensky 1978). This allows for the analysis of vector natural-territorial structures to make wider use of modeling methods developed for technical objects. An example is the experience of using the conceptual apparatus of reliability theory to calculate partial measures of the stability of a geosystem (Grodzinsky 1987). Stability is evaluated by the behavior of a natural component and is associated with the concept of failure – events where the state of this component leaves the area of homeostasis.

Ecotonal biogeosystems are the most dynamic structural–functional units of the biosphere, characterized by increased intensity of matter–energy and clear manifestations of new evolutionary tendencies in the environment (Odum 1971; Ricklefs 1976; Sochava 1979; Kolomyts 2005). Lateral fluxes and corresponding landscape connections in geo-ecotones account for the spatiotemporal ordering of natural complexes at all hierarchical levels, from a continental or oceanic sector to a biogeocoenosis (Hansen and di Castry 1992; Zaletaev 1997).

The overwhelming majority of studies, both in Russia and abroad, focus on local phytoecotones (Odum 1971; Armand 1992; Solov'eva and Rozenberg 2006) due to issues of discreteness and continuum of vegetation cover, biodiversity, as well as the distribution of species and communities along the gradients of abiotic factors and at the interface between contrasting natural environments. Various ecotone classification schemes developed in this context were considered in the work (Solov'eva and Rozenberg 2006). It is also known that studies of ecotones of higher taxonomic rank, e.g., zonal geo-ecotones and oro-ecotones, have been conducted (Ricklefs 1976; Mil'kov 1981; Kolomyts 2005; di Castri and Hansen 1992; Bailey 1995; Zaletaev 1997).

The challenge presented in this book is the study of a more complex ecotonal object: the hierarchical system of geo-ecotones at the regional and topological levels under different morphostructural and macroclimatic conditions in both flat and mountainous areas. This represents a new stage in the development of the theory and methods of ecotone research.

The central place in the global theory of zonality is occupied by the conception of climatic dependence of the main zonal types of plant and soil covers. To date, this situation has been embodied in the already traditional generalized concept of bioclimatic structural patterns of the landscape geosphere and its parts, accompanied by a productive ecological interpretation of this concept (Walter 1968; Sukachev 1975; Whittaker 1975; Sochava 1979; Bazilevich et al. 1986; Isakov et al. 1986). Some general properties of the bioclimatic system – modern climatic areas of plant species and whole communities, hydrothermal indications of palynological (spores and pollen) spectra – are widely used for global and super-regional paleogeographical reconstructions (Velichko et al. 1983; Velichko and Klimanov 1990) and ecological–geographical prognoses (Tchebakova et al. 2009; Kolomyts 2005; Kolomyts 2003; Velichko 2010). At the same time, the use of global bioclimatic connections becomes ineffective at the regional level, where researchers face an unusually high discreteness of geographical space and significant territorial diversity of geo(eco)systems. These connections lack the necessary spatial resolution and therefore fail to reflect the diversity of regional responses to global climatic signals. It would be advisable to develop a system of bioclimatic connections with more detailed parameters of climatic niches of vegetation and soils, which could serve as a basis for paleo-prognostic constructions.

We consider the regional bioclimatic system as an ordered combination of spatial connections of the soil–phytocoenological cover with landscape–geophysical factors. Elements of the system are climatic niches of phytobiotic and soil units, distributed by hydrothermal gradients and representing a hyperspatial aspect of ecological niche, according to Odum (1971) and Svirezhev (1977). The regional bioclimatic system determines the boundary conditions for the formation of phytocoenotic units of the topological level (associations, their groups, and classes) under the refractive influence of local geomorphological and edaphic factors.

Geo-ecotones, as the most sensitive fragments of the natural-territorial mosaic to external signals, should serve as primary objects of study of the human impact on the environment. The impetus for the transformation of geo(eco)systems and their transition to a qualitatively new state is the aggravation, first of all, of inter-complex (horizontal) landscape connections (Zlotin 1987), which are most characteristic of geographic ecotones (Kolomyts 1987). The activation of lateral natural interactions during periods of violation of the previously achieved (to one degree or another) stability of natural ecosystems under the influence of new climatic signals is the primary cause and driving force of those landscape–ecological rearrangements that occurred in the past and which are expected in the future. Thus, lateral relations and connections become the subject of an independent paleo-predictive study in geo-ecotones; the parameters of these links should be directly included in the corresponding calculation models.

The problems of studying geo-ecotones are closely related to the development and degradation of landscapes, the driving forces and rates of the evolutionary process, taxonomic and age relationships of the structure and functioning of geosystems, their ontogeny, and phylogenetic changes, and finally, the regularities of the formation of landscape connections on land. In essence, those scientific and methodological problems that relate to the very origins of the natural-territorial organization of the continental part of the geographic shell of the Earth should be considered.

## 3.2 BOREAL ECOTONE OF THE EAST-EUROPEAN PLAIN AS A LANDSCAPE–ECOLOGICAL SYSTEM

In Northern Eurasia, the largest bioclimatic and landscape–zonal transition zone is a broad boundary frontier between the boreal (mainly taiga-forest) and sub-boreal (forest–steppe and steppe) belts of plant formations. The frontier stretches from the Baltic Sea to the East Sayan and Lake Baikal (Bazilevich et al. 1986) and then, after a break, to Inner Mongolia. This *transcontinental boreal ecotone* is a vector (connection) geosystem of the highest, or belt, rank. The objects of our study were coniferous–forest and coniferous–broadleaf natural complexes close to the zonal boundary between sub-taiga and southern forest–steppe in the center of the East-European (Russian) Plain (Figures 3.1 and 3.2). Being the *southern outpost of the boreal belt*, these geo(eco)systems are in conditions close to critical and therefore very unstable, which advances the problem of preservation and reproduction of forest resources of this territory (industrial "core" of European Russia) among the ecological problems of high priority.

As shown by hydrothermal ordination of the zonal types of landscapes (Bazilevich et al. 1986; Isachenko 1988), northern (meadow) steppe, forest–steppe, broadleaf, mixed, and south-taiga forests coexist on the territory of the boreal ecotone in rather narrow ranges of biologically active accumulated temperatures (1,700°–2,300°), annual precipitation (600–750 mm), and Vysotskyv–Ivanov's annual atmospheric humidity factor (0.7–1.0, see above). Under these conditions, the frontier forest ecosystems, with significant mutual overlapping of their climatic areals (Bazilevich et al. 1986), inevitably acquire the features of trigger systems with hysteresis properties, according to the definition (Armand and Vedyushkin 1989). They are in near-critical states, determining their enhanced susceptibility to the influence of the global climatic system. Special investigations have shown (Armand and Kaydanova 1976) that the vast hysteresis belt, beginning close to the southern boundary of the forest zone on the Russian Plain, encompasses forest–steppe and northern steppe and is characterized by a set of metastable states of natural including forest ecosystems. Within this "hysteresis loop", even an insignificant deterioration of forest site climatic conditions can induce irreversible degradation of forest formations.

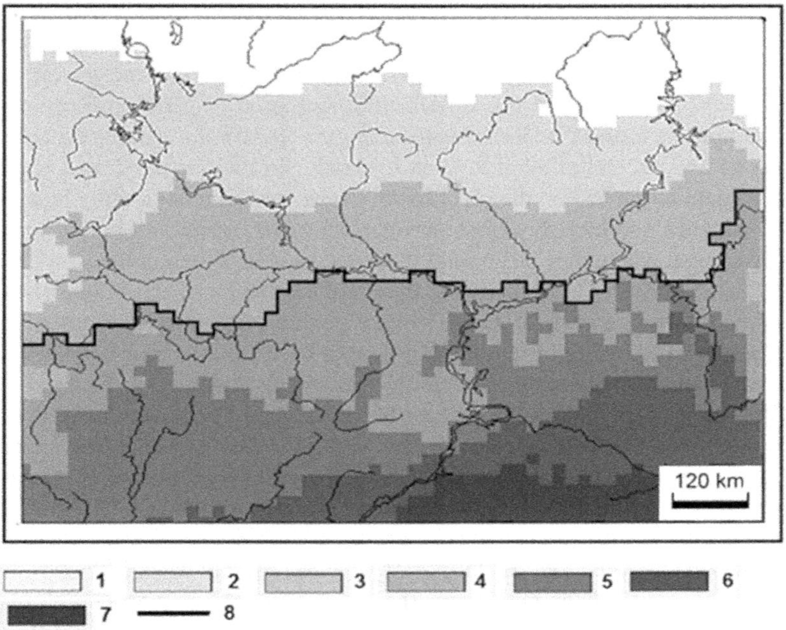


**FIGURE 3.1** Natural zonality of the East-European Plain – Boreal ecotone. Nature zones/sub-zones: 1 – middle taiga; 2 – south taiga; 3 – mixed forests; 4 – broadleaf forests; 5 – typical and south forest–steppe; 6 – north steppe; 7 – south steppe. 8 – Main Landscape Border of the East-European Plain, by Mil'kov (1981).

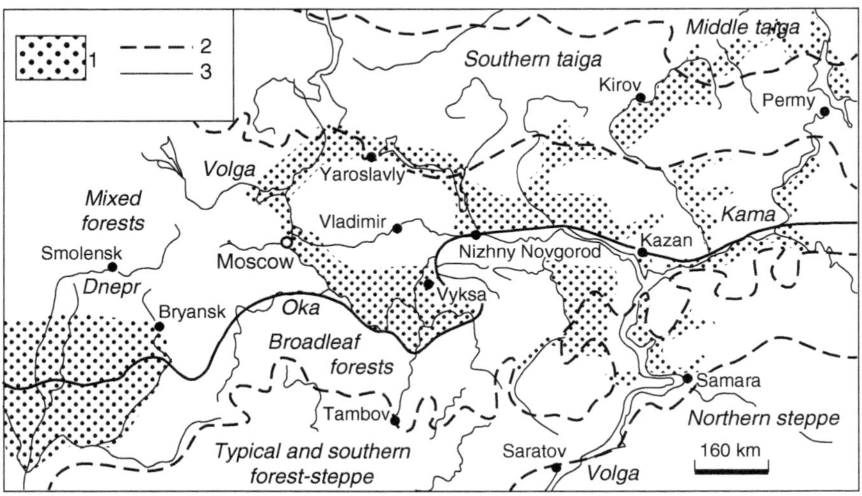


**FIGURE 3.2** Landscape–zonal structure of Volga River basin headwaters and its encirclement. 1 – forested outwash lowland; 2 – boundaries of nature zones and sub-zones; 3 – Main Landscape Border of the East-European Plain, by Mil'kov (1981).

This is enabled by the forefront factors of landscape lithogenic basis – relief and composition of soil-forming rocks. The lithogenic basis brings about the territorial interpenetration of frontier geo(eco)systems, thus creating a more mosaic character of the trigger belt. As a result of the vertical differentiation of plain landscapes, a two-member complex is formed, underlining the territorial polarization of the ecotone: sandy fluvio-glacial and ancient alluvial low-cut lowlands, on the one hand, and high loamy erosion–denudation plains, on the other hand. Boreal ecosystems (mainly pine-forested lowlands) advance southwards across outwash lowlands, while sub-boreal (oak grove and forest–steppe, where they still remain) advance across loamy uplands.

The East-European Plain, including the main catchment of the Volga River basin, has a rather pronounced "polarization" of its geomorphological skeleton (Kolomyts 2022). The morph structure of modern landscapes here finally formed in the Neogene, when, as a result of differentiated tectonic movements against the background of a general elevation of the Paleogene surface of the Russian Plain alignment, the Middle Russian and Volga Uplands arose, and the Verkhne–Volzhsky elevated ridges were formed. Active exogenous processes divided the increases into a series of reservoir elevations and ridges, while the depressions were deepened and widened (Zonn et al. 1967; Dedkov 1972). In the Pleistocene, due to the accumulation of thick (up to 150–200 m and more) strata of glacial and fluvio-glacial deposits, mainly in the lowlands, and further denudation of reservoir elevations, the main elements of morphosculpture took shape, emphasizing the vector ("polarity") of the modern geomorphological framework of the territory and thereby predetermining the system of regional and local landscape conjugations (watershed–slope–valley).

Being inherited from the geological past, these morphostructural contrasts are also supported by modern tectonic movements throughout the sub-latitudinal strip of the Oka–Volga–Kama river valley system. The northern left-bank low-lying taiga forest areas descend at a rate of up to 1.5–2.0 mm/year, while the forest–steppe and steppe regions of the elevated right bank tend to rise (up to 2.0–2.5 mm/year). Thus, the indicated system of river valleys undergoes a sharp inversion change in tectonic processes, and this change is reflected in the depth of groundwater, which characterizes the degree of drainage of the territory. As a result, the regimes of soil and soil waterlogging are maintained throughout the low-lying forest of the Oka–Volga left bank, on the one hand, and natural drainage is intensified, with erosive dissection of the surface areas of the elevated forest–steppe right bank, thereby increasing their aforementioned "predisposition" to anthropogenic warming, on the other hand.

Thus, the cardinal bioclimatic changes on the boreal ecotone of the Russian Plain are accompanied by the drawing together of several component boundaries; their combination resulted in distinguishing the *Main Landscape Border of the Russian Plain* (Mil'kov 1981), which is the landscape "core" of the boreal ecotone (see Figures 3.2 and 3.3). This boundary passes through the valleys of the Oka–Volga–Kama rivers and separates the zone of mixed forests from the sub-zone of broadleaf forests of the forest–steppe zone (Figure 2.3). Here it is actually the landscape "core" of the boreal ecotone. Ecosystems of the sub-taiga coniferous forest and broadleaf forests, being the southern outpost of the boreal belt, are in conditions close to critical and therefore very unstable, which advances the problem of preservation and reproduction of forest resources in this territory.

**FIGURE 3.3** Fragment of the Main Landscape Border of the East-European Plain. Oka River Valley in the By-Oka-Terrace Nature Reserve Area. In the foreground (right bank of the Oka River) is broadleaf and small-leaf forest; in the background is a mixed forest of the By-Oka-Terrace Reserve.

The system of chain reactions in inter-component interactions associated with the transition from north to south through the main landscape boundary of the Russian Plain is shown in Figure 3.4, which does not require special explanation. It is only worth noting the main point in these reactions: the change of some landscape–zonal systems by others on the boreal ecotone is not a purely climatic phenomenon; it occurs as a result of a rather diverse oro-climatic interaction, where hydrothermal and litho-dynamic flows are closely interconnected, and background hydro-climatic conditions and topography with surface deposits turn out to be equally powerful environmental factors. Therefore, it seems inappropriate to highlight the priority role of any of them on a regional scale.

Ecotone is known to be quite an independent subunit of the biosphere, often differing from adjacent homogeneous systems in a high complexity of coenotic structure and higher bioproductivity. Such structure–function features are fully characteristic of the boreal ecotone (Bazilevich et al. 1986). One should stress the interchange of the roles of warmth and humidity as limiting factors of bioproductivity on the ecotone. The increase in the production of zonal plant formations in the boreal belt is due mainly to the increase in warmth, and in the sub-boreal one – to the increase in humidity. In the first case, the growth of humidification results in a lower productivity of ecosystems, while in the second case, on the contrary, it results in an increase.

Consequently, in the transition zone, there must be a reconstruction of the principal scheme of territorial distribution of zonal (flat interfluve) and intra-zonal natural complexes at the local level. Therefore, the study of the topological structure of the boreal ecotone is of great importance for a differentiated assessment and forecast of the resistance of geo(eco)systems to natural and anthropogenic factors.

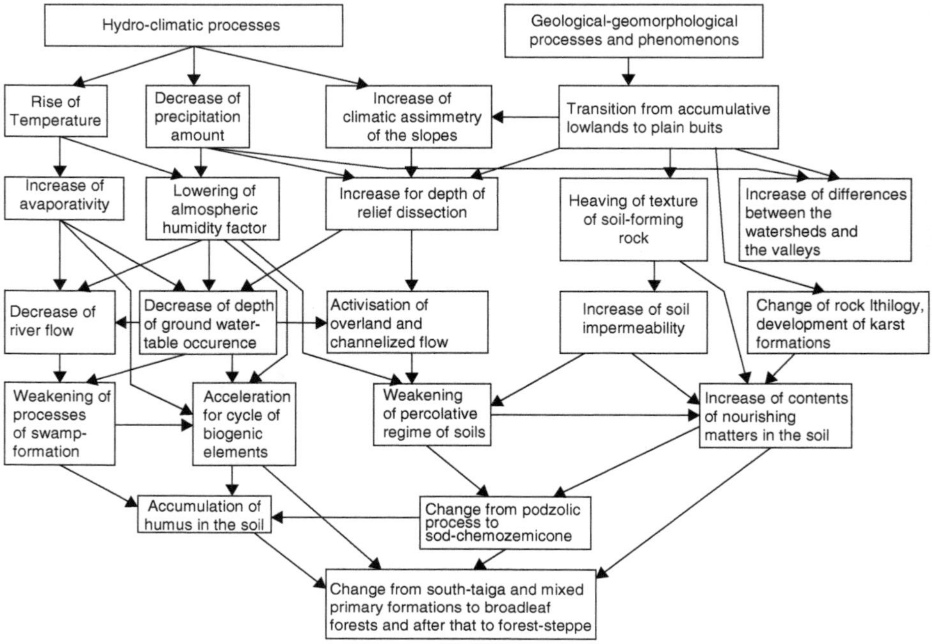


**FIGURE 3.4** System of chain reactions in geo-component correlation under the crossing through the Main Landscape Border of the East-European Plain.

On the boreal ecotone, the restructuring of landscape, phytocoenological, and soil areas begins, and new evolutionary trends in the natural environment arise here. Changes in the ratio of heat and moisture in this transitional band, caused by background climatic shifts, can entail a significant environmental change – a change in the sign of the relationship between the structure of natural ecosystems and their productivity with the main landscape and geophysical parameters. Figuratively speaking, considering the behavior of landscape–zonal systems on the boreal ecotone in the past, present, and future, the researcher has in hand a "fast-flowing" model of the regional manifestation of global changes in the natural environment.

## 3.3 THE BOREAL ECOTONE AS A NATURAL "FAST-FLOWING" MODEL OF GLOBAL CHANGE

As already mentioned, the boreal ecotone is determined first of all by the most important climatic border – the transfer of the warmth-to-humidity ratio across 1, which radically changes the energy potential of main natural processes: weathering and soil formation, transformation and migration of substances, formation, and development of biocoenoses (Glazovskaya 1992). Within broadleaf forests and a typical forest–steppe, a spectrum of contours of the annual radiation balance of 1450–1850 MJ/m$^2$ (35–45 kilocal/cm$^2$) passes. It serves as an important boundary dividing two thermal zones: northern (boreal) with insufficient heat supply, with sufficient and excessive moisture, and southern (sub-boreal), with sufficient heat supply but with a lack of moisture. Especially, it should be emphasized the change in the boreal ecotone of the roles of heat and moisture as background limiting factors of the structure and productivity of phytocoenoses. In the boreal zone, an increase in atmospheric moisture leads to a decrease in the productivity of ecosystems; in the sub-boreal zone, on the contrary, it leads to an increase (Bazilevich et al. 1986).

Because of this, the basic watershed of the Volga River basin must show a rather high sensitivity to global climatic changes. As far as is known, the ecotones are the most sensitive and fragile parts of the biosphere in reference to external influences. The dynamics of natural zones and zonal boundaries in the Holocene and historical time in the Middle Belt of the Russian Plain illustrate a rather high sensitivity of the vegetation cover and landscapes of this territory to climate fluctuations and anthropogenic influence (Budyko 1980; Gerasimov and Velichko 1982; Khotinsky and Savina 1985; Klimanov et al. 1995).

The overlapping of potential ecological niches of zonal vegetation types on the boreal ecotone causes inevitable competition between plant species and communities (Sukache 1972; Bazilevich et al. 1986). As a result, each zonal boundary becomes a "trigger belt – hysteresis loop", where both members of the competing pair of zonal communities may be quite stable. This is enabled by the forefront factors of the landscape lithogenic basis – relief and composition of soil-forming rocks. The lithogenic basis brings about the territorial interpenetration of frontier geo(eco)systems, thus creating a more mosaic character of the trigger belt. As a result of the vertical differentiation of plain landscapes, a two-member complex is formed, underlining the territorial polarization of the ecotone: sandy fluvioglacial and ancient alluvial low-cut lowlands (e.g., Pripyatsko-Desninskaya, Oksko-Moshinskaya, Nizmennoye Zavolzhye) on the one hand, and high loamy erosion–denudation plains (Central Russian Upland, Privolzhskaya, Vysokoye Zavolzhye) on the other hand. Boreal ecosystems (mainly pine-forested lowlands) advance southwards across outwash lowlands (see Figure 3.2), while sub-boreal (oak grove and forest–steppe, where they remain) advance across loamy uplands.

The increased sensitivity of the boreal ecotone to external influences is caused, first of all, by the very narrow framework of the ecological space within which a relatively quick and sharp change of natural zones occurs. As shown by the hydrothermal ordination of zonal types of landscapes (Isachenko 1988), within the temperate zone, steppe, forest–steppe, broadleaf, mixed, and coniferous forests coexist in the ranges of sums of biologically active temperatures of 1500–2000 and the Vysotsky–Ivanov's annual atmospheric humidity factor (see Chapter 5) of 0.7–1.0, i.e., in rather narrow ranges of hydrothermal space.

Hence, the inevitable mutual overlap of the climatic niches of most zonal systems on the boreal ecotone. In terms of temperature sums, this overlap is usually the smallest, which indicates the thermal regime as the leading spatially differentiating factor, while niches overlap much more in annual precipitation. Therefore, changes in the zonal structure of the boreal ecotone should be associated primarily with background thermal signals.

Thus, the boreal ecotone of the Russian Plain may be referred to the category of critical ecological zones, according to the classification (Glazovskaya et al. 1993); therefore, it is a rather favorable model territory for scientific research into a number of key ecological and geographical problems, including the problem of regional–local response to global climatic changes. Besides, it seems very promising to develop, based on this territory, the topological and regional aspects of the new global ecological concept of stability of the continental biosphere (Gorshkov 1994; Utkin 1995; Arsky et al. 1997; Kotlyakov 1997): the concept of *biotic regulation of terrestrial ecosystems* and the determinative (landscape–geophysical, hydrological, geochemical, etc.) natural processes. Further development of this concept is associated, in our opinion, with the study of local–regional regularities of the small biological cycle in natural ecosystems providing environmental stability in accordance with Le Chatelier's principle (Gorshkov 1994). As far as we know, the above concept has not yet been considered in the given statement of the problem, whereas the primary steps of practical implementation of the measures proposed in the concept and expert evaluation of their efficiency should be expected at the local and regional levels of the biosphere.

This problem has a significant practical aspect as well. Ecological safety of large territorial subunits of the continental biosphere significantly depends on the state of the zonal–regional types of natural ecosystems, first of all forest cover, and on the interrelations primarily of two competitive plant formations: forest and grassy (Figure 3.5). This determines the importance of making the boreal ecotone as a sub-planetary boundary between forest and steppe an object of regional geo-ecological investigation. Therefore,

**FIGURE 3.5** Southern strip of the mixed forest zone. Bison in dark coniferous–broadleaf forest of the By-Oka-Terrace Reserve (photo by the author).

the problem of maintaining forest ecosystems and reproducing forest resources on the southern boundary of the temperate forest zone, where forest communities are present in states close to critical, is among the fundamental ecological problems. It has always been of high priority for East-European countries, where the wide transitional zone from forest to steppe, i.e., the zonal forest–steppe ecotone, is the industrial and demographic core of this large region. At the moment, the questions of stability of natural ecosystems as a natural–historical basis for the stable development of the region are gaining more and more significance in the solution of this problem.

The boreal biogeographic ecotone of the Volga River basin is used as an example for consideration of theoretical and scientific–methodical problems of geographical zonality – a fundamental ecological–geographical conception at the present-day stage of biosphere evolution associated with the global anthropogenic impact on the climate.

## 3.4 FUNCTIONING OF GEO(ECO)SYSTEMS AS AN OBJECT OF THE ECOLOGICAL MONITORING

In monitoring the environmental effects of climate change, an important place should be given to monitoring and predicting the *functional state of natural ecosystems* – discrete parameters of the small biological cycle that determine the role of ecosystems in the *biotic regulation of the carbon cycle* and in the corresponding interactive effects of soil–plant cover on climatic processes. The minor biological turnover in forest ecosystems is known (Houghton et al. 1996; Isaev and Korovin 1999; Tarko 2005) as one of the mechanisms that provide stability to the natural environment according to Le Chatelier's principle. In respect to the carbon cycle in the biosphere, this principle is expressed by the following postulate: the state of the environment will be stable if any spontaneous increase in $CO_2$ content in the atmosphere is accompanied by an equal enhancement of carbon utilization by terrestrial and ocean biota. An apparent violation of Le Chatelier's principle is the transformation of biota from a carbon sink into a carbon source, i.e., when biota release carbon dioxide into the atmosphere under external influence.

In modeling, high emphasis was placed on the phytocoenotyic functional block – the parameters of annual and perennial cycles of plant matter turnover, which are especially important for prognostic ecological investigation. Such turnovers are the most important indicators of ecosystem resistance to external disturbance (Odum 1971; Gerasimov 1985). A multidimensional ecological analysis of various phytomass categories (live and dead, foliage and skeletal, on separate biogeohorizons, etc.) was performed, and the integral indices of mass exchange were determined.

From the standpoint of geographical and functional ecology (Odum 1983; Gerasimov 1985), the mechanisms of biotic regulation of the carbon cycle under global climatic changes can be described by the dynamics of internal turnover and balance of plant matter in ecosystems. We have performed empirical statistical modeling of functional parameters of topogeosystems based on the *bio-chorological approach* proposed in due time by Timofeev-Resovsky (1970): the principle of the spatial organization by living matter of the matter and energy cycle in biogeocoenoses as discrete elementary structural units of the biosphere.

The solution to the complex of ecological problems induced by forthcoming climatic changes is associated in many respects with the study of the causative mechanisms of the carbon cycle, which, firstly, drives all other biogeochemical cycles and, secondly, determines the content of greenhouse gases in the atmosphere and thus has a regulating effect on global climatic processes (Kobak 1988; Bonan et al. 1992; Printice et al. 1992; Gorshkov 1994; Houghton et al. 1996; Kondrat'ev et al. 2003; Tarko 2005). Further development of the concept of biotic regulation is connected with the study of local–regional regularities of the minor biological turnover, which provides stability to the natural environment in accordance with Le Chatelier's principle. Spatial diversity of the biological cycles is created by the structural heterogeneity of biogeocoenoses. As far as we know, the concept of biotic regulation of the carbon cycle in the biosphere has not yet been developed in such a statement of the problem.

The ways to solve this problem were outlined in the course of previous regional and local environmental forecast studies based on a landscape–ecological approach (see Chapters 10 and 11). Let us remind you that the dominant transmission channel of connections between regional and especially local geo(eco)systems with the background climatic system passes through the summer moisture content of soil, which is, on the one hand, quite a reliable geophysical indicator of the state of natural complexes and, on the other hand, the most powerful ecological factor that predetermines their territorial organization. It has also been revealed that the main factor of climate-genic transformation of natural ecosystems in the region is the hydrothermal conditions of the vegetation period. These are the initial ideological positions for functional ordination analysis of forest ecosystems.

It has also been accepted that shifts in the biogeocoenosis biological exchange (phytomass production and decomposition), which take the first several years in the boreal zone and less than a year in broadleaf forests, correspond to the temporal step of changes in individual links of the carbon cycle in forest phytomasses and the mobile humus of the soil. These typical times of functional relaxation approximately correspond to the carbon cycle duration in forest (live and dead) phytomasses and in mobile soil humus.

The problems of predictive dynamics of forest ecosystems under changing climate have been developed quite poorly so far. The known simulation models of forest responses to climatic impacts (Bonan et al. 1992; Kellomäki 1995; Komarov et al. 2003; Tarko 2005) meet the narrow preset framework of habitat conditions; therefore, the results of such modeling are insufficient for predicting the state of the forest community as a whole (Shugart et al. 1989) and do not cover the spatial diversity of succession changes under the same background influence. The local landscape–ecological prediction, which we have developed, is based on the construction of discrete empirical–statistical models of natural ecosystems. These models make it possible to operate with a small number of the most informative characters and to obtain results that are probably not as accurate as, for instance, in simulation modeling but with a higher level of territorial generalizations.

## 3.5 ENVIRONMENTALLY SUSTAINABILITY

The notion of sustainability (in other words, "sustainable development") does not have a clear definition, but it has already undergone various discussions among specialists in various fields and is widely used in the scientific literature (Kondrat'ev et al. 2003). In general terms, this concept refers to the "nature–society" system. In our study, we will consider the first part of the NSS – directly the state of the natural environment, with an analysis, however, of the system-forming influence of not only natural but also anthropogenic factors. Thus, we will discuss "environmental sustainability", in the interpretation of K.S. Losev (2001).

> According to the ecological approach, sustainable development is such a development that does not take the system beyond the limits of the economic capacity of the biosphere. It does not cause destruction or degradation processes in the biosphere, the result of which may be the emergence of conditions that are fundamentally unacceptable for humans.
>
> *(Danilov-Danilyan and Losev 2000, p. 43)*

The main goals of the sustainable development of the biosphere are to ensure its ecological functions, among which the preservation of biodiversity and the productivity of the vegetation cover – two of the most striking indicators of anthropogenic "desertification" of degraded areas – are of paramount importance (Glazovsky 2009).

The modern biosphere is a highly correlated, self-regulating system that ensures the stability of its constituent elements, and if that stability is violated, it "extinguishes" the violation or begins to collapse itself (Lovelock 1989), leading to an ecological crisis. One of these global crises is currently brewing due to negative climate change, which is caused by anthropogenic impacts on the chemical composition of the atmosphere, in particular on the concentration of carbon dioxide (Budyko 1980).

As the history of the development of the biosphere shows (Vernadsky 1994), life has been a powerful factor that shapes the environment – the lithosphere, atmosphere, ocean, and soil. Therefore, biotic regulation and stabilization of the environment serve as the main mechanisms for maintaining the stability of the biosphere and its parts, preserving the habitability of the environment in conditions of increasing human impact on nature (Gorshkov 1994; Steffen and Tyson 2001). The stability, vitality, and integrity of the biosphere and its elements ensure the ecological safety of humanity itself (Kondrat'ev et al. 2003). According to the decisions of the United Nations Conference on Environment and Development (in Rio de Janeiro 1992), the sustainable development strategy of open-source software is to preserve the natural resource potential of ecosystems, primarily their biotic components, to meet the vital needs of present and future generations (Koptyug 1994).

## 3.6 EMPIRICAL SIMULATION OF THE DEVELOPMENT OF ADAPTIVE STRUCTURES IN PLANT FORMATIONS

The presented empirical data and conclusions make it possible to estimate the *degree of development of adaptive structures* of the soil–phytocoenotic core of natural complexes in the Volga River basin in terms of typification of their modern climatic niches. Based on the known thesis of R. Whittaker (1975), as well as the propositions of modern phytocoenology (Mirkin et al. 1989), it may be supposed that natural ecosystems evolve toward differentiation of their climatic niches, increases in their volume, and spread of their ecological optimums (adaptive centers) in the niche space. All these ecological processes take place against the backdrop of the growth of geo-component coupling in the respective natural complexes and seem to be determined by an increase in the level of monosystem landscape organization. Both groups of processes lead to a total increase in the resistance of geo(eco)systems to external effects, including climatic changes. In this case, evolution is interpreted as a form of adaptation.

Thus, climatic niches with a distinct ecological optimum (i.e., with great power and low volume of hydrothermal niche) are characteristic of natural ecosystems that are at the lowest stage of development of their adaptive mechanisms, i.e., they have the lowest level of adaptation. At the same time, a high volume of niche with a "fuzzy" optimum (the minimal power of niche) or its complete absence is evidence of rather developed mechanisms of ecosystem adaptation to environmental fluctuations. These mechanisms provide certain climatic tolerance.

The described regularities may be observed in the column diagrams of the distribution of plant formations by the levels of their adaptation to total precipitation per year and the mean July temperature, as illustrated by two partial landscape structures: thermo-edaphic and edaphic-productive (Figures 3.6 and 3.7).

The maximum tolerance to changes in temperatures and precipitation is typical of the formations of south taiga (both dark coniferous and pine) and the eastern (more continental) sectors of middle taiga (adaptation levels IV and III). These formations have adapted to the widest range of hydrothermal space, which is demonstrated by the lower powers and high volumes of their climatic niches. Thermo-edaphic and edaphic-productive structures are characterized by high levels of adaptation of formations against rather significant (classes III–IV) equivalence of geo-component blocks. At the same time, vegetation has adapted well to both thermal and pluvial factors. It has acquired the maximally possible reserve of hydrothermal homeostaticity. It may be supposed that south taiga as a whole, the continental variants of middle taiga, and typical forest–steppe were the most stable landscape–zonal systems in the geological and historical past and must preserve this stability in the future.

Broadleaf forests and typical forest–steppe have the same high level of development of their adaptive structures, so they can also be considered rather stable phytocoenological formations.

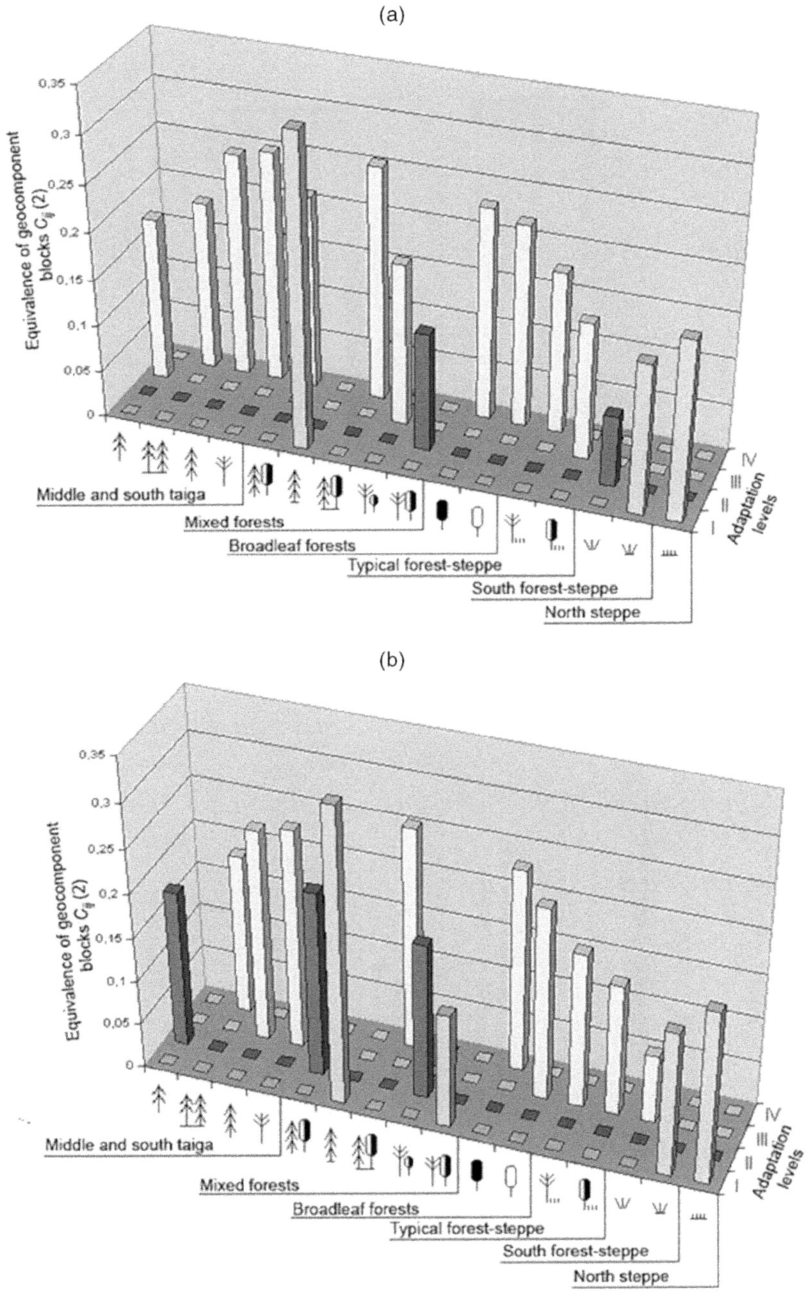


**FIGURE 3.6** Thermo-edaphic partial landscape–zonal structure for the boreal ecotone of the Volga River basin. Distribution of zonal types and plant formation groups by levels of their adaptation to total precipitation per year (a) and to average July temperature (b) under various values of the equivalence of geo-component blocks.

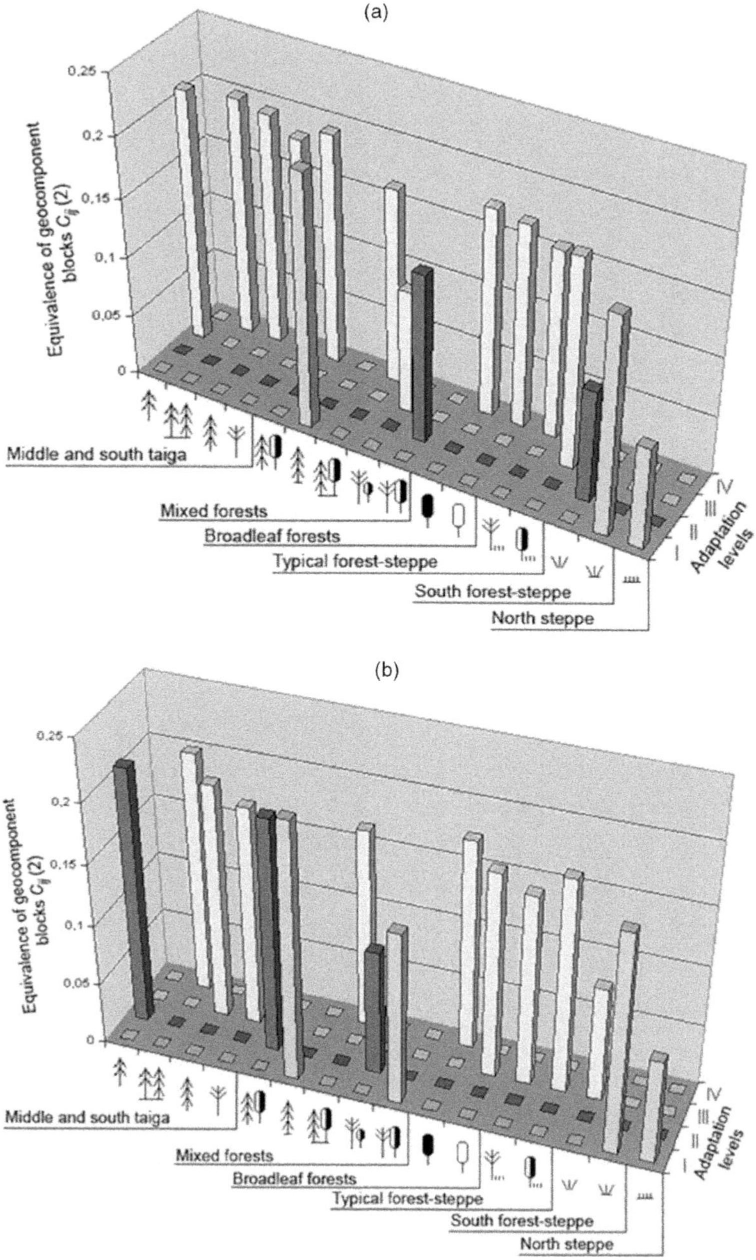


**FIGURE 3.7** Edaphic-productive partial landscape–zonal structure for the boreal ecotone of the Volga River basin. Distribution of zonal types and plant formation groups by levels of their adaptation to total precipitation per year (a) and to average July temperature (b) under various values of the equivalence of geo-component blocks.

Much lower levels of adaptation (I and II) are typical of mixed forests, as well as of the southern steppe of the western (Pontic) type and northern steppe. It seems that the three groups of landscape–zonal systems were characterized by the highest dynamicity and transgressive expansion in the past and should exhibit the same properties in the future, which is confirmed by the results of our paleo-prognostic constructions (see Kolomyts 2024). At the same time, while the first (sub-taiga) group is most sensitive to variations in summer temperature conditions, the second (southern forest–steppe and steppe) group is most sensitive to changes in annual precipitation. Complex nemorose–herbaceous spruce forests of the by-Kama region and formations of pine–broadleaf forests are especially poorly adapted to hydrothermal conditions. The latter forests demonstrate this against the relative imbalance between geo-components in the partial structures under consideration (see Figures 3.6 and 3.7).

So, the transition band from the boreal to sub-boreal belt on the Russian Plain is characterized by a general imbalance of spatial interdependence of abiotic geo-component blocks relative to each other and relative (for each of them) to the integral indicator of geophysical functioning of landscapes: their primary bioproductivity. Accordingly, the level of landscape–geophysical integrity of regional natural complexes should decrease in the same direction. An opposite picture is observed for soil–geochemical structures. The properties of landscapes' integrity determined by these structures should increase from the north southwards. Thus, one may speak about the territorial regional polymorphism of geosystems of the first order. It is demonstrated by the presence of two fundamentally different megatypes of natural–territorial organization, one of them typical of the boreal belt and the other typical of the sub-boreal belt. So, there are two spatial models of formation of landscape structures of the highest (belt) rank.

Territorial polymorphism of the second order manifests itself through longitude sector variations of the general latitude–zonal tendency of the changing megatypes of landscape organization. These variations are determined mainly by the winter continentality of the climate: one of the strongest landscape-forming factors. Under the conditions of higher climate continentality, monosystem landscape structures prove to be more ordered (more integral). Thus, the effect of the landscape-forming factor is shown not only in the qualitative differentiation of natural complexes or in the sharpening of contrasts between them, but also in the increase of their component interrelationship and in the general growth of monosystem integrity of landscapes. At the same time, the territorial orderliness of functional partial structure increases as the leading factor that forms this structure becomes limiting. Due to this reason, the landscape–geophysical (thermo-productive) structure and the group of landscape–geochemical structures prove to be higher ordered in the northern and southern bands of the main watershed of the Volga River basin, respectively.

## 3.7 SOME INFERENCES

The study of the structural–functional organization of natural ecosystems on the zonal ecotones has also scientific and methodological significance, which is common with geo-ecology and is, besides, most important for regional and local landscape–ecological forecasts. Geographical ecotones are the most sensitive (and in this sense the least stable) fragments of the natural–territorial mosaic, and the investigation of the state and dynamics of geo(eco)systems on the zonal ecotone provides the researcher with a kind of "quick-running model" of possible local responses of natural complexes to background disturbances of the global climatic system or to large-scale hydro-technical measures taken by man.

# 4 Methods of Local and Regional Landscape–Ecological Modeling

## 4.1 INTRODUCTORY REMARKS

If, to monitor general climate-genic changes in the state of zonal–regional types of the natural environment, remote radiometric and laser sensing methods using geoinformation technologies are widely used (Bartalev and Isaev 2004; Franklin et al. 2017; Santoro et al. 2021; etc.), then "the methodology of traditional field observations ... remains the best for monitoring the gradual transformation of phytocenoses, including exogenous successions" (Israel et al. 1993, p. 11). Being geographically quite limited, terrestrial research provides disproportionately more diverse and deeper content compared to the data from scanning radiometers with satellite remote sensing of the forest cover of vast territories. These data are described in detail in the works (Bartholome and Belward 2005; Bartalev et al. 2010; Bartalev et al. 2016; Santoro et al. 2021). However, remote sensing maps of the global forest biomass pool of the temperate belt were found to contain deviations from inventory data of between 25% and 120% (Rodriguez-Veiga et al. 2016; Santoro et al. 2021). It is no coincidence, for example, that the results of repeated remote sensing of boreal forests in Central Canada from 1990–2009 were verified by ground-based research materials (Gamon et al. 2004; Stinson et al. 2011). Only ground-based correction of forest phytomass in key areas made it possible to increase the resolution accuracy of satellite data for the Mari Trans-Volga region to 57%–87%. It is hoped that the results of the ground-based monitoring studies presented in this monograph will be used to verify remote sensing data from the forests of the Volga River basin.

On the other hand, at each stage of monitoring, when analyzing the materials of ground works, it is planned to use the results of remote sensing of the forests of the studied territory. A good example is the system of long-term landscape and ecological studies functioning in the Central Forest Biosphere Reserve, combining field experiments and remote information processing (Zheltukhin et al. 2021). We have conducted an experience of ecological and geographical analysis using satellite data of the normalized difference vegetation index of boreal forests of the Volga basin as an index of the physiological basis of their productivity (Kolomyts 2022).

## 4.2 GENERAL FEATURES OF LANDSCAPE-ECOLOGICAL MODELING

A systematic approach involves the use of a very broad concept in the research process – organization, organization of landscapes or ecosystems (Preobrazhensky 1986; Armand 1988b). This concept includes two parts: (1) internal orderliness, consistency, and interaction between the parts of the whole; (2) the totality of the processes of functioning, management, and self-regulation at each structural level.

Landscape organization includes two interconnected but opposite processes – the integration of various geo-components into integral natural units (landscapes) and the differentiation of these units into relatively separate territorial parts of different scales, i.e., the subordination of their structural levels – local, regional and planetary. Such duality of the organization of geographical space (integration–differentiation) is reflected in two branches of the hierarchy of geosystems – component and areal (Gettner 1930). Accordingly, the two main types of landscape organization models are distinguished – monosystem and polysystem (Preobrazhensky 1969).

The landscape ecological analysis of the spatial organization of natural complexes, their anthropogenic dynamics, and carbon balance is based on the construction of d*iscrete empirical-statistical models* of their structure and function, according to Rozenberg (1984). These models can be classified as self-organizing models. They are substantially different from the known simulation models (Printice et al. 1992; Smith et al. 1992; Holten et al. 1993; Kudeyarov 2007; Zavarzin 2007), which are more or less a priori. Empirical statistical models are a posteriori, i.e., the results of field studies are used not for verification of some calculation data but as a factual basis for modeling per se. It yields results of much higher spatial resolution and the empirically substantiated, rather extensive geographic generalizations. In contrast to simulation models, the proposed and implemented method of local discrete modeling of natural ecosystems can be much more reasonably considered a biogeocoenotic method – in the spirit of V.N. Sukachev's theory (1972). It is important to emphasize that such models allow predictive analysis with wide use of the results of traditional field landscape ecological surveys as well as various cartographic materials.

At the same time, it should be noted that simulation and empirical statistical models cannot be considered interchangeable. Each of these classes of models has its pros and cons, similar to any type of modeling. As shown by a comparative analysis of the models of plant responses to climatic changes (Lischke et al. 1998), simulation (dynamic) modeling provides a detailed "time scan" of the studied structural or functional ecosystem parameter but has no opportunity for reasonable spatial extrapolation of the results. In the meantime, empirical statistical models, being static, can describe the territorial diversity of predicted situations but suffer from uncertainty over time. In this context, we accept that the priority object in landscape ecological prediction within a century must not be the structural evolution of ecosystems but the directed change of their function, i.e., the shifts in the small biological cycle (phytomass production and decomposition, as well as humidification), which take the first few years in the taiga zone and occur during the year in the subzone of broadleaf forests (Kolomyts 2005). These characteristic times of functional relaxation are proportionate to exogenous changes in the carbon content of forest phytomasses (living and dead) and mobile humus of the soil (Osipov 2004). This approach is in complete agreement with the subject of scientific research.

From the standpoint of geographical and functional ecology (Odum 1983; Gerasimov 1985), the mechanisms of biotic regulation of the carbon cycle under global climatic changes can be described by the dynamics of internal turnover and balance of plant matter in ecosystems. We have performed empirical statistical modeling of functional parameters of topo-geosystems based on the bio-chorological approach proposed in due time by N.V. Timofeev-Ressovsky (Timoffeev-Resovsky and Tyuryukonov 1966; Timoffeev-Resovsky 1970): the principle of spatial organization by living matter of the matter and energy cycle in biogeocoenoses as discrete elementary structural units of the biosphere.

Further development of the concept of biotic regulation is seen as a study of local–regional mechanisms of the biological cycle, which provides stability of the natural environment in accordance with Les Chatelier's principle (Gorshkov 1994). The sphere of *biogeocoenoses* (*landscape facies*) as the elementary bio-chorological units, i.e., ecosystems of the local (topological) level, is the most complex and active part of the natural environment, its functional "core" (Sochava 1974). This elementary unit of geographical ecology accepted in Russia corresponds to the categories "Site", "Eco-element", and "Land-type Phase" by classifications of Australia-Britain, Canada, and the USA, respectively (Klijn and Udo de Haes 1994). The spatial diversity of the biological cycles is created in the first instance by the structural heterogeneity of biogeocoenoses. As far as we know, the concept of biotic regulation of the carbon cycle in the biosphere has not yet been developed in such a statement of the problem.

According to the known theoretical works (Krapivin et al. 1982; Gorshkov 1994; Kondrat'ev et al. 2003), each ecoregion is considered a statistical ensemble of biogeocoenoses (landscape facies): elementary bio-chorological units that weakly interact with each other but are highly ordered internally (on the basis of stabilizing selection). Each facies-level ecosystem acquires particular functional properties associated with the structure of the local flows of matter and energy. This approach makes it possible to reveal the behavioral patterns of biotic communities under different geomorphological and edaphic conditions based on massive empirical information collected on test grounds, while maintaining the statistical methods of its analysis.

The techniques of quantitative basic and predictive ecological and geographical modeling are based on the methods of quantitative information theory (Ashby 1956; Kustler 1957; A.D. Armand 1975; Puzachenko and Skulkin 1981), as well as descriptive ("fuzzy") sets (Syomkin 1973; Andreev 1980; Odeh Inakwu et al. 1992). Both of these methodological apparatuses, which have already firmly entered the instrumental arsenal of geography and ecology, complement each other well. The first of them allows for revealing the laws of geo-component relations in a given territory, and the second addresses the features of inter-complex relations and relations between its individual sections. In the first case, each object under consideration in the lateral (horizontal) direction is assumed to be homogeneous within a given territory, while in the second, this territory is preliminarily divided into relatively homogeneous sections, which are then compared according to the state parameters of the entire (or maximum possible) set of objects.

Thus, with the help of information theoretical analysis, monosystem models of the landscape are built, and by applying the methods of set theory, polysystem models are developed. The ideological basis of the models of the first group is the idea of the continuity of the landscape sphere, while the second group of models is based on the concept of its discreteness and mosaic. Based on information modeling, a quantitative assessment is given of the spatial or temporal conjugation between various natural components and factors, each of which can be represented by several features. The basic principles and methods of information statistical and set theoretic modeling are described. Complex interconnections were revealed, and the territorial ecological space of regional and local geo(eco)systems was described based on information statistical and fuzzy set theoretical models.

## 4.3 MONOSYSTEM INFORMATION STATISTICAL MODELS

The information statistical analysis of empirical material was employed to uncover the interconnections of landscape components (see above). The main parameters of these connections are as follows: (1) the *normalized coefficient of interrelation* C(A;B) of two attributes taken in pairs: considered object – phenomenon A, and the effects on it factor B, and (2) the *partial coefficient of connection* $C(a_i/b_j)$ characterizing the closeness of interrelationship of individual classes (gradations) of these attributes. These gradations are marked in the form of elements of the next sets: $A = \{a_1, a_2, \ldots, a_i, \ldots, a_n\}$ and $B = \{b_1, b_2, \ldots, b_j, \ldots, b_m\}$. Each phenomenon and each factor has a certain multitude of states.

The first parameter indicates the degree of general closeness of spatial interrelations between the attributes of forest ecosystem states. In the initial working matrix, the vector of states (gradations) of factor B forms the rows, and the vector of states of phenomenon *A forms* the *columns*. The matrix cells contain probabilities, or frequencies, $p_{ij}$ of co-occurrence (probability) of the given states of *A* and *B*. The sums by the columns and rows yield a priori (independent) probabilities $p(a_i)$ and $p(b_j)$, respectively; they are used to calculate the values of conditional probabilities $p(a_i/b_j)$. At the same time, a priori probabilities $p(a_i)$ characterize phenomenon *A* under the assumption of its absolute independence of factor *B*, when $p(a_i/b_j) = p(a_i)$. The measure of the difference between the conditional and a priori distribution of *A* is the function $I(A/b_j)$ equal to

$$I(A/b_j) = H(A) - H(A/b_j), \tag{4.1}$$

where

$$H(A) = -\sum_{i=1}^{N} p(a_i) \log_2 p(a_i), \tag{4.2}$$

$$H(A/b_j) = -\sum_{i=1}^{N} p(a_i/b_j) \log_2 \rightleftarrows p(a_i/b_j). \tag{4.3}$$

Parameter $I(A/b_j)$ is the measure of the force of the general effect of this factor on the phenomenon. The higher the $I(A/b_j)$ value, the stronger their connection. Parameters $H(A/b_j)$ and $I(A/b_j)$ are calculated for each row of the working matrix. The total amount of information, which is transferred from factor to phenomenon and characterizes the closeness of their connection, is equal to

$$T(AB) = \sum_{j=1}^{m} p(A/b_k) I(A/b_k). \tag{4.4}$$

The general rule for the conjugation of two objects is (Kustler 1957) if a complex event *AB* consists of two independent events *A* and *B*, then the diversity (negentropy) of a complex event is equal to the sum of the varieties that characterize separately *A* and *B*, excluding the measure of their joint diversity, i.e.,

$$T(AB) = H(A) + H(B) - H(AB), \tag{4.5}$$

were $$H(AB) = -\sum_{ij=1}^{N} p_{ij} \cdot \log_2 p_{ij}. \tag{4.6}$$

Graphically, this rule is shown in Figure 3.1. Parameter $p_{ij}$ means that when the state $a_i$ is realized, the state $b_j$ occurs only with a certain degree of probability – due to the influence of other unaccounted factors, which play the role of "noise", distorting the tightness of the bonds between *A* and *B*. This "noise" must be subtracted from the general variety of conditions.

The variety $H(A/B)$ denotes one part of "noise" due to the uncertainty ai with the known $b_j$, and $H(B/A)$ denotes the second part of "noise" associated with the uncertainty of $b_j$ with the known $a_i$. If you push the figures in Figure 4.1, then $T(AB)$ will decrease and eventually turn to zero, which will mean the absence of communication between the objects. If the figures are combined, then $H(A) = H(B) = T(AB)$, i.e., the connection becomes functional, without extraneous distorting influences.

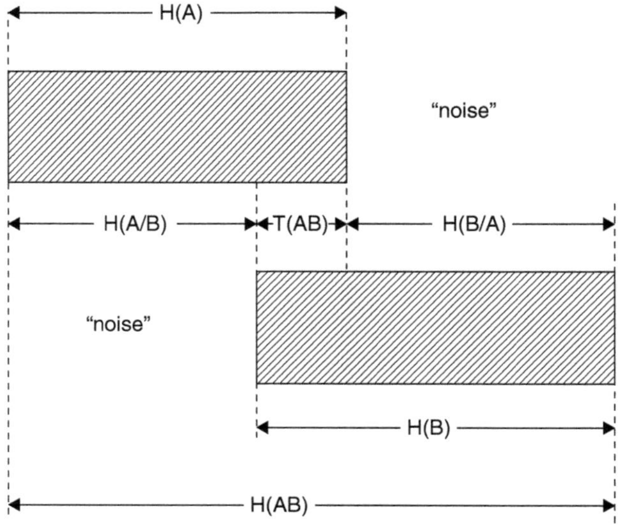


**FIGURE 4.1** Diagram of relationships between information functions, after (Kustler 1957). Explanations in the text.

The amount of information is estimated in binary units (bits); however, it is more convenient to use the relative measures of connection expressed in quotas of 1. In the present study, the normalized coefficient of interrelation has been used:

$$C(A;B) = \frac{2^{T(AB)} - 1}{2^{H(\min A,B)} - 1}. \tag{4.7}$$

Here, *2T(AB)* is the number of general states of *A* and *B*, while *H(A)* and *H(B)* are the common measures of the diversity of characters *A* and *B*, respectively. Based on parameter *C(A;B)*, a generalized information statistical model of landscape connections in ecosystems is constructed. The coefficient of information receipt *C(A/B)* has a certain meaning (A.D. Armand 1975):

$$C(A/B) = \frac{T(AB)}{H(A)}. \tag{4.8}$$

Parameter *C(A;B)* was the basis for the construction of generalized information statistical models of landscape–ecological connections in regional and local ecosystems. Note that $C(A;B) \approx 0.19$ also corresponds to a correlation coefficient of 0.65–0.70 (Puzachenko and Skulkin 1981).

Parameter *H(AB)* is used as Shannon's measure of the diversity (complexity) of object structure as well (Ashby 1956). This measure is maximal when all morphological parts are of equal size (equally possible). This maximum depends only on the number of morphological parts and is determined by Hartley's formula:

$$H_{\max} = \log_2 N. \tag{4.9}$$

The actual value of the diversity measure is usually below the maximum due to the different dimensions of morphological parts. The ratio of parameters $H_i$ and $H_{\max}$ gives insight into the degree of geo-space territorial order (or, on the contrary, disorder). The index of relative organization of biological systems (Antomonov 1977) was used as a measure *C(d/o)* of geographical space disorder:

$$C(d/o) = 1 - H_i/H_{\max}. \tag{4.10}$$

The next parameter $C(a_i/b_j)$, the partial coefficient of connection, made it possible to reveal the informational structure of relations in the binary "factor–phenomenon" system and thus to find some kind of *ecological niches* of each value (gradation) $a_i$ of phenomenon *A* in the space of values $b_j$ of factor *B*. Parameter $C(a_i/b_j)$ allowed for the identification of *ecological niches* of each value (gradation) $a_i$ of phenomenon *A* in the space of values $b_j$ of factor *B*. In ecology, it is quite permissible to apply two of the three known aspects of the conception of the ecological niche (Odum 1971; Begon et al. 1986): (1) spatial, close to the conception of site inhabitance, and (2) resource, determined by the position of an object relative to the gradients of some or other substance–energy factor (temperature, moisture content, physicochemical properties of soil, etc.). Taken together, both these aspects characterize the niche as a certain space of vital environmental factors (Svirezhev 1982).

Nevertheless, the conception of niche applies not only to individual plants or animals but also to entire biotic communities, ecosystems, and their selected structure–function attributes. According to Whittaker (1975), we may generalize the notion and space of a niche, but the principles of the concept remain the same. The essence of such a generalization in the case of plant formations considered below is that attention is focused on the populations of coenosis-forming edificatory species (after which the formations are named), rather than on the entire variety of constituent plant species.

The ecological niche of landscape systems is, first of all, their phytocoenotic unit, the location of which in geographical space corresponds to the climatic niches of dominant kinds – indicators and the communities they form. The conception of a niche applies not only to the geo(eco)system as a whole but also to its structure–function attributes. The series of ecological niches for each phenomenon describes *a phase space* of the factor–phenomenon system. According to the statistical conception of ecosystem stability (Puzachenko 1990), the most probable value of the ecological niche of the given gradation of the phenomenon is taken as stable, corresponding with the equilibrium process of changes in the phenomenon's state by the gradient of factor conditions.

Each niche is presented in the form of a row (descriptive) vector – a "fuzzy" set of the states of phenomenon *A* caused by the given factor *B*. The row vector consists of standardized partial association coefficients:

$$C(a_i/b_j) = \frac{p(a_i/b_j)}{p(a_i)}. \tag{4.11}$$

It is agreed that the association between $a_i$ and $b_i$ is significant at $C(a_i/b_j) > 1$. The higher the probability, the greater the coefficient; therefore, the components of the row vector $C(a_i/b_j)$ are considered as "weighting" coefficients.

The reliability of analysis results significantly depends on the correctness of using the mathematical apparatus in calculations. Following the requirements of informational analysis, the statistical excerpt of a limited number of variables (in this case, less than 100 points) can be divided into no more than 5–6 classes (gradations). Our verification by the criterion $\chi^2$ (Mitropol'sky 1971) showed that even at such a comparatively coarse quantization, the normalized coefficient of interrelationship, below 0.07, often appears to be inexact (interrelationship of the attributes below the 90% level of significance). Therefore, all connections with coefficients lower than the above threshold were excluded from the analysis.

The matrix of partial association coefficients describes a system of ecological niches of either the ecosystem or its individual attributes, defined as "zones of consumption", in the space of changes of the given factor (Odum 1971; Begon et al. 1986). It is accepted that each value $C(x_i/y_j)$ is a certain equivalent of probability (finding) of any gradation of phenomenon *i* by the given gradation of a factor *j*. The higher the probability, the greater the coefficient; therefore, the components of the row vector $C(x_i/y_j)$ coefficients. On the basis of such matrices, the vector graphs of binary ordination *X* and *Y* have been created. The series of ecological niches for each phenomenon describes a *phase space* of the system of factor–phenomenon (Puzachenko and Skulkin 1981).

In the matrix of significant (>1) partial coefficients of connection, the phenomenon and the factor are graded by columns and rows, respectively. Factor gradations with the maximum values of the coupling coefficient form a certain region of phenomenon dominance (hereinafter *ecological dominant*), while other gradations belong to the "fuzzy" part of the niche. If we define ecological domination in each vector column of a niche by symbols "×" or "+", and "fuzzy" links of the niche by the symbol "•", then we will obtain a formalized matrix of the climatic niche. In this form, the matrix demonstrates the character and closeness of connections. Many characteristics of forest ecosystems have not only the fundamental ecological niche for some factor or another (with the respective ecological dominant and the "fuzzy" part of the niche) but also a kind of *enclave* separated from the niche on the graph of partial coefficients by more than one gradation of the factor. Attention was also paid to the cells of binary ordination of the phenomenon with the factor, where $C(x_i/y_j) \leq 1$. These were the regions of *sporadic distribution* of the given phenomenon by the gradient of changes in the factor under consideration, frequently with the emergence of an enclave.

It is accepted that each value $C(a_i/b_j)$ is a certain equivalent of the probability of the zonal–regional biogeocoenosis group *i* by the given graduation *j* of a geophysical parameter. The coefficient is higher at a high-level probability; therefore, the components of row vector $C(a_i/b_j)$ are considered "weighting" coefficients. Such matrices were used to obtain taxonomic base norms of the above hydrothermal parameters: the values averaged by the territory. The central values of all gradations for the given geophysical parameter and matrices of normalized (by the columns) partial association coefficients have been used for the calculation of taxonomic norms.

The most probable (dominant) value of the ecological niche at the given gradation of the phenomenon is taken as stable, corresponding to the equilibrium process of changes in the phenomenon's state by the gradient of factor conditions. The series of ecological niches by each phenomenon describes *a phase space* of the system of factor–phenomenon (Ashby 1956).

The phase space of ecological niches reflects the dynamics of their volume under changing factor states. While moving along the lines of the matrix or along the line of ecological determinants (which is the same), one can follow the variation of the "homeostasis reserve" of the phenomenon and, accordingly, its sensitivity to changes in factor states. Abrupt narrowing or broadening of the niche demonstrates that with each subsequent state, the phenomenon changes the character of its response to the variation of factor states, becoming more or, on the contrary, less sensitive to this factor. It appears that irreversible changes occur in the mechanisms of interaction between the components of the phenomenon system, and its behavior becomes nonlinear, according to Zeidis et al. (2001). The nonlinear character of the transition of the states of phytocoenotic, soil, and landscape complexes as a whole by the arid gradient is typical of their binary relations with most of the abiotic factors under consideration. It adds complicating uncertainty to the landscape–ecological prediction based on the principle of actuality (Puzachenko and Skulkin 1981).

Analysis of the information statistical model begins by highlighting three categories of signs: edificators (environment), indicators, and repeaters (Kolomyts 2005). First, the output arrows on the digraph are predominantly second, while the input arrows are fewer. Repeaters do the same thing: they are intermediate information nodes, so they have both input and output arrows. The more output arrows there are and the larger they are, on the one hand, and the fewer input arrows there are and the smaller they are, on the other hand, the more pronounced are the edificatory properties of this sign.

## 4.4 TYPICAL SCHEMES OF BINARY ORDINATION FOR THE COMPONENT ATTRIBUTES

Systems of ecological niches are presented graphically, allowing one to judge not only the connection channels between factor and phenomenon but also to follow implicitly the dynamics of this connection, to find critical (transitional) points on the trajectory of connection, and, which is no less important, to reveal in certain regions of the phase space attributes of the mechanisms of positive or negative feedback in the system "factor–phenomenon", as well as the moments of direct connection changeover. Thereby, we manage to leave the scope of binary relations of the phenomenon with the given factor to estimate, first, the role of other factors and, second, the processes of autoregulation of the phenomenon that are immediately related to the stability of its certain conditions.

Simple regulation includes six typical cases of an unambiguous connection between phenomenon and factor (Figure 4.2, *a–f*). Among them, the first two are referred to as linear (or related) connections, in the absence of any noticeable autoregulation of the phenomenon. The switching-on of feedback at a certain stage of the phenomenon variation trajectory results in a concave, convex, or broken curve of this trajectory (see Figure 4.2, *c–f*). Simple regulation is typical of the marked binary connection of the phenomenon with a given factor, in the absence of or under weak influence from other,

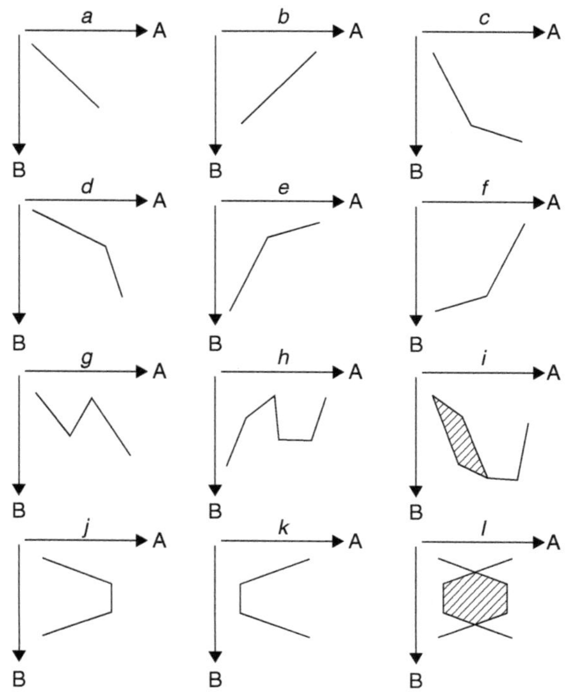


**FIGURE 4.2** Typical trajectories of behavior of phenomenon *A* under the changing factor *B*, with corresponding regulation types of direct and reverse connection. The "zone" of tolerance is shaded (see in the text).

unaccounted factors. Therefore, the direct connection is unambiguous, though reverse impact may change the gradient of this relation.

Complex regulation is typical in cases of substantial "noisy" effects, when the binary connection between A and B is disturbed by the influence of other cause-and-effect relations and becomes ambiguous. There are one or several switchings of the connection (Figure 4.2, *g–i*). In the presence of two trajectories simultaneously at a certain segment of the binary connection, a "zone" of tolerance appears in the phase space, where the given state of the phenomenon may be observed under different states of the factor (in Figure 3.2, this "zone" is shaded).

Parabolic connections often occur under complex regulation (see Figure 4.2, *g–k*). Here, the negative connection switching-on in the region of extreme (maximum or minimum) values of the phenomenon is so strong that the regulator of direct connection switches from "+" to "–" or vice versa. This turning point of the phenomenon trajectory implies nothing but the presence of competitive relationships in the factor–phenomenon structural–functional system (Armand 1989). One of the two competing subsystems dominates at each branch of the parabolic trajectory. Competition may result, first of all, from the appearance of a new factor having an opposite effect on the state of the phenomenon compared to the previous factor. This effect becomes dominant from a particular moment, and that is when the regulator switches.

Second, competitive relations are also possible among the characteristics of the phenomenon itself when dealing with a rather complex receptor system, both structural (e.g., groups of forest types or ecological groups of herbage) and functional (e.g., total above-ground phytomass or total annual productivity of phytocoenosis). Such a system (receptor of influence) consists of two competing subsystems, each responding differently to the influence of the same factor and becoming dominant in a certain interval of factor values. The appearance of a strong negative feedback may indicate just the change of the dominant: the receptor of influence in the phenomenon itself.

The same state of the phenomenon may be observed with the highest probability at two rather distant factor values, thereby determining the potential width of the ecological niche of the given gradation of the phenomenon, with two ecological optima. In the presence of several (two or more) pairs of competing subsystems, a complex harmonic trajectory appears, most often S-shaped, with two or more cases of feedback switching-on and the respective switching-over of direct regulation.

Finally, a quite real case is the parallel existence of two subsystems in the factor–phenomenon system with obvious intensification of competitive relations between them. The diagram of ecological niches shows it as an intercrossing of two parabolic curves (see Figure 4.2, *l*) with the formation of a partially or completely closed tolerant "zone" of unstable equilibrium. Within this region, the equally possible existence of several states of the phenomenon at a given factor value is admissible. However, such states are unstable, and each of them, having an insufficient "stability reserve", may be irreversibly disturbed by outside influence, which will signify the "noise" changeover of the regulator. Such properties are typical of so-called trigger systems (Armand and Vedyushkin 1989).

The system of ecological niches was used to calculate the *taxonomic norms* of functional characteristics of the given object. For this purpose, the central values of all gradations of the factor were used, as well as the matrix of the normalized partial coupling coefficients of the object introduced into the calculations as "weight" coefficients. The normal value, or in other words, the weighted average value of the functional feature (Ramenskiy 1971), was determined as a sum of products of central values of all its gradations by the "weight" coefficient corresponding to the given gradation of the phenomenon (object).

## 4.5 POLYSYSTEM FUZZY SET THEORETICAL MODELS

The second aspect of the natural–territorial organization describes polysystem models (Preobrazhensky 1969; Syomkin 1973; Andreev 1980; Kolomyts 2022). Most generally, these models are based on the insight into the similarities and differences between elements of the finite set of attributes adequate to some or other state of the ecosystem. The models of system states expressed in the measures of ordering and similarity are, in any case, discrete. Their advantage is that they can represent the diversity of qualitative properties of initial structural elements of geo(eco)systems.

The most common structural–functional models of complex relationships can serve as models based on the idea of the similarity and difference between elements of a finite set of features, adequate to a particular state of a geo(eco)system. Therefore, the analysis of pair relations between natural complexes should be carried out using set theory methods. The state models of systems expressed by these methods are somewhat discrete. Their advantage is that they can display the variety of qualitative properties of the initial elements of the structure. Recall that a set is any combination of definite and quite distinguishable objects considered as a whole. The objects themselves are called the elements, or members, of this set. A set of structural and/or functional signs of an ecosystem (values of its variables) that fit into the designated boundaries characterizes a certain state of the system (Cherkashin 1997).

According to set theory (Stoll 1968), relations between objects *A* and *B* are revealed using the following operations (Figure 4.3): (1) associations – (*AUB*) = *C*; (x 2) intersections (product of sets) – ($A \cap B$) = *C*; (x 3) the inclusion of the set *A* into the set *B*, i.e., $C = A|A \ni B|$. The measure of similarity of two sets *N*(*A*, *B*) is determined by the number of their common elements *n* – *N*(*A*, *B*) = *n* ($A \cap B$). The third operation gives full inclusion, and the second one gives incomplete inclusion when set *C* characterizes the degree of mutual inclusion of the sets *A* and *B* into each other. Since in the general case $A \neq B$, the relative measure of inclusion *C* turns out to be asymmetric.

The description of the natural complex as a composite system provides the researcher with a set of attributes usually forming indistinct, "fuzzy" sets. Here, similarities and differences can be revealed statistically using the "weight" of each attribute. We have used two measures of the pairwise ordering of the objects: the measure of similarity (or difference) and the measure of inclusion of one object into another. The binary relations of biogeocoenoses (*A*, *B*, *C*, etc.) were estimated by the set of their functioning parameters acting as their elements. If there are two topo-ecosystems *A* ($A_1$) and *B* ($A_2$) consisting of the same number *n* of elements $a_i$ and $b_i$, respectively, which have certain numerical values, then the measure of similarity $N(a_i,b_i)$ of these elements taken in pairs, as well as the inclusion measures $J(a_i;b_i)$ and $J(b_i;a_i)$ of element $a_i$ into $b_i$ and vice versa, can be found from the ratios:

$$N(a_i,b_i) = \frac{2(a_i \cap b_i)}{a_i + b_i}; J(a_i;b_i) = \frac{a_i \cap b_i}{a_i}; J(b_i;\ a_i) = \frac{a_i \cap b_i}{b_i}. \tag{4.12}$$

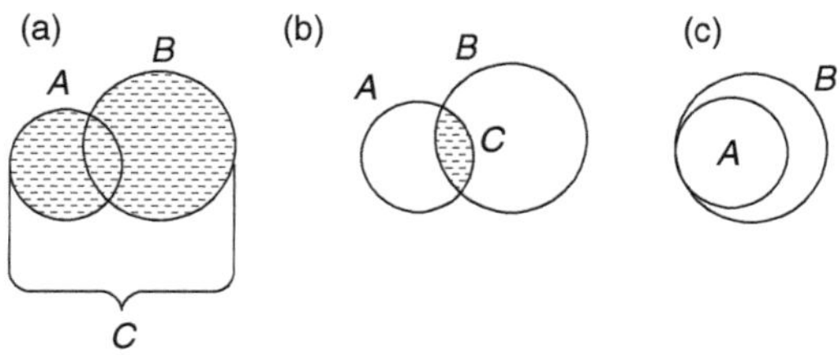


**FIGURE 4.3** Typical trajectories of behavior of phenomenon A under the changing factor *B*, with corresponding regulation types of direct and reverse connection. The "zone" of tolerance is shaded (see in the text).

Here, $a_i \cap b_i$ is the intersection of the values of elements, determined as the minimum value of the two elements $a_i$ and $b_i$. The measures of difference, or distance, between the elements is $D = 1 - N$. In the general case, $J(a_i;b_i) \neq J(b_i;a_i)$. If both of these measures exceed a certain threshold (critical) value $J_{crit}$ of the inclusion measure (a certain "banality" threshold $Z_{crit}$), then it is possible to speak about the tolerance of elements (objects), i.e., about their relative similarity at a given $J_{crit}$ (Andreev 1980); we believe that $Z_{crit} \equiv J_{crit}$. For each pair $A_1$ and $A_2$ of the vectors of ecological niches or any other descriptive sets, each containing *n* elements (quantitative attributes), the above measures of binary relations are calculated by the formulas:

$$N(a_1,a_2)=\frac{2\sum_{j=1}^{N}\sum_{k=1}^{S_j}\min\left\{P_j^k(a_1),P_j^k(a_2)\right\}}{\sum_{j=1}^{N}\sum_{k=1}^{S_j}P_j^k(a_1)+\sum_{j=1}^{N}\sum_{k=1}^{S_j}P_j^k(a_2)}; \tag{4.13}$$

$$J_1(a_1;a_2)=\frac{\sum_{j=1}^{N}\sum_{k=1}^{S_j}\min\left\{P_j^k(a_1),P_j^k(a_2)\right\}}{\sum_{j=1}^{N}\sum_{k=1}^{S_j}P_j^k(a_1)} \tag{4.14}$$

$$J_2(a_2;a_1)=\frac{\sum_{j=1}^{N}\sum_{k=1}^{S_j}\min\left\{P_j^k(a_1),\right\}}{\sum_{j=1}^{N}\sum_{k=1}^{S_j}P_j^k(a_2)}. \tag{4.15}$$

Here, $P_j^k(a_j)$ is the "weight" *j*, the *k*-th attribute at $a_i$ element; *r* is the common number of attributes, each of them having $S_j$ values. The respective graphs of similarity measures and orientation graphs (or-graphs) of inclusion measures for the objects under consideration are plotted by matrices $N(A_1, A_2)$ and $J(A_1; A_2)$.

Then simple operations were performed with the matrices of inclusion measures stated in the works of Andreev (1980) to obtain re-sorted single adjacency matrices. These matrices were used to construct the *structural schemes of the dominance* of the objects, which we interpret as the generalized models of partial *polysystem geo-ecological spaces*. These schemes are constructed at the critical values $Z_{crit}$ of the inclusion measure when the continuous chain of overlapping ecological niches has not yet been broken and the oriented graph (or graph) preserves its integrity. The system of inclusion relations contained the connections of tolerance (one-to-one correspondence), strongly and weakly connected objects, and the critical connections, the disappearance of which leads to the degradation of the entire or graph of inclusion relations.

The measure $N(A,B)$ of similarity between two objects essentially characterizes *the degree of their isomorphism* by a given factor. At the same time, their inclusion measures $J(A;B)$ allow the ordering of geo(eco)systems into definite rows, or some *ecotypes*, as stated by Bykov (1973). The ecotypes of landscape units in the system of their hydroclimatic ordination characterize the level of structural–functional development and the level of efficiency of using the resources of the abiotic environment. At $J(A;B) > J(B;A)$, one may state that geo(eco)system *A* is more widely distributed in the space of a given factor and more developed in its changes compared to system *B*; at $J(B;A) > J(A;B)$, on the contrary, it proves to be more pessimistic and, accordingly, less stable. Thus, the case in point is the phenomenon of homomorphism in the binary relations between the objects. Object *A* is homomorphically represented by object *B* in the former case and vice versa in the latter case.

The ecotypes, or *ecological groups*, of landscape faces (see below) are revealed using simple operations on the matrices of inclusion measures (Andreev 1980; Kolomyts 2022) to obtain re-sorted single matrices of adjacency of objects (see Kolomyts et al. 2012, pp. 89–92). These matrices are used for the creation of structural schemes of domination of biogeocoenoses at some or other thresholds of banality *Z* and the determination of critical values $Z_{crit}$ of this threshold.

## 4.6 INTEGRAL MODELS OF THE POLYSYSTEM ORGANIZATION OF NATURAL COMPLEXES

The combination of information and set theoretical methods of analysis makes it possible to obtain complex discrete models of the state of geographical space. Let us present the scheme of the further calculation procedure by the example of elemental landforms (geotopes). More than 20 structural and functional indicators of the state of the respective forest biogeocoenoses were taken for each of them. They included: (1) the characteristics of "structural metabolism", according to Odum (1971), i.e., soil morphological and phytocoenotic structural units, as well as phytomass reserves; (2) the parameters of the functioning of plant communities (green mass productivity); and (3) the hydrothermal indicators of the state of edaphic-topes. The following three complex parameters were determined for each geotope based on the matrices of similarity and inclusion relations between its biogeocoenosis and the coenoses of other geotopes: (1) the measures of biogeocoenosis isomorphism (invariance) $I_k$; (2) the metabolic index of biogeocoenosis $S^m{}_k$, characterizing the general level of its structural and functional development; and (3) the significance index of biogeocoenosis in the given altitudinal–zonal geo-space $G_k$, representing the reserve of its homeostatic (elastic resistance). The ideology and methods for calculating the above parameters are given below.

Ecological objects belong to an extensive group of natural systems with relatively independent behavior of their components; therefore, geo(eco)systems are rather weakly connected natural formations (Armand 1988a), which is one of the basic conditions for the manifestation of polystructural properties, or polymorphism in a broader sense (Solntsev 1997).

Polymorphism is expressed, in particular, by the relative mutual independence (autonomy) of structural and functional characteristics of natural ecosystems. Under varying external conditions, a geo(eco)system, first of all, changes its structure to maintain the previous level of functioning, including the level of bioproduction as its most important invariant property (Timofeev-Resovsky and Tyuryukanov 1966; Sochava 1978; Svirezhev 1982; Kerzhentsev 1999).

In cases when two ecosystems under consideration are similar in their structural and/or functional characteristics, we can say that they are *isomorphic* to each other. The concept of the isomorphism of two systems suggests the interchangeability of their elements, providing the similarity of their structure and function (Paffengolts 1973). If an ecosystem shows isomorphism concerning quite a few other systems, it may be considered to possess certain invariance of its properties in a given geo-ecological space.

This property of polymorphism allows the information theory method to be used for calculations of landscape component interconnection in the territory, which is considered an indivisible system (Andreev 1980). The main parameter of these connections is the normalized coefficient of interrelation *C(A;B)* between the two attributes taken in pairs. One of them is considered a phenomenon A and the other is considered a factor B. Each phenomenon and each factor has a certain multitude of states, and the parameter *C(A;B)* is a measure of diversity for these spatial couplings. Parameter *C(A;B)* also expresses the diversity of microstates of object *A* ($a_1, a_2, a_3, \ldots$) or object *B* ($b_1, b_2, b_3, \ldots$) themselves.

We have used the matrices of similarity measures to calculate the measure of structural–functional isomorphism (invariance) of biogeocoenoses $I_k$. For each *k* geotope with the respective biogeocoenosis, the Shannon information measure $I_k$ was defined as

$$I_k = H_k \,/\, H_{\max}, \tag{4.16}$$

where

$$H_k = -\sum_{i=1}^{n} N(k,i) \cdot \log_2 N(k,i), \ldots H_{\max} = \log_2 N \tag{4.17}$$

Here, *N(k,i)* is the measure of similarity between the *i* and *k* geotopes; *n* and *N* are the numbers of connections of the *k* geotope and the number of geotopes, respectively. In our case, $N = 6$ and $n = 5$. A geotope can be maximally invariant if its elemental ecosystem is maximally similar to other systems of the same rank and, in this sense, maximally isomorphic to them.

The $I_k$ parameter is used for comparative assessment of the degree of isomorphism between the systems with a different number of differentiated gradations. It characterizes the *variety of similarities* between biogeocoenoses of the given geotope and all biogeocoenoses of other geotopes by a different set of their structural and functional characters. The more diverse such a set is, the higher the degree of complexity of similarity relations; i.e., this geotope can be considered to a greater extent as a structural and functional invariant of the given altitudinal zone. An invariant geotope is not necessarily a geotope with the most developed biogeocoenosis. The invariance of a geotope characterizes the degree of structural–functional variety of similarity between its biogeocoenosis and the coenoses of all other geotopes in the ecoregion of interest.

High $I_k$ values indicate the most equal levels of similarity between this object and all other objects, which are close to a certain optimal value of similarity ($N(k,i) \rightarrow N(k)_{opt} \approx 0.38$ when the product $N(k) \cdot \log_2 N(k)$ is found to be maximal (0.5307 bit). If this object has a very high similarity (0.90–0.95 and more) with some objects and, on the contrary, low similarity (≤ 0.65–0.70) with other objects, then the final measure of the variety of similarity will be minimal (Σ $[N(k,i)\cdot\log_2 N(k,i)] \rightarrow \min$). In addition to geotopes, analogous measures of a variety of similarities were calculated also for phytocoenotic groups per se, irrespective of their location. In this case, the degree of invariance of each $k$ meadow community $I_k$ by the set of its structural–functional characteristics was considered.

Two other complex parameters of the state of meadow biogeocoenoses were determined using the initial matrices and/or or-graphs of inclusion relations. Here, we will also give necessary explanations.

Both parameters are calculated based on the structural scheme of geo-space dominance. For each altitudinal–zonal geo(eco)system, it is necessary first of all to determine the degree of structural–functional development of its components: the groups of biogeocoenoses or phytocoenotic groups. Thereby, it is possible to assess the general level of organization of an altitudinal–zonal system, as well as the field of its spatial biogeocoenotic or phytocoenotic contrasts. As follows from the factorial-based dynamic analysis (Kolomyts 2022), the degree of development of a mountain forest biogeocoenosis is greater in the case of higher soil temperatures in the vegetative season (at rather high humidity of the edapho-topes), larger herbage projective cover and average height (as well as floristic and phytocoenotic diversity), a thicker organo-mineral soil layer with a lower content of lithomass, and finally, larger total and fractional reserves of living phytomass and higher productivity of green mass.

Eventually, the general structural and functional state of each biogeocoenosis is expressed in a certain value of its metabolic index $S_k^m$. On the structural schemes of altitudinal–zonal geo-spaces, the resultant pattern of the level of biogeocoenosis organization is fixed by the edges of or-graphs of inclusion relations. The connection arrows are directed from less developed local ecosystems towards more developed ones.

The number of input ($n_i$) and output ($n_j$) connections was determined for each geotope (or each phytocoenotic group) on the structural schemes of dominance. The total number of connections in the system ($N$) was also calculated. Based on the above, the metabolic index $S_k^m$ of the biogeocoenotic group of the $k$ geotope, or the $k$ group of phytocoenosis, was found from a simple ratio:

$$S_k^m = P_i - P_j + 1, \tag{4.18}$$

where

$P_i = n_i / N; P_j = n_j / N$. Obviously, $0 \le S_k^m \le 2$.

At the same time, it is accepted that the incoming and outgoing connections are mutually independent. Hence, it is possible to use the property of additivity of individual information (Kustler 1957).

The value of the parameter *Smk* may be considered a generalized structural–functional index of the above *ecological groups* of forest or meadow phytocoenoses on some geotopes or others. Hence, it is possible to differentiate certain *metabolic ecotypes of biogeocoenoses.* It is known (Bykov 1973; Odum 1983) that the concept of ecotypes is used to single out ecosystems adapted to local soil climatic conditions. At $S_k^m >> 1$, the comparative level of structural–functional development of a biogeocoenotic group at the considered geotope is sufficiently high. Such a group falls into the category of *subdominant* and can be named a *sub-macrobolic ecotype*. If $n_j = 0$, it becomes *a dominant*, i.e., *a macrobolite*. As the metabolic index decreases, the group of biogeocoenoses first becomes *a rank-and-file member* of the altitudinal–zonal set of such groups, i.e., *a mesobolic ecotype*. Finally, at $S_k^m \ll 1$, meadow ecosystems are characterized by the least developed structure and low intensity of functioning; hence, they are classified into the categories of *reduced ecotypes: sub-microbolites* and *microbolites*. In the latter case, the respective geotope, or a phytocoenotic group, has no input connections in the structural scheme of dominance ($n_i = 0$).

Finally, the third complex parameter $G_k$ of polysystemic organization of high mountain meadow natural complexes characterizes the significance of each geotope in the representation of the complete diversity of the states of input ecological factors by its biogeocoenosis (i.e., true altitude, slope exposure and steepness, and hydrothermal conditions). The value of significance was used for the ranking (a kind of classification) of topo-ecosystems, i.e., the determination of their position in the structure of a given altitudinal–zonal geo-ecological space.

The $G_k$ parameter represents one of the aspects of geo(eco)system resistance to external perturbations. According to the most common definition (Ricklefs 1976), system resistance is a ratio between variability in the environment and variability in the system per se. Here, the matter in question is the ability of the system to withstand environmental perturbations or to rather quickly neutralize the consequences of disturbances in its structure and function caused by this perturbation. The so-called *resistant* and *elastic stability* is demonstrated in the former and latter cases, respectively (Armand 1989).

When we consider the response of the geo(eco)system to the external hydrothermal signals that have disturbed it from the previous dynamic equilibrium, we are concerned, first of all, with the resistant stability of this system. When determining the significance of one or another biogeocoenosis in a given altitudinal–zonal geo-space, we estimate the "reserve of its homeostatic", according to Armand (1989), within which an elemental natural complex can vary in its phytocoenotic structure while preserving its qualitative definiteness and the level of the small biological cycle in accordance with Le Chatelier's principle (Gorshkov 1994).

The Shannon function of diversity was used to assess the measure of significance $G_k$. The measure of significance of the $k$ topo-ecosystem in a given altitudinal–zonal space was calculated based on the probabilities of input and output connections in its matrices $J_1(A;B)$ and $J_2(B;A)$. At the same time, it was assumed that the input and output connections were mutually independent. This allowed us to use the property of additivity of individual information (Kustler 1957). The general significance of connections in a given biogeocoenosis, which characterizes the diversity of altitudinal–zonal space properties represented in this biogeocoenosis, comprises the negentropy (Brillouin 1956) of in-degrees (incoming arrows on the or-graph) and the entropy of out-degrees (outgoing arrows). The relative reserve of homeostaticity $G_k$ for the group of biogeocoenoses at the $k$ geotope was calculated by the formula:

$$G_k = \left[-P_i \cdot \log_2 P_i + P_j \cdot \log_2 P_j + 1\right] / \left[\log_2 N + 1\right], \quad (4.19)$$

where $P_i$ and $P_j$ are the probabilities (in unit fractions) of input and output connections, respectively, and $N$ is the total number of connections. The $G_k$ parameter is used for the direct assessment of the degree of susceptibility of a biogeocoenosis developed at a given geotope to changing hydrothermal conditions.

The $S_k^m$ and $G_k$ parameters are closely interrelated. By differentiating the metabolic ecotypes of elemental geo(eco)systems based on the generalized measure of their structural–functional development, we simultaneously determine their homeostatic potential, contributing to their stability under certain variations of climatic and geomorphological conditions of the given geo-space. The higher level of phyto- or biogeocoenosis organization corresponds to a more intensive biological cycle, which serves as a mechanism of elastic stability, neutralizing external perturbations and returning the ecosystem to the previous state after the "elimination" of the influence. In general, the elastic stability of biogeocoenoses under variations of the abiotic environment must decrease from their macrobolic ecotypes to mesobolites and further to microbolic ecotypes.

One should mention the significance of the critical threshold of banality $Z_{crit}$. This threshold denotes the lower limit of the measures of inclusion of system elements when the structural and functional integrity of the system persists. Lower $Z_{crit}$ values denote a relatively high degree of inter-complex interactions in a given ecological space. It provides the territorial integrity of the latter even at relatively low values of inclusion measures. Here, one may expect the greatest diversity of spatial couplings with a predominance of critical couplings and the presence of tolerance couplings forming the typological "cores" of ecological space.

At a relatively high banality threshold, one should discuss the weaker total degree of inter-complex interactions, despite significant mutual overlapping of the ecological niches of system elements. The system preserves its structural–functional integrity only under a substantial crossing of these niches, which is manifested by the high threshold of $Z_{crit}$ values. Weak connections are predominant in the structural scheme, in the absence of a strong and significant number of critical couplings. The ecological space appears to be disintegrated into separate structural elements, indicating a total low level of system organization.

By distinguishing the ecological groups of landscapes, biogeocoenoses, or separate geo-components based on the measure of their significance in the given ecological space, we simultaneously determine the fields of their *homeostasis*: the state of dynamic equilibrium, potential stability under some variations of parameters of this space. The value of significance of some geo(eco)system characterizes, in essence, the "reserve of homeostasis", within which a natural complex, following Le Chatelier's principle, may vary in its phytocoenotic structure while preserving the qualitative distinctness and the level of the minor biological cycle. Stability expressed by the reserve of homeostasis of a geo(eco)system will be further termed as *the potential stability* of the geo(eco)system. The total reserve of homeostasis as a measure of potential stability of a geo(eco)system must enhance with an increase in the volume of its ecological niche.

# 4.7 DISTINGUISHING FOREST BIOGEOCOENOSIS ECOTYPES

The measure $N(A,B)$ of similarity between two objects (see above) essentially characterizes the *degree of their isomorphism* by a given factor. At the same time, their inclusion measures $J(A;B)$ allow the ordering of geo(eco)systems into definite rows, or some *ecotypes*, by Bykov (1973). The ecotypes of landscape units in the system of their hydro-climatic ordination characterize the level of structural-functional development and the level of efficiency of using the resources of the abiotic environment. At $J(A;B) > J(B;A)$, one may state that geo(eco)system $A$ is more widely distributed in the space of a given factor and more developed to its changes compared to system $B$; at $J(B;A) > J(A;B)$, on the contrary, it proves to be more pessimal and, accordingly, less stable. Thus, the case in point is the phenomenon of homomorphism in the binary relations between the objects. Object $A$ is homomorphically represented by object $B$ in the former case and vice versa in the latter case.

The ecotypes, or *ecological groups*, of landscape faces are revealed by means of simple operations on the matrices of inclusion measures (Andreev 1980; Kolomyts 1998) with the obtaining of re-sorted single matrices of adjacency of objects. These matrices are used for the creation of structural schemes of domination of biogeocoenoses at various thresholds of banality $Z$ and for the determination of critical values $Z_{crit}$ of this threshold (see below). The system of inclusion relations comprises strongly and weakly coupled objects and critical couplings, the disappearance of which leads to the disintegration of the entire or-graph of inclusion relations. The whole set of objects was divided into four ecological groups:

1. dominant objects with the maximum number of ingoing arrows on or-graphs, i.e., having maximally stable ecological niches most similar to the niches of all other objects; according to Odum (1971), this ecological group may be named as *eurytopic*;
2. sub-dominant objects with an average number of ingoing arrows (often in the presence of outgoing arrows) showing enhanced stability; that is, the *sub-eurytopic* group;
3. ordinary objects (*mezzo-topic* group) fixed with a proportionate number of ingoing and outgoing arrows on the or-graph;
4. and (5) sub-reduced (*sub-stenotopic*) and reduced, or "banal" (*stenotopic*), objects with the maximum number of outgoing arrows in a single number or almost total absence of ingoing arrows; they take the lowest step of the hierarchy of inclusion relations and are characterized by narrow (reduced) ecological niches, which gives them the minimum stability. This row essentially characterizes the different *levels of ecosystem functioning.*

One should mention the meaningful content of the critical threshold of banality $Z_{crit}$. This threshold denotes the lower limit of the measures of inclusion of system elements, when the structural and functional integrity of the system still persists. Lower $Z_{crit}$ values must denote a relatively high degree of inter-complex interactions in a given ecological space. It provides the territorial integrity of the latter even at relatively low values of inclusion measures. Here, one may expect the greatest diversity of spatial couplings with a predominance of critical couplings and the presence of *tolerance couplings* forming the typological "cores" of ecological space.

At a relatively high banality threshold, one should speak about the weaker total degree of inter-complex interactions, in spite of significant mutual overlapping of the ecological niches of system elements. The system preserves its structural–functional integrity only under rather substantial crossing of these niches, which is manifested by the high threshold of $Z_{crit}$ values. Weak connections are predominant in the structural scheme, in the absence of strong and with an insignificant number of critical couplings. Ecological space seems to be disintegrated into separate structural elements, indicating the total low level of system organization.

By distinguishing the ecological groups of landscapes, biogeocoenoses, or separate geo-components based on the measure of their significance in the given ecological space, we simultaneously determine the fields of their *homeostasis*: the state of dynamic equilibrium, potential stability under some variations of parameters of this space. The value of significance of some geo(eco)system characterizes, in essence, the "reserve of homeostaticity", within which a natural complex, in accordance with Le Chatelier's principle, may vary in its phytocoenotic structure while preserving the qualitative distinctness and the level of the minor biological cycle. Stability expressed by the reserve of homeostaticity of the geo(eco)system will be further termed as *potential stability* of the geo(eco)system. The total reserve of homeostaticity as a measure of potential stability of the geo(eco)system must enhance with the increase in the volume of its ecological niche.

The quantitative assessment of ecotypes of biogeocoenological systems includes (1) their metabolic index as a measure of significance in the space of regional or local relations $S_k^m$; (2) a reserve of homeostativity as a measure of potential stability of forest communities $G_k$. So, with a theoretical analysis of the reserves of living forest phytomass on the eluvial types of locations for four experimental landfills of the Volga River basin, the following indicators of their functional ecotypes were obtained:

| Experimental test sites | $S_k^m$ | $G_k$ |
|---|---|---|
| Kerzhenets reserve | 0.67 (reducent) | 0.125 |
| By-Oka-Terrace reserve | 0.84 (sub-reducent) | 0.126 |
| Green Town (Nizhny Novgorod) | 1.00 (mesotop) | 0.231 |
| By-Sura ("Chavash Varmane") | 1.50 (dominant) | 0.338 |

The measure of reflection of a given regional or local ecological space is expressed by the metabolic index $S_k^m$ of the biogeocoenotic group, calculated using Equation (3.18). The structural schemes of temporal states of regional and local geo(eco)systems, as well as schemes of their climate-genic transformations, have been created based on the parameter $S_k^m$ values (see Chapters 13 and 14).

# 4.8 REPRESENTATION OF CALCULATING CLIMATIC MODELS

All known developments of global and regional ecological predictions are based on deviations of the main climatic system parameters anticipated in the nearest century from their values in a certain base period under the effect of anthropogenic changes in the chemical composition of the atmosphere, with some consideration of the role of social and economic factors in civilization development (Budyko 1980; Budyko et al. 1992; Houghton et al. 1996; Izrael 2001; Kondrat'ev et al. 2003). As a base period, we have taken the 100-year interval (1881–1980) of instrumental observations from the hydro-meteorological network, with the results published as multiyear climate norms. This interval covers the entire so-called industrial period. During this time, the European territory of Russia experienced five significant climatic cycles, each lasting about 15 years (Izrael 2001). The upper limit of fluctuations of 11-year moving averages, which occurred around 1985, was taken as the beginning of the prognostic period characterizing modern global warming. Starting from that time, the annual temperature progressively increased during the 80s and 90s of the 20th century, leaving the previous extreme points of the base period far behind.

We will consider four main prognostic intervals (steps): 1985–2050, 2050–2100, 2100–2150, and 2150–2200. The period of meteorological observation from 1881 to 1985 serves as the basis. Recent global warming began at the end of this period (Houghton et al. 1996). Prognosis until the fifth step (from the year 2100) has an extrapolative character.

The first step is sometimes combined with the second. A considerable part of the first predictive step (1985–2025) has already passed at present. The following values of the main climatic parameters for the end of each interval will be used as ecological estimates: the mean January and July temperatures, annual precipitation, and precipitation during the cold and warm periods. The 10-year average values of these parameters are taken for each term, including 5 years before and after this date. Averaging by decades is extensively used during the analysis of multiyear climatic fluctuations and reveals significant connections between the regional spectra of temperature courses and the changes in atmospheric circulation forms (Rubinstein and Polozova 1966), which supports the validity of such averaging when using climate prediction models of general atmospheric circulation.

The hydrothermal trends for the period up to 2150 have been taken from two global coupled atmosphere–ocean general circulation models (AOGCMs): (1) the moderate global prognostic climatic E GISS model from the Goddard Institute of Space Studies, USA (Hansen et al. 2007), with its earlier version, GISS – more moderate (Hansen et al. 1988); and (2) the more extreme HadCM3 model, Version A2 (Hadley Centre of Climate Research, UK; Gordon et al. 2000; Pope et al. 2000).

The first model predicts relatively temperate global warming. This is one of the most preferred general circulation models, according to the results of testing the statistical significance of the data obtained and the simulation of the state of contemporary climate, including its seasonal characteristics (Santer 1985). It has been successfully used in a number of prognostic–ecological studies (Izrael 1992; Holten et al. 1993; Smith & Shugart 1993). The initial global model GISS-1988 global model GISS-1988 was transformed for a regional level (territory of the Russian Plain) in the model GISS-1993 by G.V. Menzhulin and S.P Savvateev (1993) at our request. The transformed model GISS-93 has proven to be more realistic compared to the initial model GISS-88 (Kolomyts 2005).

The primary GISS climatic scenario was based on the assumption that the doubling of $CO_2$ concentration in the atmosphere compared with the "preindustrial period" (1886–1905) would be reached by the year 2050 (Budyko 1980). More recent investigations have pushed this date to the end of the 21st century (Albritton et al. 2003; Kislov 2006). Consequently, we will adjust the dates of the GISS scenarios as follows: 2010 → 2050 and 2050 → 2100.

The E GISS model provides limits on climate change that correspond to the goals of the Paris (2015) Agreement – to prevent average annual warming of more than 1.5°–2.0° by the middle of this century, i.e., by the time of doubling CO2 concentration in the atmosphere.

The second model represents a range of the latest AOGCM models. It is a combined atmosphere–ocean general circulation model that adequately reflects minor climate changes. The A2 scenario was developed for conditions of more rapid population growth but relatively slow technical progress, when the influx of greenhouse gases into the atmosphere from human activities decreases only slightly (Albritton et al. 2003). This feature differentiates it from the B2 scenario and thus results in stronger anthropogenic warming.

The HadCM3 model is one of the most extreme models predicting the maximal warming for the recent 10,000 years. The results of numerical simulation (Albritton et al. 2003) envisage a much greater increase in the mean annual temperature than the mean global increase according to version A2, for 5°–6° over the entire vast area of East Europe (including the whole Volga River basin), and West and East Siberia in the period of 1971–2100 compared to the period of 1961–1990. The ecological prognosis by the second model permits an evaluation of the forest ecosystem's reaction to such global climatic changes, which may exceed the level of tree endurance and cause the disintegration of both nemoral and boreal forests over large areas.

According to the extreme HadCM3 model, the increase in July temperatures in the Middle Volga region will be 2°–3° by 2050, and at the end of the 21st century, will reach 5.5°–7.0°. The annual precipitation amount will change slightly. According to this thermo-arid scenario, the entire south of the boreal zone on the Russian Plain will experience sub-boreal climatic conditions. The latter can already exceed the endurance threshold of the main forest-forming species and cause the disintegration of not only nemoral but also boreal forests over vast areas. It should be noted that the climate forecast according to the HadCM3 model is adequate to the current warming trend, which can lead to an increase in the average global temperature by 4° by 2100 (Le Quere et al. 2015), and in Russia by 6°–11° (Shvidenko et al. 2014). We considered all forecast periods regardless of their timing, i.e., they are used as models to answer the question "what will happen if ...".

It should be noted that the prognostic mapping assesses only a few equifinal states to which these systems will tend during the course of first functional (metabolic) and then structural transformations under the action of climatic signals. Prognostic models are non-equilibrium, and the zonal areas presented on them, as well as probabilities of transitions, indicate the landscape–zonal conditions in which these ecosystems will function in the coming decades. This primarily concerns changes in the rate of small biological turnover and the transition of an ecosystem to a new level of balance between the productive and detrital branches of metabolism. This process is quite adequate in duration to the total shift of landscape–ecological conditions. Thus, the object of priority in landscape–ecological prognosis is functional relaxation as a primary reaction of ecosystems to external impacts. In ecological assessments of anticipated global warming for the coming decades, the focus may concern only functional but not structural–morphological prognosis.

The spatial resolution of the above models, with an interval of 1° latitude and 1° longitude, allowed for computer mapping (software package "Arc View GIS, Vers. 3.0") of the main hydro-climatic parameters: temperature, precipitation, evaporation, annual atmospheric humidity factor, etc., on a working scale of 1:2,500,000 (Kolomyts 2005), i.e., much larger than it had been done previously in ecological prognoses and paleo-reconstructions.

# Regional and Local Geosystems as the Objects of Predictive Research

# 5

## 5.1 INTRODUCTORY NOTES

In world literature, we are not aware of forest models that use an integral natural elementary unit – biogeocoenosis – as a basis. Biomes are often considered in models (Gil'manov and Bazilevich 1983; Aber et al. 2001), as well as forest plant communities (Komarov et al. 2003), individual tree species in forests, etc. The landscape variables in these models are climatic, soil, and topographic characteristics, but they are most often treated as a mechanistic set of factors that are not organized into any functional system. Even when an attempt is made to distinguish landscape types (land types, sites, and habitats) for forest areas, it is very formal, without considering the system inter-component and inter-complex landscape communications.

The studies of the behavior of local ecosystems in a changing climate are very few. Among them, the so-called island patch models are known – "patch models" (Forman 1995; King et al. 1999) – which consider relatively small areas of forest cover that are distinguished by the functional features of structurally homogeneous areas. These models require a detailed description of the "island" site, but geographically they are very limited because they describe only "typical sites in typical biomes (bottom–up approach) but not the entire topological diversity of ecosystems inherent in some ecoregion or another.


DOI: 10.1201/9781003733119-5

## 5.2 ECOLOGICAL CONCEPT IN LANDSCAPE STUDIES AND FOREST TYPOLOGY

In the main directions of bio-ecological and geosystem monitoring (Izrael 1984; Gerasimov 1985), the priority tasks are the study of the internal turnover of substances in various natural ecosystems and the assessment of the balance of the biological cycle as a necessary condition for their sustainability. These provisions fully apply to predictive assessments of the environmental consequences of global warming, but they need significant clarification. The experiments of the stationary method of research (Krauklis 1969; Krauklis 1974; Green et al. 1980; Beruchashvili 1990) show that the most favorable objects for understanding the mechanisms of the structural and functional organization of the natural environment are nature complexes with small spatio-temporal gradations: topographic geosystems and ecosystems. They represent the structural and morphological parts of the landscape or regional ecosystem, which form a natural category of greater dimension.

Back in 1968, N.V. Timofeev-Resovsky defined biogeocenosis as an elementary unit of the biological cycle, which is a very complex self-regulating system (Tyuryukanv and Fedorov 1996). The absolute dominance of vertical (inter-component) connections in the landscape facies (Sochava 1974) predetermines a high degree of correlation of its constituent geo-components and a rather rigid closure of the biogenic cycle of matter–energy metabolism (Kondrat'ev et al. 2003). "The material and energy cycles of biogeocenoses are interconnected and form a giant cycle of the Earth's biosphere" (Timofeev-Resovsky 1970, p. 90).

The mechanisms of biospheric responses to background climatic impacts are supposed to originate also at the local (topological) level because the sphere of topes is known (Sochava 1974) as the most complex and active part of the natural environment, its functional "core". The morphological parts of the landscape, i.e., terrains, stows, and facies (biogeocenoses), can be described by the route and stationary methods with the use of exact sciences more completely and deeply than the larger (e.g., regional) units of the biosphere. In this regard, the study of the structure and function of geo(eco)systems of topological order has long been associated with the concept of an experiment developed in physical geography and biogeocoenology (Sochava 1969; Sochava 1974; Dylis 1978; Krauklis 1979). This makes no sense because small natural complexes can be described using modeling methods much more completely and deeply than the larger (e.g., regional) units of the biosphere. This also includes the study of the response of local geo(eco)systems to global climate change.

The absolute domination of inter-component (vertical) connections in landscape facies (Mil'kov 1981) predetermines the high degree of correlation between its geo-components and a rather strict closure of the biogenic cycle of matter and energy exchange. "The matter and energy cycles of biogeocoenoses are interdependent and form a giant cycle of biosphere on the Earth" (Timofeev-Resovsky and Tyuryukanov 1966, p. 90). The spatial diversity of the matter and energy cycles in the biosphere is created by the structural heterogeneity of biogeocoenoses.

Biological sciences, including forest science, have made a significant contribution to the formation and development of geo-topology as a modern stage of landscape science. The term "ecotope" proposed by Vysotskiy (1909) and developed by Ramensky (1971), is used as a biological interpretation of the concept "location occupied by organisms". The concept of "forest type", by Kolesnikov (1956), is distinguished by its wide geographical content and includes not only the phytocoenotic (geo-botanical) aspect but also the ecological one, which is expressed in the detailed description of the types of forest conditions, i.e., conditions of the location of forest areas, combined in one type.

As defined by Kolesnikov (1956), *forest type* is an ensemble of tree species, which is a qualitative element in the genetic series of forest vegetation development in some landscape–geographical area or another, a stage of a particular forest-forming process typical of the given forest growth conditions. Forest type combines plots with different stages of age-related and short-term restorative successions, i.e., primary and derivative communities. It is characterized by the commonality of the principal and admixed tree species, as well as the types of undergrowth and soil cover. Finally, each forest type is characterized by its own productivity expressed in terms of quality classes. The type of forest growing conditions is defined as an abstract set of plots of the Earth's surface that produce the same forest growing effect – forest productivity (Sukachev 1972) – and have similar ecological regimes, first of all, hydrothermal and geochemical.

As a biotic core of an elementary natural ecosystem, we consider (with rare exceptions) phytocoenosis (plant community) – a qualitatively peculiar and homogeneous plot of plant cover, not further divided, with a homogeneous relationship between plants, as well as between the plants and the environment, according to Nitsenko (1971). In the context of the principles of geo(eco)system organization, the well-known interpretation of phytocoenosis as a historically organized apparatus adapted to the optimal use of the productive forces of habitat for the accumulation and transformation of matter and energy and for the construction of living matter (Sukachev 1972) is very important.

A landscape facies (biogeocoenoses) occupying an elementary homogeneous spatio-temporal range – the type of location site (LS) or physiotope, according to Neef (1968), is the limit of the geographical division of the territory. In our studies, we usually operate with facies groups. Each facies group, which is a kind of typological generalization of related types of locations (facies), corresponds not only to the root type of phytocoenosis, which is in an untouched pristine state, but also to a certain set of its derivatives that characterize the stages of spontaneous or anthropogenic changes in plant communities. The physiotope, to some extent, determines the rate of the forest-forming process, the general set of its intermediate stages, as well as their relative duration.

The group of landscape facies (biogeocoenoses) is a set of elementary geosystems located within a genetically uniform surface and close to each other in terms of their vertical structure, regime parameters, dynamic attributes of the past, and trends of future development (Sochava 1962; Mikheev 1974). The combination of facies into groups, as an essential element of ecosystem analysis, is accompanied by an inevitable generalization of their secondary features; therefore, it is necessary to distinguish the groups that would best characterize the local contrasts of location types associated with morphogenetic, lithological, exposure, hydro-edaphic, and other intra-landscape patterns. Contrasts can be revealed on the conjugate rows of facies and their groups in the *system of local landscape couplings*, i.e., on the catena (Gerrard 1981).

Examination of the catenary organization of geo(eco)systems allows for a deeper understanding of the natural laws governing their connections with relief, climate, and historical factors (Tishkov 2005). The most objective indicators of the structure and function of natural complexes are revealed when their catenary organization is taken into account, i.e., in the analysis of landscape conjugations of a particular taxonomic rank.

## 5.3 GROUND EMPIRICAL MATERIAL FOR MODELING

### 5.3.1 Landscape Structure of the Nature Zones

Landscape–zonal geo-space is cartographically expressed by its landscape texture as a form of territorial placement and positional ordering of typological landscape units, according to Simonov (1982). Landscape classification and the respective legends to the landscape map of the boreal ecotone of the Volga River basin (Figure 5.1), constructed at a working scale of 1:2,500,000, are based on the classification system proposed by A.G. Isachenko (1965, 1985). The classification criteria include latitudinal zonality, longitudinal sectorality, altitudinal layering of landscapes, and lithogenic factors (the geological foundation of a landscape with inherent tectonics and relief). The zonal groups corresponding to zonal subdivisions of the terrestrial parts of the world were accepted as classification associations of the highest rank. The zonal groups corresponding to zonal subdivisions of the terrestrial parts of the world were accepted as classification associations of the highest rank. The *zonal–sectoral types* and *sub-types of landscapes* can be distinguished by combining the zonal and sectoral criteria (associated with the degree of continentality of the climate). The types include landscapes with common bioclimatic characteristics, demonstrating the most general features of the hydrothermal regime that determine the development of a certain class of plant formations and types of soils. The groups of plant formations and soil subtypes correspond to the subtypes of landscapes (Sochava 1979; Glazovskaya 1981).

Sub-regional and local attributes of the lithogenic factor are determinative criteria at the lower steps of landscape classification. The genesis, common features, and age of the morphogenetic complexes of relief are used to distinguish *landscape genera*, i.e., genetic groups (erosion–denudation, moraine, outwash, etc.). Lithological and mechanical compositions of soil-forming rocks and the forms of meso-relief predetermine the discrimination of *landscape kinds*; however, soil and plant characteristics are widely used here as diagnostic attributes.

For the purpose of environmental forecasting related to climate change, the landscape map was "greened" so that the zonal–regional features of the background bioclimatic system were more clearly reflected in the appearance of the main operational units (landscape types). It was absolutely clear that it was necessary to increase the rank of the initial landscape units, unifying landscape types into larger categories. To designate such categories, we used the definition of "landscape groups" as interpreted by V.A.

**FIGURE 5.1** Fragment of the landscape map of the main catchment area of the Volga basin. I – designations of landscape types; in the numerator – the number of the zone or sub-zone (1 – middle taiga, 2 – southern taiga, 3 – sub-taiga, 4 – broadleaf forests); in the denominator – the number of the landscape kind (see below for the explanation). Landscapes: II – river valleys, III – swamp. Boundaries: IV – natural zones and sub-zones, V – types of landscapes.[1]

Nikolaev (1978). This definition is based on the same Polynov scheme on the scale of regional types of locations. These are fairly high-ranking unities that follow the historical–genetic classes and sub-classes of landscapes and stand above their zonal types. According to Nikolaev, landscape groups are distinguished by the types of water and geochemical regimes: the ratio of atmospheric, ground, and drip moisture, the degree of drainage of the territory, and the prevalence of removal, transit, or accumulation of mobile chemical elements.

For the purposes of background landscape–ecological prediction associated with the changes in the global–regional climate, it was necessary to "saturate" the landscape legend with ecological attributes, providing a more distinct representation of zonal and regional peculiarities of the bioclimatic system in the aspects of landscape kinds. The following two attributes (unconventional in the practice of regional landscape mapping) were added to the legend: (1) the type of sites (eluvial, trans-eluvial, transit, etc.) and (2) the degree of lithomorphicity–hydromorphicity.

On the generalized meso-catena for all zones and sub-zones, we identified four types of sub-regional locations: eluvial, trans-eluvial, transit, and accumulative (together with trans-accumulative location). All types of landscapes of the Volga basin (61 types) were combined into 20 typological landscape groups. The type of location, or group of landscapes, according to V.A. Nikolaev, is a "cross-cutting" taxonomic unit not only for the landscape types and sub-types themselves but also for the corresponding groups and classes of plant formations and types (sub-types) of soils. In order to bring landscapes, vegetation, and soils to a single taxonomic rank, we combined the types of landscapes into higher-ranking groupings, taking into account their belonging to both the natural zone (sub-zone) and the type of location, and obtained 20 typological groups of landscapes (Figure 5.2). This was an "ecologized landscape map" for the headwater of the Volga River basin.

Each typological group of landscapes is diagnosed by its inclusion in a particular natural zone (sub-zone) and location type, as well as by a property derived from these initial features – the ratio of the factors "lithomorphism–hydromorphism", which replace each other when changing types of locations on meso-catenas. As a result, it was possible to briefly characterize the most important features of each typological group of landscapes and bring closer the ranks of the landscape units we are considering with the units of plant and soil covers.

## 5.3.2 Regional Bioclimatic System

The main catchment area of the Volga River basin is to serve as an object of local and regional monitoring of the sustainable development of forests in the southern margin of the boreal belt of the East-European sub-continent under the conditions of modern global climate change. This territory is included in three world centers of extreme anthropogenic destabilization of the natural environment, with the loss of up to 80%–90% of the forest cover area (Table 5.1; Figure 5.3) and with the dominance of secondary associations (Losev 2001). Even in the taiga zone of European Russia, by the beginning of the 1990s, slightly more than 60% of forest areas remained undisturbed. In the remaining Upper Volga-Oka forests, more than 50% loss of phytomass occurred.

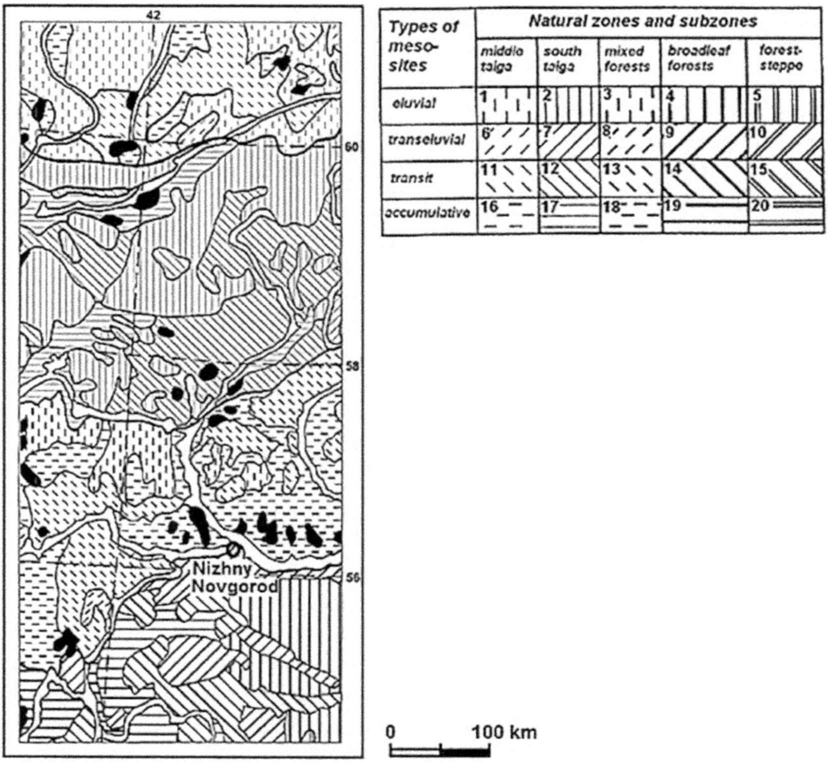


**FIGURE 5.2** Map of landscape groups for the headwater of the Volga River basin. Conventional meanings: 1–20 – numbers of landscape groups. The rest of the meanings are as in Figure 5.1.

Based on processing the recent satellite data from the digital Forest Map (Bartalev and Isaev 2004), it has been established that no more than 43% of the Oka River basin is covered with continuous and open forests. Crown density varies from 38%–45% to 56%–63% in the basin, with an average of 50.7%,

General thinning of the stands and defoliation of the crowns have resulted from fungal diseases of the trees, infestation by entomophytes, and desiccation of some species (in particular, spruce and oak), as well as from forest management stresses (Smirnova 2004). Negative processes of forest which brings a considerable part of forest areas in the Oka basin to a thin hemi-xerophytic state according to Gulisashvili (1980).

General thinning of the stands and defoliation of the crowns have resulted from fungal diseases of the trees, infestation by entomorphytes and desiccation of some species (in particular spruce and oak), as well as from forest management stresses (Smirnova 2004). Negative processes of forest ageing and degradation are developing everywhere; e.g., "the forests of Moscow region are currently losing their protective properties and their environmental, sanitary and recreational potential is decreasing" (Korotkov et al. 2020, p. 4).

**TABLE 5.1** Legend for the forest map of the Oka-Volga River basin (Figure 4.3)

| | |
|---|---|
| **Primary Plant Formations** | |
| 1 | Middle and southern taiga spruce forests and broadleaf spruce sub-taiga forests (44 a, 48 b, 49, 52 b, 53, 54), southern taiga fir–spruce (65), broadleaf–fir–spruce sub-taiga forests (68) |
| 2 | Middle and southern taiga pine forests (76, 76-1, 76-2, 76-3, 77, 78) |
| 3 | Broadleaf–pine sub-taiga forests (81, 81-1, 81-2, 81-3, 82, 84) |
| 4 | East-European forest–steppe and steppe–pine forests (87) |
| 5 | Middle and southern taiga pine forests with an admixture of larch (92) |
| 6 | Broadleaf oak and linden–oak from pedunculate oak with an admixture of other broadleaf forest species (118 a, b, c, d), broadleaf lime trees with an admixture of other broadleaf forest species (120 a) |
| 7 | Sphagnum southern taiga and sub-taiga raised bogs, pine–dwarf shrub–sphagnum (203, 204) |
| 8 | Willow forests – swampy deciduous and dark coniferous forests (233 b), nemoral floodplains – willow forests, black alder, elm, willow, sedge and oak forests (234, 236 b) |
| **Birch and Aspen–Birch Forests in Place** | |
| 9 | Southern taiga fir–spruce forests (50, 66) |
| 10 | Middle and southern taiga pine forests (79) |
| 11 | Sub-taiga fir–spruce forests (55) |
| 12 | Sub-taiga pine forests (85) |
| 13 | Forest–steppe and steppe–pine forests (88) |

*Note:* The numbers in brackets are the numbers of groups of forest formations, after (Gribova et al. 1980).

The cartographic materials we use (Gribova et al. 1980; Kotova 1987) have not lost their relevance to the present day. It has been established (Isaev 2012) that during the period from 1968 to 2008, the forest cover of the European part of Russia changed by only 6%–8%. The species structure of forests also changed insignificantly: the share of coniferous species decreased by 4%, and the share of broadleaf trees increased by 5%, which meant a slight "decrease in the share of indigenous, economically valuable coniferous and hard-leaved species" (p. 39). The ratio of areas with different age structures of forests also changed slightly.

Generally, there was an overall increase in forest cover across European Russia between 1961 and 2000 (Shvidenko et al. 2017); however, in the boreal belt, the area of small-leaf forests increased twice as much as that of coniferous forests (Smirnova 2004). The replacement of indigenous (climax) zonal forest formations of the Oka River basin (dark coniferous, mixed, broadleaf) by secondary small-leaf communities (birch and aspen forests) caused by centuries-long intensive forest and agricultural activities has led to a decrease in the efficiency of solar energy use for the production of living phytomass (Kolomyts 2022) and, consequently, to a significant loss of environmental resources by the forest cover in its feedback to the climate through the mechanisms of carbon cycle regulation, according to the concept (Gorshkov 1994; Utkin 1995; Losev 2001). At the same time, the dynamics of forest communities have increased – their sensitivity to perturbing hydrothermal signals has increased, and successional shifts

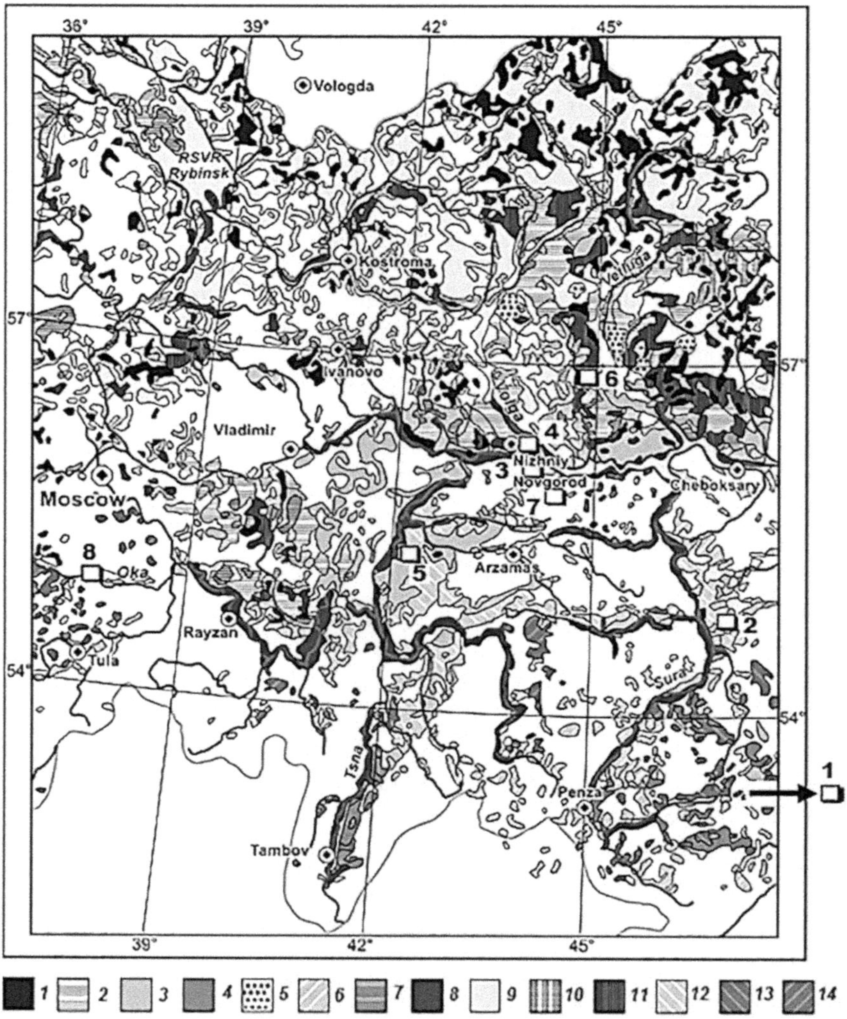


**FIGURE 5.3** Map of forests of the Oka-Volga basin, according to Isachenko and Lavrenko (1974) and location of experimental test sites. For symbols of plant formations, see Table. 5.1. Experimental test sites: 1 – Zhiguli; 2 – By-Sura; 3 – Green Town; 4 – Shchelokovsky Khutor; 5 – Vyksa; 6 – Kerzhenets; 7 – Kudma; 8 – By-Oka-Terrace Reserve.

have accelerated, seeking to return them to their previous (or new) sustainable state (Kolomyts 2022).

According to Sukachev (1975), the main representatives of natural zones and sub-zones are primary plant formations, the classification scheme of which for the territory of the East-European Plain is given in Table 5.2. According to the rules of classical phytocoenology (Sochava 1979), the phytocoenological unity identified here refers to the classes (and sub-classes) of plant formations that are regional variants (e.g., East-European or Kama-Pechersk) of the types and sub-types of vegetation (middle taiga, sub-taiga, etc.). The Volga River basin accounts for 13 classes of indigenous plant formations.

Figure 5.4 shows a basic raster map of zonal–provincial groups of plant formations of the Volga River basin. The map is based on the "Vegetation Map of the European Part of the USSR and the Caucasus" (Kotova 1987). The ranges of each phytocoenological group included indigenous forest communities (dark and light coniferous, mixed) – both modern and restored on the site of long-derivative (small-leaf – birch and aspen) forests, as well as agro-landscape complexes. Thus, this map is a model of

**TABLE 5.2** The classification scheme of primary plant formations of the natural zones of the East-European (Russian) plain

| *PLANT FORMATIONS, BY GRIBOVA ET AL. (1980)* | | | *GROUPS OF PLANT ASSOCIATIONS* | |
|---|---|---|---|---|
| *ZONAL TYPES AND CLASSES* | *REGIONAL VERSIONS* | *SUB-ZONAL SUB-TYPES* | *BRIEF CHARACTERISTICS* | *NUMBER AND SYMBOL* |
| Dark conifer and broadleaf–dark conifer forests (secondary aspen–birch) | East-European (Upper Volga region) | Middle taiga | Spruce green mosses with small shrubs | 1  |
| | | South taiga | Spruce small shrub–grass | 2  |
| | | Sub-taiga | Broadleaf–spruce complex nemorose–herbal | 3  |
| | Kama – Pechora – West Ural region | Middle and south taiga | Fir–spruce and spruce–fir grass–small shrub, with green mosses, and grass | 4  |
| | | Sub-taiga | Fir–spruce complex nemorose–herbal | 5  |
| | | | Broadleaf–fir–spruce nemorose–herbal | 6  |
| Pine and broadleaf–pine forests (secondary aspen–birch) | East-European (Upper Volga Region) | Middle and south taiga | Pine, with spruce, green mosses with small shrubs | 7  |
| | | Sub-taiga | Pine (with oak in undergrowth) small shrub–grass | 8  |
| | | | Broadleaf–pine and pine complex, with spruce | 9  |
| | | Forest–steppe and steppe | Pine and broadleaf–pine, with steppe undergrowth, and herbs–cereals | 10  |

(*Continued*)

**TABLE 5.2 (*Continued*)** The classification scheme of primary plant formations of the natural zones of the East-European (Russian) plain

| *PLANT FORMATIONS, BY GRIBOVA ET AL. (1980)* | | | *GROUPS OF PLANT ASSOCIATIONS* | |
|---|---|---|---|---|
| *ZONAL TYPES AND CLASSES* | *REGIONAL VERSIONS* | *SUB-ZONAL SUB-TYPES* | *BRIEF CHARACTERISTICS* | *NUMBER AND SYMBOL* |
| Broadleaf forest | East-European | Northern forest–steppe | Lime–oak and oak | 11a |
| | | | Lime with admixture of other broadleaf kinds | 11b |
| Typical and southern forest–steppe | of the Pontic type | Typical forest–steppe | Meadow steppes with combination of oak forests | 12 |
| | | Southern forest–steppe | Rich herb–sheep's fescue–feather grass steppes, with oak copses | 13 |

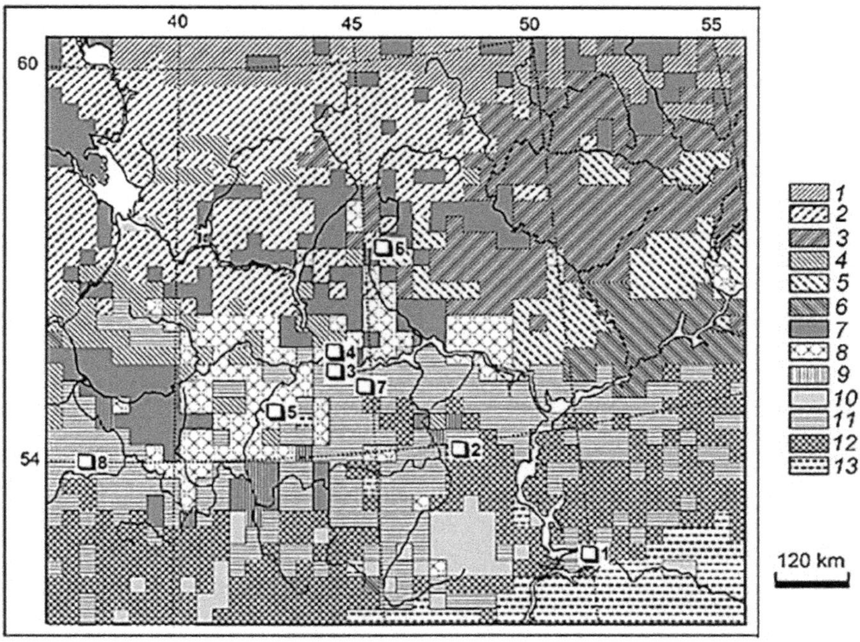


**FIGURE 5.4** Raster base map of zonal–provincial groups of indigenous plant formations (modern+restored) on the territory of the main drainage basin of the Volga River basin. The map is designed and compiled with the participation of L.S. Sharaya. Symbols see Table 5.2.

the potential bioclimatic forest system of the Volga River basin. It reflects the *zonal–provincial phyto-climatic structure* of this territory and can serve as a basis for constructing background bioclimatic forecast maps. Based on it, forecast estimates of the carbon balance of forest cover hypothetically represented only by indigenous formations will be given, which is very important for comparison with the predicted carbon balance of real forests. It will also be shown how the regulatory role of the boreal and nemoral forests of the Volga River basin in the carbon exchange between the Earth's surface and the atmosphere has changed as a result of their centuries-old anthropogenic transformation and general deforestation. Thus, we can speak about the full temporal correspondence of the analytical material obtained by us on the state of the forest formations of the Volga River basin with their distribution areas presented on the indicated geo-botanical maps. The content of these maps was used as the initial factual material for basic and predictive environmental calculations, including the carbon balance of forest formations.

The initial information for statistical data analysis was taken from any of the landscape maps, as well as from 25 maps of landscape–geophysical conditions of the headwaters of the Volga River basin (Kolomyts 2005). For this purpose, the well-known method of biogeographic grids was used (Kobeleva 1975; Puzachenko and Skulkin 1981). The step between the nodes of the square grid (points) was usually less than the average cross-section of the landscape contour. The entire territory was covered by 1,467 points. Various published and cartographic fund materials were also used (see Kolomyts 2005).

## 5.3.3 Experimental Test Sites

It is planned to carry out initial (basic) and repeated (in 25, 50, and more years) landscape–ecological surveys according to the full program on the experimental sites of the study area. Each such survey is a spatio-temporal cut of the state of forest biogeocoenoses of a particular region at a given point in time. The list of structural and functional biogeocoenotic features necessary for the construction of empirical and predictive models is given in Supplement 1.

Our local–regional basic and forecast analysis was based on the materials of large-scale landscape–ecological surveys carried out by the authors and their colleagues (1987–1998) at six experimental test sites (ranges) of the Middle Volga Region (see Figures 5.3). These test sites were chosen on the zonal ecotones where the phytocoenoses are most sensitive to climate changes. They embrace a wide range of zonal transitions from south forest–steppe to mixed forests (southern taiga appearance) and characterize the main diversity of zonal transitions on the boreal ecotone of the eastern sector of the Russian Plain. Let us enumerate these test sites:

1. The zonal ecotone of forest and steppe – oak forests and steppified pine forests of the Zhiguli low-mountain range (the region of "Samarskaya Luka" Nature Park);

2. The sub-taiga/forest–steppe ecotone – pine, mixed, and broadleaf forests of the high right-bank By-Sura (the region near the Sura River, the territory of "Chavash Varmane" National Nature Park);
3. The boreal–nemoral forest ecotone – coniferous, mixed and broadleaf forests of high-plain by-Volga region (the right-bank region of the Volga River) in the vicinity of Nizhny Novgorod; this compound ecotone system consists of two subsystems: (3) erosion loamy plain with maple–lime–oak and aspen forest (stow "Shchelokovsky Farmstead") and (4) moraine–outwash plain with spruce, pine and spruce–lime–oak forests (stow "Green Town");
4. Outwash forested lowland of the Oka-Tesha region (the right-bank strip of the Oka River) – pine forests and subors near the boundary between the mixed forest zone and the nemoral–forest sub-zone (surroundings of the town of Vyksa, Nizhny Novgorod region);
5. Moraine–outwash lowland on the trans-Volga Region (north of Nizhny Novgorod) – spruce and pine forests in the southern strip of the zone of mixed forests ("Kerzhensky" Nature Reserve).
6. Sub-zone of northern forest–steppe in the high-plain by-Volga region; lake-alluvial sandy and loamy plain on the right bank of the Volga, near its small lateral inflow – Kud'ma river; the oak–lime and pine, with spruce forest.
7. South boundary of the sub-taiga zone; lowland moraine–outwash plain on the middle left bank of the Oka River; mixed, conifer, broadleaf, and small-leaved forests (By-Oka-Terrace biosphere reserve – OTR).

Each of these test sites characterizes a particular regional ecosystem, with the corresponding conventional names: Zhiguli, By-Sura, Shchelokovsky Farmstead (Hutor), Green Town (Zeleny Gorod), Vyksa, Kerzhenets, OTR, Kud'ma.

On each test site, the empirical information (statistical sample of geo-component signs) included about 80 characteristics of the structural–functional state of forest topo-ecosystems. These characteristics consisted of 5 blocks: (1) hydro-geomorphological; (2) soil morphological; (3–4) phytocoenotic structural and functional; and (5) landscape–geophysical. The collection of factual material at each point, its processing, and subsequent calculations were made in accordance with the known methodological regulations (Glazovskaya 1964; Bazilevich et al. 1978; Anuchin 1982; Beruchashvili and Zhuchkova 1997; Shvidenko et al. 2008).

## 5.3.4 System of Local and Regional Landscape Coupling

Six facies groups, which covered the major diversity of forest types and *types of location sites (geotopes)*, have been formed based on 45–65 sample areas in each of the test sites (Table 5.3). Facies groups settled down in the vector system of local landscape couplings (micro-catenas) in the direction of increasing edaphic moistening (Figure 5.5): from eluvial (E) and trans-eluvial (TE) types of local sites (LS) through slope transit (T) and trans-accumulative (TA) to riverine accumulative (A) and bed super-aqual (Saq),

**TABLE 5.3** Groups biogeocoenoses (landscape facies) of different Middle Volga ecoregions represented by corresponding experimental sites

| *NO. OF FACIAL GROUPS* | *ECOREGIONS (EXPERIMENTAL TEST SITES)* | | | |
|---|---|---|---|---|
| | *ZHYGULI LOW-MOUNTAIN RANGE (ZHYGULI)* | | *HIGH RIGHT-BANK BY-SURA REGION (BY-SURA)* | |
| 1 | Light steppified pine forests on the steep sunny slopes | 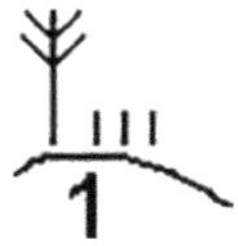 | Pine and spruce–pine forests on the upper parts of slopes of sabulous ancient alluvial plain |  |
| 2 | Subors and pine–broadleaf forests on the steep neutral and shadowy slopes |  | Pine–broadleaf and aspen forests on the flat watershed of second moraine plain |  |
| 3 | Maple–lime–oak and aspen forests with nut grove on the shadowy slopes of middle steepness |  | Oak–lime and aspen forest with nut grove on the loamy high watersheds (ecosystems of the flat-interfluves) | 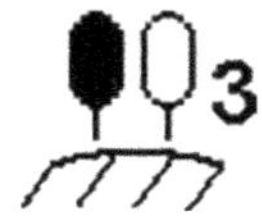 |
| 4 | Elm–maple–lime forests on the gently sloping watersheds (ecosystems of the flat-interfluves) | 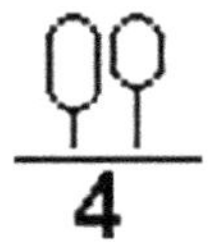 | Spruce–pine forests on the lower parts of slopes of sabulous–loamy ancient alluvial plain and high floodplain |  |
| 5 | Shadowy broadleaf forests on the lower parts of steep slopes |  | Spruce–lime–oak and pine–birch–aspen forests on the gently concave slopes of high sabulous–loamy interfluves | 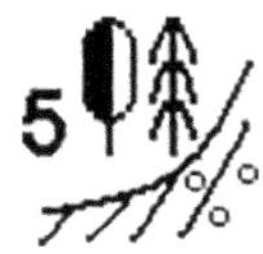 |
| 6 | Lime, maple, and aspen forests of the bottoms of small deep valleys |  | Swamped spruce–pine forests on the interfluve depressions of moraine and outwash sabulous–loamy plain |  |

*(Continued)*

**TABLE 5.3 (*Continued*)** Groups biogeocoenoses (landscape facies) of different Middle Volga ecoregions represented by corresponding experimental sites

| | *ECOREGIONS (EXPERIMENTAL TEST SITES)* | | | |
|---|---|---|---|---|
| *NO. OF FACIAL GROUPS* | *HIGH RIGHT BANK REGION OF BY-VOLGA RIVER (GREEN TOWN)* | | *HIGH RIGHT BANK REGION OF BY-VOLGA RIVER (SHCHELOKOVSKY FARMSTEAD)* | |
| 1 | Pine forests and subors on the sandy hilly watersheds |  | Maple–lime–oak forests on the near-watershed sunny loamy slopes | 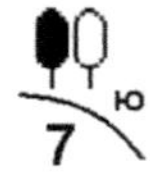 |
| 2 | Spruce forests and spruce–lime–oak forests on the flat sandy interfluves |  | Lime and aspen forests on the shadowy and neutral loamy slopes; | 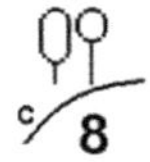 |
| 3 | Spruce–lime–oak forests and birch forests on the flat sabulous–loamy watersheds (ecosystems of the flat-interfluves) |  | Maple–lime–oak forests on the gently sloping loamy watersheds (ecosystems of the flat-interfluves |  |
| 4 | Spruce–pine forests on the gentle sandy slopes |  | Maple–lime–oak forests on the transit sunny loamy slopes | 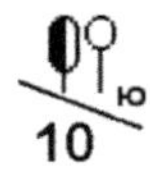 |
| 5 | Pine and spruce forests on the depressions of sandy interfluves and on above-floodplain terraces |  | Oak, lime, and aspen forests on the transit shadowy and neutral loamy slopes |  |
| 6 | Swamped floodplain small-leaf forests of the valleys of small rivers |  | Oak and lime forests on the lower parts and bottoms of loamy slopes |  |

(*Continued*)

**TABLE 5.3 (*Continued*)** Groups biogeocoenoses (landscape facies) of different Middle Volga ecoregions represented by corresponding experimental sites

| *NO. OF FACIAL GROUPS* | *ECOREGIONS (EXPERIMENTAL TEST SITES)* | | | |
|---|---|---|---|---|
| | *OKA-TESHA FORESTED LOWLAND (VYKSA)* | | *TRANS-VOLGA OUTWASH FORESTED LOWLAND (KERZHENETS)* | |
| 1 | Pine and spruce–pine forest with lichen and green moos on the apexes of sandy ridges |  | Pine forests with spruce and birch on sandy ridges |  |
| 2 | Spruce–pine forests with red bilberries on the loamy–sandy gently wavy watersheds and slopes |  | Pine forests with birch and spruce on slightly undulating interfluves of sandy ancient alluvial plain |  |
| 3 | Spruce and spruce–pine forest with wood sorrel on the sandy flat-interfluve and gently slopes (with small depth of loam occurrence) |  | Spruce and spruce–birch forests on high interfluves of sabulous–loamy moraine–outwash plain |  |
| 4 | Oak–pine–spruce forests with nemoral herbage the well-drained sandy–loamy slopes and their foots |  | Spruce–oak–lime and small-leaf forests on high floodplain and above-floodplain terraces | 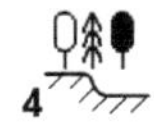 |
| 5 | Pine and spruce–pine forests with bilberries and long moos on the sandy–loamy gently concave slopes |  | Coniferous and small-leaf forests on the depressions of moraine–outwash plain |  |
| 6 | Pine and black alder forests with sphagnum and long moos on the loamy small stream valleys and floodplain depressions |  | Swamped mixed forests on the floodplain depressions and small stream valleys |  |

(*Continued*)

**TABLE 5.3 (*Continued*)** Groups biogeocoenoses (landscape facies) of different Middle Volga ecoregions represented by corresponding experimental sites

| *NO. OF FACIAL GROUPS* | *ECOREGION (EXPERIMENTAL TEST SITE)* | | | |
|---|---|---|---|---|
| | *BY-OKA-TERRACE BIOSPHERE RESERVE (BOTR)* | | *LAKE-ALLUVIAL PLAIN ON THE RIGHT BANK OF VOLGA (KUD'MA)* | |
| 1 | Pine–birch forests, with aspen and lime, on gently bulging sandy interfluves and high parts of slopes |  1 | Oak–lime primary forests on the flat loamy interfluves and adjoining slopes |   |
| 2 | Pine–spruce and spruce–pine forests on the gently sandy–loamy watersheds |  2 | Pine and pine–oak–lime, with spruce, primary forests on gently rolling sandy interfluves and their slopes |   |
| 3 | Pine–lime–oak and pine–lime forests on watersheds (with small depth of carbon eluvium) |  3 | Aspen–oak–lime and birch forests on sandy–loamy and loamy watersheds and their slopes |   |
| 4 | Lime–aspen–birch forests on high and middle parts of sandy–loamy slopes (with small depth of carbon eluvium) |  4 | Birch–aspen forests, with oak and lime, on the flat sandy watersheds and their slopes |   |
| 5 | Spruce and spruce–pine forests on middle and lower parts of sandy–loamy slopes |  5 | Oak–lime and aspen–birch forests on sandy–loamy above-floodplain terraces and watershed depressions |   |
| 6 | Swamped coniferous and small-leaf forests on the floodplain depressions and small stream valleys |  6 | Swamped alder and willow forests on the floodplain depressions and small stream valleys |   |

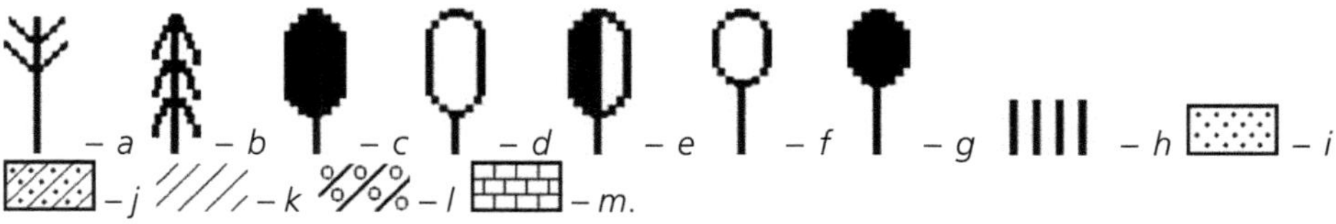


Convention meanings to symbols of biogeocoenosis groups (for all tables and figures). Dominant forest stand: *a* – pine; *b* – spruce; *c* – oak; *d* – lime, elm; *e* – broadleaf kinds without the division; *f* – birch, aspen; *g* – black alder. Grass cover: *h* – forb–meadow steppe. Soil-formatting rocks: *i* – sandy; *j* – sandy loam; *k* – middle and heavy loam; *l* – sandy–loamy moraine with small boulders; *m* – limestone bedrock.

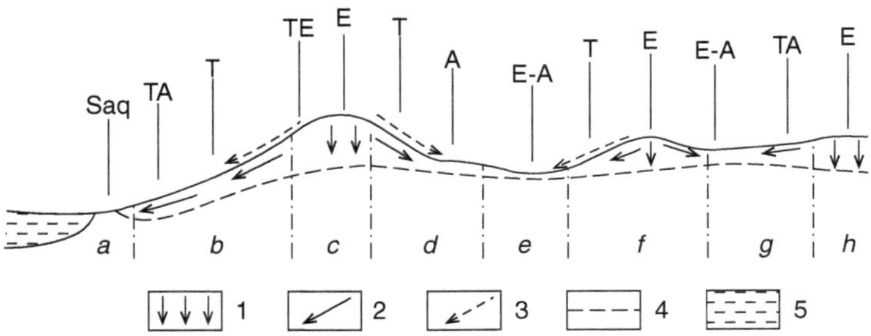


**FIGURE 5.5** Distribution of the local site types (geotopes) on the catenas under conditions of plain relief. Moisture and substance transit in soil and friable deposits: 1 – vertical; 2 – along a slope; 3 – litho-dynamic and air flows; 4 – level of the groundwater table; 5 – water body. The local site types: E – eluvial; TE – trans-eluvial; T – transit; TA – trans-accumulative; A – accumulative; E-A – eluvial–accumulative; Saq – super-aqual *a* – *h* – stows (landtype associations), or urotshistshes, in Russian transcription, as the meso-topological units; each of them may spread to several site types and consist of certain facial series.

according to classification (Polynov 1956; Glazovskaya 1964). The eluvial–accumulative sites on interfluve padings have excessive edaphic moistening as well. We unite these sites together with super-aqual ones. Each type of LS corresponded to a specific set of forest types.

*Biogeocoenosis (facies) groups* were distinguished by the combinations of local sites (geotopes) and forest type groups. According to Vysotsky's conception (1909, 1960), a group of *flat-interfluve (placor) facies* was distinguished among them. They usually occupy the eluvial sites and reflect the zonal–regional background of the given territory. The facies groups of the remaining geotopes have been considered as *extra-zonal* categories (Mil'kov 1981) as representatives of other zonal types of geographic environment, not only neighboring but also rather distant.

The collection of factual material at each test site, its processing, and the subsequent calculations were made following the known methodological regulations (see above). An example of such an experimental survey with a network of sample areas is shown in Figure 5.6.

Examination of catenary organization geo(eco)systems permits a more profound understanding of the natural laws of their connections with relief, climate, and historical factors (Tishkov 2005). The meso-site types of regional level have the same system of landscape couplings (Polynov 1956). The methodological construction "Empirical Simulation of regional Bioclimatic Trend by Local Ecosystems" (see below, Section 7.1) is grounded on the demonstrated landscape couplings of topological level.

At the same time, certain *forms of micro-relief* (minor surface landforms) have been distinguished by the degree of their dissipating or, on the contrary, accumulating influence on water and ground masses. This catenary system includes six geotopes – forms of micro-relief.

For the conventional meaning of the biogeocoenosis groups, see Table 5.3.

Among the local factors refracting this zonal–regional physical–geographical background, the following two are of crucial importance (Kolesnikov 1956): (1)

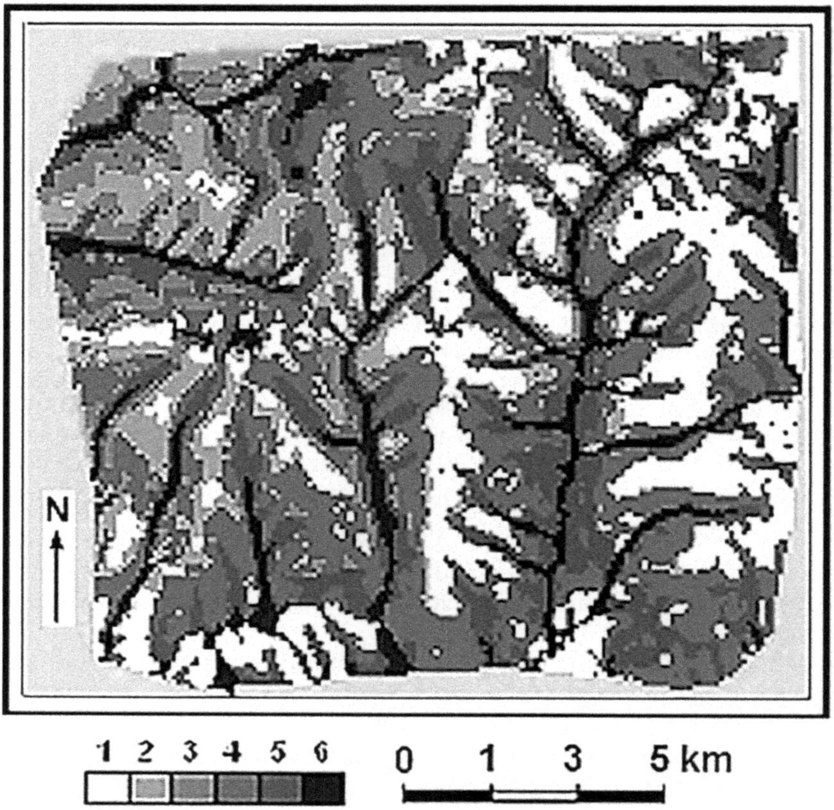


**FIGURE 5.6** A raster map of the distribution of forest biogeocoenosis groups (1–6) over the territory of the By-Oka-Terrace biosphere reserve. The map was calculated and compiled by L.S. Sharaya.

lithomorphization – updating the mineral substrate, mainly associated with the removal of dusty–clay particles from soil or their accumulation; (2) hydromorphization – the improvement or weakening of drainage, determining the degree of concentration of slope runoff. Lithomorphism and hydromorphism are the two local environmental factors that successively replace each other with the respective change in the elements and shapes of the meso-relief. The near-divide facies of the coupled paragenetic series (catenas) will primarily belong to the lithomorphic category, while the valley facies will belong to the hydromorphic and super-hydromorphic categories. Various sub-lithomorphic, mesomorphic, and sub-hydromorphic facies will take an intermediate position. We used these signs as a priority for identifying various biogeocoenotic groups in each regional ecosystem.

The type of habitat conditions is ultimately determined by the nature and intensity of biological absorption of substances obtained from the underlying water–mineral substrate, as well as the radiation energy supplied to the active surfaces of the geo(eco) system. This is the leading role of edaphic factors in the formation of the intra-landscape structure of the territory (Vysotskiy 1909; Ramenskiy 1971; Gerrard 1981).

The forms of meso- and micro-relief, together with bedrock lithology, create the genetic principle in many relationships between intra-landscape dynamic factors; therefore, the lithogenic basis is key factor in the formation of not only the landscape per se but also its parts up to facies (Solntsev 2001). Dynamically stable natural boundaries are also associated with geomorphological boundaries, at the intersection of which there is a sharp change in phytocoenosis and soil differences, indicating significant changes in metabolic processes and results. The shape of the relief is a synthetic expression of the sum of most environmental factors that determine the habitat of the studied grouping of vegetation (Kolesnikov 1956). This is the essence of the ecological multifactorial nature of the lithogenic basis of landscapes.

## 5.3.5 Large-Scale Landscape–Ecological Surveys

The first step toward understanding the local mechanisms of global changes is being implemented with the help of a methodological construction with the working title "Empirical imitation of a regional bioclimatic trend by ecosystems at the local level". In this regard, regularities of refraction of the zonal–regional bioclimatic background by local geomorphological and hydro-edaphic factors are revealed, according to the materials of the basic large-scale landscape–ecological survey. Thanks to these regularities, spatially ordered biogeocoenotic systems arise. The latter are adequate to the vector of predicted climate changes and are therefore able to simulate the main directions and scales of ecosystem rearrangements (see Table 5.3), thereby creating an empirical basis for predictions. The entire spectrum of sample plots should cover the main variety of geomorphological conditions and soil–phytocoenotic structures inherent in this ecoregion.

The following conditions must also be observed in the collection of the actual material:

1. Each geographic point (trial area) characterizing a certain homogeneous area (landscape facies is biogeocoenosis) is described by many features that are simultaneously fixed and tied to the same local location and to a certain shape of the micro-relief (the size of the trial area in our studies was usually 20×20 m, although the standard size of 50×50 m is accepted in forestry).
2. Each trial area of 20×20 m in size characterized a specific forest phytocoenosis as a stable system of phytobiota elements coexisting on a certain part of the Earth's surface and the phytocoenotic environment created by them and their predecessors (Bykov 1973). Phytocoenosis has a specific component composition and structure, a certain biological cycle, and productivity. This is the lowest topological category of the phytosphere, defined as a site, or a highlight, of a plant association (Walter 1976; Sochava 1979).
3. The points should cover, if possible, the entire territorial diversity of local geo(eco)systems according to their independent and partially dependent attributes to more fully identify the "environmental amplitude" of dependent

variables. This provides the necessary factual basis for the subsequent analysis of the spatial contiguity of geo-components and the relationships between natural complexes.

4. All signs must be "cross-cutting", i.e., be present to one degree or another at all points and have a minimum number of zero values. Thus, there is pairwise comparability of all points among themselves over the entire set of geo-component features.
5. To build models, all signs must be expressed by numerical indicators or ordered qualitative characteristics estimated at the points. The number of gradations (classes) of the attribute should not be excessively large. The experience of statistical analysis (Kolomyts 1995) shows that no more than five to seven gradations should be distinguished in fewer than 100 test plots with the modern accuracy of field observations and measurements.
6. According to views on the fundamentally probable nature of landscape patterns (Armand et al. 1969), the initial empirical material should be massive and suitable for statistical processing. The number of trial plots should be proportionate to the number of characteristics recorded in the field (A.D. Armand 1975). Only in this case will the statistical sample size reflect the general set of variations of each characteristic for a given territory.

   In our studies, 40–70 points were described at each experimental training ground; information was collected on each of them according to 60 parameters of the state of forest biogeocoenoses.
7. In a single landscape–ecological survey, the studied features characterize the "instantaneous" state of each geo-component and, in their totality, provide a certain spatiotemporal section of the natural complex with the general trajectory of its different-frequency changes (primarily seasonal and perennial), which ensures the comparability of data for all trial plots. In this regard, the entire cycle of descriptions and measurements is carried out as soon as possible so that the dynamic indicators, such as the temperature and humidity of the middle and lower layers of the soil, as well as the composition and mass of grassy plants, do not undergo significant changes. Moreover, the values of these indicators are reduced to a certain uniform term with observations at the control site. High-frequency indicators, such as the temperature and humidity of the surface air layer, are not recorded.

Each trial (sample) area characterizes a specific homogeneous range (facies isolation or biogeocoenosis) and is considered a specific micro-state of higher-ranking geosystems (stows, landscapes). It is important to emphasize that the set of structural and functional features is fixed on each trial area at the same time and is tied to the same local type of location (see below). All geo-component features are "cross-cutting", i.e., taken over the totality of biogeocenoses. This makes it possible, first, to find spatial relationships between different geo-component features, and second, to carry out a paired analysis of the topographic ecosystems by a given set of geo-components.

Information on the state of indigenous forest biogeocoenosis and their succession series is being collected according to 47 initial and 25 additional calculated parameters (see Supplement 1). Detailed information about the methods of surveys is provided in (Kolomyts 2005; Kolomyts 2022; Sharaya 2009; Kolomyts et al. 2012). The following types of work will be carried out on each sample area of size $20 \times 20\,m^2$:

- determination of the local sites type and micro-relief parameters;
- forest inventory and calculation of phytomass of undergrowth, dead stand, and dead wood;
- mowing of grass on three micro-sites, measuring on them the mass of plant litter, dead wood, and forest litter mass, as well as the mass of the roots of the ground cover and low shrubs;
- phytocoenotic descriptions of the entire sample area, with measurement and calculation of numerical parameters of the structure of the ground cover by micro-transects;
- three-time drilling, with the determination of soil temperature at standard depth and with the collection of samples for determination of moisture and chemical composition of the soil;
- schematic descriptions of the vertical soil profile according to the drilling data;
- measurements of total and absorbed solar radiation beneath the forest canopy and in a nearby forest opening; measurements of canopy density.

The materials from such surveys allow us to carry out the following analysis:

1. The description of the spatial organization of forest ecosystems will allow (1) identifying the most effective channels of external influences on the hydrothermal regime, structure, and functioning of biogeocoenoses; (2) establishing causal mechanisms for the formation of their various edaphic climaxes; (3) assess the role of the catenary organization of forest communities in the diversity of their successional dynamics and response to climatic influences; (4) to identifying forest phytocoenoses that are structurally most similar to zonal plant climaxes and, according to the parameters of the biological cycle, can serve as priority objects of global biosphere monitoring. Methods of empirical simulation of the regional climate trend by local-level ecosystems may be used, based on the properties of poly-zonality of these ecosystems as their response to global climate change. This will enrich monitoring with a more diverse range of probabilities of direct and inverse connections.
2. Based on the data from re-surveys at intervals of 25, 50, or more years, a comparative statistical analysis of catenary biogeocoenotic structures in the given territory may be carried out, along with the construction of analytical models of inter-component connections, productivity, and sustainability of forest biogeocoenoses for these spatial–temporal cuts. According to these models, it is necessary to find the main trends of changes in the conditions of forest biogeocoenoses over the past period of modern global warming.

When analyzing the materials of the main landscape–ecological surveys, the paramount prognostic objective will be to extract the signs of climate-related functional and structural exogenesis in forest biogeocoenoses over the past period and to relate these signs statistically to the trends of the main climatic parameters. The technique for this operation is described in sufficient detail in Section 3.3.

Further analysis of the materials from the first landscape survey will be performed using methods of experimental forest biogeocoenology, which are based on the study of the mechanisms of spatial organization of geo(eco)systems of topological dimension and use the ergodic properties of these systems for predictions of their climate-related dynamics. In particular, it is planned (1) to identify the causal mechanisms of the territorial organization and functioning of forest ecosystems in the given zonal–regional conditions; these mechanisms can be considered as disseminators of local spread and transformation of global–regional climatic signals; (2) to reveal the ways of transmission and transformation of these signals according to the system of inter-component and inter-complex landscape connections, with the construction of appropriate empirical models.

In the intervals between the main landscape–ecological surveys, annual data collection can be carried out according to a reduced program. These survey objects will be limited to the composition of the herbage, the productivity of the ground vegetation cover, and the temperature and humidity of the soil in the upper 50-cm soil layer. These are the functional parameters of the forest biogeocoenoses that respond almost instantly to the signals of the background climate system. As our follow-up research on the ground in the Oka-Mokshino forested lowland revealed (test site Vyksa), it is precisely such parameters that can be used to calculate scenarios for changes in forest productivity and carbon stock with upcoming inter-annual fluctuations in meteorological conditions that can simulate long-term (inter-decadal) climate changes. Thus, we are talking about empirical simulation of the functional response of forest biogeocoenoses to long-period climate fluctuations constructed with the help of vector series of placor–extra-zonal rows of topological units.

Inter-annual fluctuations in temperatures and precipitation are usually quite pronounced. Weather meteorological anomalies give us an opportunity to foresee future climatic situations when one or another anomaly may become a longstanding norm. The meteorological conditions of each year will be considered as an analog of hydrothermal conditions for a particular period of the climate forecast. Then it will be possible to verify the expected calculated parameters of the state of forest ecosystems (productivity, sustainability, carbon balance, etc.) according to the empirical material of landscape surveys of this year, which will turn out to be a hydrothermal analog of a certain interval of the predicted climatic trend. This will be one of the stages of the "work" described above (see clause 4.6) of the changing operating system of long-term geosystem monitoring.

## 5.4 PARAMETERS OF THE BIOLOGICAL CYCLE AS INDICATORS OF ECOSYSTEM STABILITY

In modeling, a high emphasis was placed on the phytocoenotic functional block – the parameters of annual and perennial cycles of plant matter turnover, which are especially important for prognostic ecological investigation. Such turnovers are the most important indicators of ecosystem resistance to external disturbance (Odum 1971; Gerasimov

1985). The multidimensional ecological analysis of various phytomass categories (live and dead, foliage and skeletal, on separate biogeohorizons, etc.) was performed, and the integral indices of mass exchange were determined. Let us enumerate the measured and calculated discrete parameters of the biological cycle, with indications of their designations.

### 5.4.1 Phytomass (t/ha)

1. wood mass, BW;
2. skeletal tree–shrub phytomass, BS;
3. total live above-ground phytomass, BL;
4. root mass, BR;
5. total green mass, BV;
6. total live phytomass (above- and underground), BC;
7. forest litter mass, ML;
8. dead skeletal mass (brushwood and dead wood), WD;
9. total dead above-ground phytomass, BD;
10. humus mass (it has been in view the labile humus in upper soil layers), HU.

### 5.4.2 Productivity (t/ha per year)

11. total annual production of forest phytocoenosis, PC (≈ $P_{prim}$; NPP);
12. annual growth of skeletal phytomass, PS;
13. annual production of green mass, PV.

### 5.4.3 Complex Indices of Biological Cycle

14. productivity to phytomass ratio, CE = PC/BC;
15. allometric coefficient, CA = log(PC)/log(BC);
16. coefficient of annual phytomass turnover (rotation) – the ratio of green mass production to the total above-ground live phytomass, CT = PV/BL;
17. coefficient of annual destruction (known as the litter-fall index as well) – the ratio of forest litter mass to the mass of green substance produced annually, CY = PV/ML, years;
18. coefficient of perennial phytomass destruction – the ratio of dead above-ground mass to the total live above-ground mass, CP = BD/BL;
19. coefficient of utilization of dead above-ground phytomass – the ratio of the difference of the two total above-ground phytomasses, live and dead, to the total live above-ground mass, CU = (BL – BD)/BL;
20. index of the rate of perennial phytomass destruction, ID = BD/(PV + PS), years.

The first four complex indices are already known, and we have proposed the rest of them.

Factual material was collected at each point (test site), processed, and subsequent calculations were made in accordance with the known methodological regulations.

It is necessary to elucidate the complex indices of the short-term biological cycle. The parameter CE is a measure of the ecological efficiency of the phytoproduction process (Odum 1971). The coefficient CT indicates the portion of live above-ground phytomass involved in its annual turnover. The coefficient CY is also known as the litter-fall index (Bazilevich and Rodin 1971), which is expressed as the period (in years) necessary for the accumulation of a given mass of forest litter (dead plants) at the existing annual rates of plant mortality and organic matter decomposition. The parameter CP characterizes the balance of the ratio of creative and reduced branches of the plant matter perennial cycle during the succession of several generations of stands. The index CU represents the portion of phytoproduction that is completely decomposed and transformed into organic–mineral soil substance in the subsequent detritus chain of turnover, with losses for gas exchange with the atmosphere and for the production of saprophages.

## 5.5 THE IMPORTANCE OF STUDYING LOCAL GEO(ECO)SYSTEMS FOR LANDSCAPE–ECOLOGICAL FORECASTS

The ecological prediction developments known to date are, with few exceptions, very schematic and aimed mainly at assessing the general future condition of the biosphere as a planetary system (Emanuel et al. 1985; Bonan et al. 1992; Holten et al. 1993; Gorshkov 1994; Tarko 2005). Accordingly, prediction maps are too small-scale, and cartographic prognosis confines itself to the schematic demonstration of changes in the boundaries of natural zones and areas of some plant formations. The regional and local levels of ecological prognoses have not been sufficiently developed yet due to the lack of factual material and methodological difficulties in transferring hydro-climatic prognoses from global to regional and local (topological) levels.

This part of the monograph report expounds on the main statements of the original ecological concept of prediction: "Global Changes on the Regional and Local Levels", as a basis for regional and local ecosystem monitoring under global anthropogenic climate changes. The objects of the research are (1) forest biogeocoenoses (landscape facies) as the elementary unit of geographical ecology (see further) and (2) landscape–zonal systems of the headwater of the Volga River basin and its surroundings. The conservation and reproduction of forest resources under changing climatic conditions at the southern boundary of the temperate forest zone, where forest communities are in conditions close to critical, is one of the current fundamental ecological problems.

As is known, global ecosystem monitoring is most up-to-date and actually realizable on the scale of individual ecological regions. However, natural processes and events on the regional and local (topological) hierarchical levels are characterized by the greatest diversity and high discreteness (Sochava 1974); therefore, the regional response to global climatic changes inevitably takes the form of multiple reactions of vegetation, soils, and landscapes as a whole to background climatic signals. So far, there has been no distinct notion of this multiplicity because the measure of the sensitivity of soil–biotic components to climatic changes in different zonal–climatic and geomorphological conditions has not yet been estimated.

While developing this concept, the author faced significant methodological difficulties. Most of them arose while assessing the transformation of global climatic signals into local ones, proportional to the scales of regional landscapes and their parts, down to biogeocoenoses. At present, there are actually no experiments on the determination of the most probable directions and rates of distribution of hydrothermal signals from global to topological (local) levels. The response of ecological niches of local ecosystems to the behavior of background climatic systems is also unclear. The regional hydrothermal trends are generally described by two initial parameters: temperature and precipitation (Budyko 1971; Sykes and Prentice 1993; Houghton et al. 1996; Aber et al. 2001). However, the connections of these parameters with landscape facies (biogeocoenoses) are very weak and often statistically unreliable.

It was necessary to establish the causal mechanisms of local responses to global and regional signals by revealing the transfer functions in landscape connections, which transform these signals when passing through the coupled ensembles of natural complexes and their components. We have established empirically (Kolomyts 2022) that micro-relief (small-scale landforms) and summer soil moisture content are the main channels of connection between local ecosystems and the background climatic system. The former factor exerts a multilateral effect on the geographical differentiation of local natural complexes, while the latter is the most powerful direct ecological factor. The solution to this problem is associated in many respects with the study of the internal cycle of plant matter in various natural ecosystems and the estimation of biological cycle equilibrium as a necessary condition for their stability (Odum 1971). These statements entirely concern forest ecosystems and need substantial specification. The relatively closed cycle of phytomass transformation, including two branches – plant mass production and decomposition – is the major part of ecosystem function.

The forthcoming global warming also makes it necessary to solve the complex ecological problems associated in many respects with the study of the mechanisms of changes in the carbon cycle in the "Earth's surface–atmosphere" system. This cycle, firstly, drives all other biogeochemical cycles and, secondly, determines the content of greenhouse gases in the atmosphere and thus has a regulating effect on global climatic processes. These problems have already been highlighted in numerous papers (Kobak 1988; Bonan et al. 1992; Smith et al. 1992; Gorshkov 1994; Gorshkov et al. 1999; Aber et al. 2001; Osipov 2004; Tselniker 2006; Zalihanov et al. 2006). High emphasis is placed on the development of a procedure for determining carbon pools in different components of phytobiota and in soil, as well as the creation of a database on the current carbon content in forests, marshes, and other ecosystems, with territorial generalizations by natural and administrative regions. Biosphere processes determining the "work" of the carbon cycle under various natural conditions and anthropogenic successions are also considered in detail.

The problems of a quantitative assessment of responses of different biotic components of the carbon cycle to global climatic changes are illustrated much less. Accordingly, there are actually no predictions of the climate-genic dynamics of factors realizing the biotic regulation of carbon dioxide content in the atmosphere. Meanwhile, a solution to these problems is necessary for revealing mechanisms that provide stability of various zonal–regional subdivisions of the biosphere under the forthcoming global warming (Izrael 1984; Moiseev et al. 1985; Tarko 2005).

The present research aims at further development of the concept of biotic regulation of the carbon cycle in the biosphere (Gorshkov 1994; Tarko 2005) by studying local regularities of the minor biological turnover, which provides environmental sustainability in accordance with Le Chatelier's principle. As has already been mentioned, the mechanisms of biospheric responses to background climatic impacts are supposed to originate at a local (topological) level. The sphere of topes is known (Sochava 1974) as the most complex and active part of the natural environment. The absolute domination of inter-component (vertical) connections in the landscape facies (Mil'kov 1981) predetermines the high degree of correlation between its geo-components and rather strict closure of the biogenic cycle of matter and energy exchange; therefore, biogeocoenosis is quite a complex self-regulating system (Sukachev 1972). "The matter and energy cycles of biogeocoenoses are interdependent and form a giant cycle of the biosphere on the Earth" (Timofeev-Resovsky and Tyuryukanov 1966, p. 90).

The spatial diversity of the matter and energy cycles in the biosphere is created by the structural heterogeneity of biogeocoenoses. As far as we know, the concept of biotic regulation of the carbon cycle has not yet been developed in such a statement of the problem.

The forest cover is one of the leading factors providing for the stability of the continental biosphere. Forest ecosystems are the most potent regulators and stabilizers of natural biogeochemical cycles (Sukachev 1972), including the carbon cycle, as well as the temperature regime of the atmosphere (Gorshkov 1994; Gorshkov et al. 1999). Global-scale deforestation is considered to be one of the main causes of the increasing greenhouse effect.

## NOTES

1 LANDSCAPE CLASS: **plain landscapes**
LANDSCAPE TYPE: **temperate continental typical boreal**
Genera and kinds of landscape
**Middle taiga**

I. *Erosion–denudation landscapes (**E-D**):* 1/1 – hydro-lithomorphic, loamy, eluvial, fir–spruce landscapes with green mosses, grasses, undershrubs, and wood sorrel.
II. *Denudation–outwash landscapes (**D-O**):* 1/3 – **hydro**-lithomorphic loamy–sandy, transit, fir–spruce landscapes with green mosses, grasses, and undershrubs.
III. *Morainic landscapes (**M**)*: 1/4 – hydro-**lithomorphic**, eluvial, spruce–pine, and spruce undershrub landscapes with loamy soil, more rarely with loamy sand; 1/5 – spruce forests eluvial, hydro-lithomorphic, grass–shrub and sorrel, with sod-podzolic residual carbonate soils.
IV. *Morainic–outwash landscapes (**M-O**)*: 1/8 – sub-hydromorphic sandy, more rarely loamy, transit, spruce-pine, and pine, with green mosses, haircap moss, and undershrubs.
V. *Outwash landscapes (**O**):* 1/10 – sub-**hydromorphic**, trans-accumulative, spruce–pine with green mosses, sphagnum, and undershrubs, on the thick sands.
VI. *Lake-glacial and lake-alluvial landscapes (**L/Al**)*: 1/11 – hydromorphic, accumulative, loamy and sand, spruce and spruce–pine undershrub, haircap moss, **and** sphagnum.

**South taiga**

I. *Erosion–denudation landscapes (**E-D**)*: 2/1 – **hydro**-lithomorphic, loamy, eluvial, fir–spruce grass–undershrub.

II. *Denudation–outwash landscapes (**D-O**)*: 2/4 – hydro-lithomorphic, transit, fir–spruce and pine grass–undershrub landscapes on the loamy sands with the **moraine**.

III. *Morainic landscapes (**M**)*: 2/5 – hydro-lithomorphic, hilly eluvial, spruce grass–undershrub forests with loamy sands, and loamy soils.

V. *Outwash landscapes (**O**)*: 2/7 – sub-hydromorphic, trans-accumulative, pine, with) green mosses on the thick sands; 2/8 – spruce–pine forests, trans-**accumulative**, hydromorphic, long-moss–sphagnum, on sands underlain by moraine, with podzolic-bog soils.

VI. *Lake-glacial and lake-alluvial landscapes (**L/Al**)*: 2/9 – hydromorphic, loamy, accumulative spruce, and pine grass–sphagnum.

VII. *Valley–outwash landscapes (**V/O**)*: 2/10 – hydromorphic, accumulative, spruce–pine, mossy, with sands and loamy sands.

**Sub-taiga landscapes/**

III. *Denudation-outwash landscapes (**D-O**)***:** 3/5 – hydro-lithomorphic, transit, broadleaf–pine undershrub–grass, with sands and loamy soils.

IV. *Morainic landscapes (**M**)*: 3/7 – hydro-lithomorphic, eluvial, complex forb spruce nemorose–herbal landscapes of the flat hilly area with loamy sands **and** loamy soils.

V. *Morainic–outwash landscapes (**M-O**)*: 3/8 – sub-hydromorphic, sandy–loamy, transit, complex forb spruce nemorose–herbal, and mossy spruce–pine landscapes.

VI. *Outwash landscapes (o)*: 3/10 – hydromorphic, sandy (on the moraine), in places marshy, trans-accumulative, spruce, and spruce–pine, with green **mosses** and sphagnum of the plane area.

VII. *Lake-alluvial and lake-glacial landscapes (**L/Al**)*: 3/11 – hydromorphic, loamy, Altransit, spruce complex forb, and grass–sphagnum landscapes.

VIII. *Valley–outwash landscapes (**V-O**)*: 3/12 – hydromorphic, marshy, accumulative, pine and spruce–pine undershrub–grass–mossy landscapes with loamy sands and sands.

**Broadleaf forest (Northern forest–steppe)**

I. *Erosion–denudation landscapes (**E-D**)*: 4/1 – lithomorphic, loamy, eluvial, oak–lime–tree nemorose–herbal landscapes of the rolling-ouval area; 4/3 – lithomorphic, heavy loamy, and clayey eluvial, oak–lime–tree nemorose–herbal landscapes with herb layer of rolling–ouval area.

II. *Erosion landscapes (**E**)*: 4/6 – lithomorphic, carbonate loamy, trans-eluvial, oak–lime–tree nemorose–herbal landscapes of rolling area, with dark-gray forest soils.

III. *Denudation–outwash landscapes (**D-O**)*: 4/5 – oak-linden, trans-eluvial, lithomorphic, nemoral–grassy, with gray forest soils; 4/7 – sub-lithomorphic, transit, broadleaf–spruce–pine undershrub–grass, and mossy landscapes of the rolling area with loamy sands; 4/8 – sub-lithomorphic, transit, broadleaf–spruce–pine undershrub–grass, and mossy landscapes of the rolling area with sands and loamy, on the rocks under karstification.

V. *Outwash landscapes (**O**)*: 4/10 – hydro-lithomorphic, trans-accumulative, spruce–pine, and pine grass-moss, lichen, and sphagnum landscapes with sands and loamy, with slight steppization; 4/11 – sub-hydromorphic, trans-accumulative, broadleaf–pine, and spruce–pine, grass–moss, and sphagnum landscapes, with sands and loamy on the moraine.

VI. *Lake-glacial and lake-alluvial landscapes (**L-Gl/Al**)*: 4/12 – sub-lithomorphic, transit, oak–lime–tree, with pine and spruce, nemorose–herbal landscapes of the flat rolling area, with loamy soils and loamy sands; 4/13 – sub-lithomorphic loamy, on the moraine and sands, transit oak–lime–tree nemorose–herbal landscapes of the flat rolling area.

# Landscape–Geophysical Foundations of Empirical-Simulation Predictive Modeling

# 6

## 6.1 CHAIN REACTIONS IN THE REGIONAL LANDSCAPE–GEOPHYSICAL RELATIONS

As provided by the maps of plant and soil covers of European Russia, bioclimatic objects are considered to be as follows: (1) the groups and classes of plant formations combined into zonal (sub-zonal) types (sub-types) of vegetation, according to Lavrenko and Isachenko (1976), and (2) the groups of soils isolated based on soil–climatic relations having a priority significance in the formation of soil cover zonal structure (Volobuev 1974).

The hydro-climate description of the territory of the Volga River basin and its surroundings was based on 25 landscape–geophysical maps of the Volga River basin (1:2,500,000), representing different links of contemporary turnover and transformation of solar energy and atmospheric moisture (Kolomyts 2005). These results were used for paleogeographical and prognostic constructions in light of global climate changes (Kolomyts et al. 2012). The taxonomic norms of hydrothermal parameters (their weighted average values for the territory) were calculated for each of the zonal–regional


DOI: 10.1201/9781003733119-6

phytocoenological units, which allowed for the comparison of landscape–ecological parameters of the past, present, and future.

The system of regional hydrothermal parameters that may be used for prognostic modeling (as well as for paleogeographical reconstructions) in connection with global climate changes is examined by the example of the Volga River basin and its surroundings. These parameters were calculated by the author for three prognostic periods by the GISS model: 2050, 2075, and 2100, as well as for the Mikulino interglacial period (125 thousand years ago) and the Holocene optimum (5–6 thousand years ago). The analysis was based on 25 landscape–geophysical maps of the Volga River basin (1:2,500,000), representing different links of contemporary turnover and transformation of solar energy and atmospheric moisture (Kolomyts 2005; Kolomyts et al. 2012). In the territory of the basin, there are about 170 weather stations, 120 agro-meteorological stations, and more than 300 observation points. The step between the nodes of the square grid (points) was usually less than the average cross-section of the landscape contour. The taxonomic norms of hydrothermal parameters (their weighted average values for the territory) were calculated for each of the zonal–regional phytocoenological units, which allowed for the comparison of landscape–ecological parameters of the past, present, and future.

The total list of landscape–geophysical parameters is given in Table 6.1. Natural vegetation is defined at the level of classes of plant formations and soils, i.e., at the level of soil species groups. The methods of information theory were used to construct a model of chain reactions in the regional bioclimatic system of the Russian Plain (Figure 6.1).

The model shows the main channels of connections that serve as conductors of external impacts (both natural and anthropogenic), for example, global warming or

**TABLE 6.1** List of landscape–geophysical factors used in modeling

| *No* | *NAME OF SIGN* | *SIMBOL* |
|---|---|---|
| 1 | Annual total radiation, MJ/m$^2$ | $Q_{sum}$ |
| 2 | Annual radiate balance, MJ/m$^2$ | $R_{ann}$ |
| 3 | Average January temperature, °C | $t_{Jan}$ |
| 4 | Average July temperature, °C | $t_{July}$ |
| 5 | Sum of the biological active temperature, °$^s$C | $\Sigma\, t_{daily} \geq 10^0$ |
| 6 | Annual potential evaporation, mm | $E_o$ |
| 7 | Duration of vegetation period, days | $T_{veg}$ |
| 8 | Totals of precipitation per year, mm | $r_{ann}$ |
| 9 | Sum of the precipitation of the cold period, mm | $r_{cold}$ |
| 10 | Maximum height of snow cover (field), cm | $h^{snow}_{max}$ |
| 11 | Sum of the precipitation of the warm period, mm | $r_{warm}$ |
| 12 | Annual evapotranspiration, mm | $E_c$ |

(*Continued*)

**TABLE 6.1 (*Continued*)** List of landscape–geophysical factors used in modeling

| *No* | *NAME OF SIGN* | *SIMBOL* |
|---|---|---|
| 13 | Annual complete flow, mm | $S_{com}$ |
| 14 | Annual surface flow, mm | $S_{ann}$ |
| 15 | Annual groundwater flow, mm | $U_{ann}$ |
| 16 | Flow coefficient (Runoff) | $C_{flow}$ |
| 17 | Total humidification | $W_{tot}$ |
| 18 | July soil moisture resources in stratum 0–20 cm | *W*–20 |
| 19 | July soil moisture resources in stratum 0–50 cm | *W*–50 |
| 20 | July soil moisture resources in stratum 0–100 cm | *W*–100 |
| 21 | Budyko's radiate index of the drought | $I_{Bud}$ |
| 22 | Bazilevich's index of aridity | $I_{Baz}$ |
| 23 | Vysocky–Ivanov's atmospheric humidity factor | $F_{hum}$ |
| 24 | Selyaninov's hydrothermal coefficient | HTC |
| 25 | Rikhter's snow–temperature coefficient | STC |
| 26 | Simonov's coefficient of continentality | $C_{contin}$ |
| 27 | January latitude continentality, by Polozova | $C_{JanC}$ |
| 28 | July latitude continentality, by Poloziva | $C_{JulC}$ |
| 29 | Annual primary productivity of natural ecosystems (net primary production – NPP), t/h | $P_{prim}$ (NPP) |

any hydro-technical construction. The normalized coefficients of binary relations of phenomenon ***A*** with factor B in these channels *C(A;B)*=0.20–0.25 and more, which corresponds to the coefficient of correlation above 0.8.

Even a cursory glance at the results of information analysis (Table 6.2; Figure 6.1) shows the leading role of not only the primary input (background frame) but also the processor material–energy parameters, which, as is known (Zonn and Chikichev 1976), clearly indicate the general zonal structure of the territory of the Russian Plain. The initial input variables with the maximum mutual independence are as follows: annual total radiation ($Q_{sum}$), annual precipitation ($r_{ann}$), types of morphostructures and morphosculptures (MST+MSC), and mechanical composition of soil-forming rocks ($MC_{soil}$).

Global climate changes are transmitted through the annual radiation balance ($R_{ann}$) and precipitation: annual ($r_{ann}$) and of the warm period ($r_{warm}$). The first level of

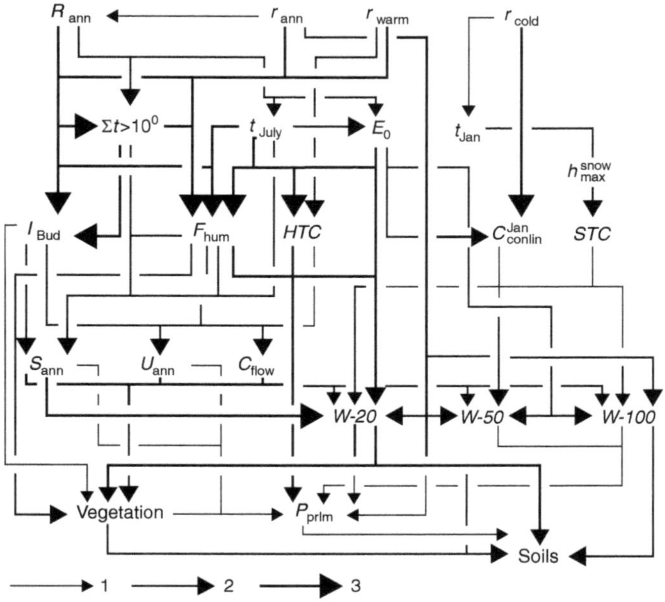


**FIGURE 6.1** The information statistical model of landscape–geophysical organization of the Volga River basin. Standardized contingency coefficients ***C(A;B)***: 1 – 0.050–0.120; 2 – 0.121–0.200; 3 – 0.201–0.300. The abbreviations are shown in the text and in Table 5.1.

**TABLE 6.2** Information indicators of relationships (see Chapter 3) of classes of plant formations and groups of soil types with landscape–geophysical characteristics of the Volga River basin

| *GEO-COMPONENT FEATURES (NAME AND DESIGNATION)* | *VEGETATION* | | *SOILS* | |
|---|---|---|---|---|
| | *C(A;B)* | *C(A/B)* | *C(A;B)* | *C(A/B)* |
| Annual radiate balance, MJ/m², $R_{ann}$ | 0.103 | 0.265 | 0.087 | 0.226 |
| **Average January temperature, $t_{jan}$** | 0.085 | 0.129 | 0.030 | 0.111 |
| January latitude continentality, $C_{Cont}^{Jan}$ | 0.142 | 0.372 | 0.101 | 0.290 |
| Average July temperature, $t_{july}$ | 0.132 | 0.355 | 0.090 | 0.264 |
| Sum of biological active temperatures, $\sum t \geq 10^0$ | 0.131 | 0.354 | 0.091 | 0.270 |
| **Length of the vegetation period, $T_{veg}$** | 0.193 | 0.374 | 0.150 | 0.326 |
| **Annual potential evaporation, $E_0$** | 0.154 | 0.350 | 0.109 | 0.266 |
| Sum of the precipitation of the cold period, $r_{cold}$ | 0.047 | 0.164 | 0.034 | 0.121 |
| Maximum height of the snow cover, $h_{max}^{snow}$ | 0.072 | 0.230 | 0.054 | 0.180 |

*(Continued)*

**TABLE 6.2 (*Continued*)** Information indicators of relationships (see Chapter 3) of classes of plant formations and groups of soil types with landscape–geophysical characteristics of the Volga River basin

| *GEO-COMPONENT FEATURES (NAME AND DESIGNATION)* | *VEGETATION* | | *SOILS* | |
|---|---|---|---|---|
| | *C(A;B)* | *C(A/B)* | *C(A;B)* | *C(A/B)* |
| Sum of the precipitation of the warm period, $r_{warm}$ | 0.121 | 0.335 | 0.079 | 0.241 |
| Annual surface flow, $S_{ann}$ | 0.115 | 0.323 | 0.077 | 0.250 |
| Annual groundwater flow, $U_{ann}$ | 0.147 | 0.381 | 0.082 | 0.263 |
| Runoff coefficient, $C_{flow}$ | 0.095 | 0.271 | 0.073 | 0.214 |
| July soil moisture in stratum 0–20 cm, *W-20* | 0.235 | 0.367 | 0.227 | 0.372 |
| July soil moisture in stratum 0–50 cm, *W-50* | 0.191 | 0.359 | 0.187 | 0.367 |
| July soil moisture in stratum 0–100 cm, *W-100* | 0.192 | 0.379 | 0.167 | 0.363 |
| Budyko's radiate index of the drought, $I_{Bud}$ | 0.117 | 0.328 | 0.076 | 0.244 |
| Atmospheric humidity factor, $F_{hum}$ | 0.138 | 0.367 | 0.082 | 0.262 |
| Hydrothermal coefficient, s | 0.142 | 0.374 | 0.093 | 0.275 |
| Snow–temperature coefficient, STC | 0.086 | 0.262 | 0.064 | 0.212 |
| Classes of the plant formations, *Vegetation* | – | – | 0.110 | 0.303 |
| Soil groups, *Soils* | 0.111 | 0.315 | – | – |
| Primary bioproductivity, $P_{prim}$ | 0.074 | 0.182 | 0.049 | 0.123 |

transformation of external signals is composed of thermal parameters of the vegetation period: the mean July temperature ($t_{July}$), total biologically active temperature ($\Sigma\, t \geq 10^0$), and annual evaporativity ($E_0$). These signals are further transmitted to complex geophysical indices: Budyko radiation index of drought ($I_{Bud}$), annual humidity factor ($F_{hum}$), and summer hydrothermal coefficient (HTC). The winter hydrothermal signal directed from rainfall of the cold period ($r_{cold}$) to the mean January temperature ($t_{Jan}$) and the height of snow cover $\left(h_{max}^{snow}\right)$ weakens on the Richter snow–temperature coefficient and on winter continentality $\left(C_{contin}^{Jan}\right)$.

We express the degree of atmospheric humidification by two coefficients (indexes): (1) atmospheric humidify factor –Vysotsky–Ivanov's index $F_{hum}(1)$ as a ratio of annual precipitation $r_{ann}$ to evaporativity $E_0$ (potential evapotranspiration):

$$F_{hum}(1) = r_{ann} / E_0; \tag{6.1}$$

and (2) Budyko's radiation dryness index ($I_{Bud}$):

$$I_{dry} = R_{ann} / r_{year}\ \gamma, \tag{6.2}$$

where $r_{year}$ is the total of precipitation per year and $\gamma$ is the latent heat of vapor formation.

The next level of signal transformation is formed by the expendable parts of the water balance: annual surface and groundwater flow ($S_{ann}$ and $U_{ann}$) and flow coefficient ($C_{flow}$). All geophysical relations are focused on hydro-edaphic parameters: July stored soil productive moisture in the 20-, 50-, and 100-cm-thick strata (*W-20*, *W-50*, and *W-100*). These attributes integrate the whole system of heat and moisture cycle in landscapes and are the most powerful ecological factors directly affecting the distribution of vegetation and soils, as well as the primary bioproductivity ($P_{prim}$). The water content in upper soil horizons is most sensitive to external impacts. Rather close relations of (*W-20*) and (*W-50*) with $F_{hum}$ and HTC ($C(A;B)$=0.168–0.213) indicate that the summer water content in soil on the Russian Plain is zonal as a whole. Consequently, soil water storage can be used for the prognosis of total reconstruction of soil vegetation cover and its productivity for certain scenarios of regional climate changes.

According to the method proposed by Puzachenko and Skulkin (1981), the relative contribution of the main geophysical factors to the distribution of the classes of plant formations ($Y_1$) and soil groups ($Y_2$) might be expressed as the following polynomials:

$$Y_1 = 0.129 \cdot t_{Jan} + 0.355 \cdot t_{July} + 0.164 \cdot r_{cold} + 0.335 \cdot r_{warm} + 0.017 \cdot X_i; \qquad (6.3)$$

$$Y_2 = 0.111 \cdot t_{Jan} + 0.274 \cdot t_{July} + 0.121 \cdot r_{cold} + 0.241 \cdot r_{warm} + 0.123 \cdot B_{prim} + 0.109 \cdot Z^{gr}_{water} + 0.021 \cdot X_i. \qquad (6.4)$$

Here, $Z^{gr}_{water}$ is the depth of occurrence of subsoil waters; the coefficients of arguments are the values of the information reception coefficient, and the residual term $X_i$ is the unaccounted factor (the total of all coefficients is equal to 6.1). As is seen, the joint effect of summer temperatures and precipitation is nearly 2.5-fold more efficient than that of the same winter characteristics, being significantly higher for vegetation than for soils (69% vs. 52%). Consequently, during the vegetation period, natural ecosystems must experience the maximal structure–function reconstructions under global warming. At the same time, the primary productivity of ecosystems as an integral parameter of their functioning quite weakly depends on the distribution of plant formations and zonal soil types. Bioproduction is directly related to the thermal conditions and duration ($T_{veg}$) of the vegetation period, as well as to the annual and summer coefficients of humidification. The coefficients of information reception by parameter $P_{prim}$ of $r_{ann}$, $t_{July}$, and $T_{veg}$ are equal to 0.439, 0.324, and 0.316, respectively. Thus, the dynamics of primary bioproductivity in the middle zone of the Russian Plain (from south taiga to grass steppe) will depend not only on the anthropogenic warming per se but even more on the concomitant changes in atmospheric humidification.

One more important pattern should be noted. The chain reaction model shows (see Figure 6.1) that the productivity of forest communities is a functional parameter, quite autonomous from their very structure, which has already been empirically proven many times (Puzachenko and Skulkin 1981; Diakonov 2002). Autotrophic biogenesis directly depends on the hydrothermal conditions of the surface atmosphere. Annual evaporativity (i.e., mainly the temperature of the growing season) and the atmospheric humidity factor affect $P_{prim}$ 3–4 times more than the structure of the vegetation cover or soil. In the first case, the parameter *C(A;B)* is 0.200–0.240, which corresponds to a correlation coefficient ≥ 0.75–0.85), and in the second, it does not exceed 0.05–0.08, i.e., it turns out to be below the confidence level (0.07).

# 6.2 METHODS FOR CALCULATING LANDSCAPE–GEOPHYSICAL CHARACTERISTICS

The initial climatic parameters of global and regional paleoreconstructions and prognoses are, as is known, $t_{Jan}$ and $t_{July}$, as well as $r_{ann}$, sometimes with the division of the latter into $r_{cold}$ and $r_{warm}$. As regards other geophysical parameters for the territory of the Volga River basin, we propose to calculate them by the empirical formulas (Table 6.3).

**TABLE 6.3** Formulas for paleo-predictive calculations of particular and complex landscape–geophysical characteristics, with corresponding correlation coefficients *R* and determination $R^2$, according to the initial hydrothermal parameters: $t_{Jan}$, $t_{July}$, and $r_{ann}$

| *CALCULATION FORMULAS* | *R* | $S^2$ |
|---|---|---|
| $R_{ann}=378.8 \cdot t_{July} - 6.667 \cdot t_{July}{}^2 - 3180$ | 0.96 | 0.92 |
| $\Sigma\, t \geq 10°C = 235.008 \cdot t_{July} - 2287C$ | 0.94 | 0.88 |
| $T_{veg}=5.6 \cdot t_{July} + 2.87 \cdot t_{Jan} + 102$ | 0.93 | 0.87 |
| $E_0 = 1384 - 161.6 \cdot t_{July} + 6.245 \cdot t_{July}{}^2$ | 0.94 | 0.88 |
| $r_{cold}=0.3339\ r_{ann} - 1.1372 \cdot t_{Jan} + 43$ | 0.86 | 0.74 |
| $E_c=0.7375 \cdot r_{ann} - 0.9761 \cdot S_{ann} + 118$ | 0.83 | 0.69 |
| $I_{Bud} = 0.0833 \cdot t_{July} - 0.0015 \cdot r_{ann} + 0.4$ | 0.92 | 0.85 |
| $F_{hum} = 0.001726 \cdot r_{ann} - 0.1954 \cdot t_{July} + 3.72$ | 0.95 | 0.90 |
| $F^*_{hum} = 0.001362 \cdot r_{ann} - 0.1069 \cdot t_{July} + 2.25$ | 0.96 | 0.92 |
| $HTC = 5.353 - 0.2621 \cdot t_{July} + 0.0038 \cdot r_{warm}$ | 0.94 | 0.88 |
| $HTC^* = 28.37 \cdot t_{July} + 0.0529 \cdot t_{July}{}^2$ | 0.95 | 0.90 |
| $(W\text{-}20) = 0.0257 \cdot r_{ann+} 0.0676 \cdot S_{ann} - 0.0136 E_0$ | 0.88 | 0.77 |
| $(W\text{-}50) = 0.0631 \cdot r_{ann+} 0.1746 \cdot S_{ann} - 0.0231 E_0$ | 0.89 | 0.79 |
| $(W\text{-}100) = 0.0964 \cdot r_{ann} + 0.3216 \cdot S_{ann} - 0.0802 \cdot E_0$ | 0.86 | 0.74 |
| $P_{prim} = 0.0139 \cdot r_{ann} - 0.2064 \cdot t_{July} + 0.055 \cdot T_{veg} - 4.22$ | 0.95 | 0.90 |
| $P_{prim} = 0.0258 \cdot r_{ann} - 0{,}00000973 \cdot r_{ann}{}^2 - 2.27$ | 0.92 | 0.85 |
| $P_{prim} = 2.001 \cdot t_{July} - 0.0716 \cdot t_{July}{}^2 - 1.92$ | 0.82 | 0.67 |
| $CY = \exp(6.453 - 0.2993 \cdot t_{July})$ | 0.84 | 0.70 |
| $CY = 7.93 - 0.19993 \cdot t_{jan} - 0.5356 \cdot t_{july} + 0.00351 \cdot r_{ann}$ | 0.78 | 0.61 |
| $h^{snow}_{max} = 0.0871 \cdot r_{ann} - 5.083 \cdot t_{jan} - 80$ | – | – |

*Notes:* *For the regions with average, $t_{July} > 19°C$. $T_{veg}$ – the length of the vegetation period (number of days)

The summer-stored soil productive moisture is directly associated with the average July temperature, but it may also be measured by $E_0$ and $S_{\text{ann}}$. The offered formulas for the calculation of soil water reserves belong only to open territories of agro-ecosystems. The transition to the same parameters for forest-covered lots ($W_f$) may be realized with the help of the following equalities (based on the data from the Near-Moscow water balance station, $R$=0.87–0.90):

$$(W-20)_f = 16.7 + 0.834(W-20); (W-50)_f = 35.6 + 0.814(W-50); \tag{6.5}$$

$$(W-100)_f = 53.0 + 0.853(W-100). \tag{6.6}$$

The assessment of past and future values of the annual surface and groundwater flow must be more differential. For the Holocene optimum and the predicted period of 2050, climate changes evidently should not go beyond contemporary variations (for the last 100 years), for which there are data from instrumental observations. Therefore, for these calculated periods, we may use contemporary relations expressed by the equation of regression:

$$S_{\text{ann}} = 0.1028 \cdot r_{\text{ann}} - 37.724 \cdot t_{\text{July}\,I} + 794;\ R = 0.90; \tag{6.7}$$

$$U_{\text{ann}} = 0.0776 \cdot r_{\text{ann}} + 0.2079 \cdot S_{\text{ann}} - 38;\ R = 0.87. \tag{6.8}$$

The calculation of the $C_{\text{flow}}$ is carried out according to the following formula:

$$C_{\text{flow}} = r_{\text{ann}} / (S_{\text{ann}} + U_{\text{ann}}). \tag{6.9}$$

For the periods of 2075 and 2100, according to the prognostic GISS model, the temperature increase will be accompanied by an increase in atmospheric humidification, which has not been observed before. A similar situation probably occurred in the optimum of the Mikulino interglacial period, when the climate was not only warmer but also much more humid than the modern one, according to (Velichko 2010; Velichko et al. 1983; Velichko and Klimanov 1990). For the calculation of the annual surface flow, the following single-factor relation can express these phenomena:

$$S_{\text{ann}} = \exp(2.402 + 0.004 \cdot r_{\text{ann}}); R = 0.82. \tag{6.10}$$

The formula for the calculation of January latitude continentality, by Polozova (1954), has the shape:

$$C_{\text{JanC}} = \left(a_{\max}^{+} - a_i\right) / \left(a_{\max}^{+} - a_{\max}^{-}\right) \cdot 100\%, \tag{6.11}$$

where $a^{+}_{max}$ and $a^{-}_{max}$ are maximal positive and negative anomalies of average January temperature at a given latitude, respectively, and $a_i$ is the real anomaly in given point.

The calculation of the snow-–temperature coefficient is carried out according to the following formula:

$$\mathrm{STC} = 10 \cdot t_{\mathrm{aver}} \,/\, h_{\mathrm{aver}}. \tag{6.12}$$

Here, $t_{\mathrm{aver}}$ and $h_{\mathrm{aver}}$ are average negative temperature of air and average height of snow for the period with stability snow cover, respectively.

Ecological prognoses need to assess the extent, to which the contemporary character of territorial relations of temperature and precipitation was typical of the past and will remain in the future (under global anthropogenic warming). We have first managed to compare calculation-prognostic scenarios of climatic conditions (Figure 6.2).

It turned out that the whole prognosis trend on the Russian Plain is characterized by an inversely proportional statistical relation of summer temperatures and annual precipitation: the warmer regions have always been and will be drier and vice versa. Consequently, in the nearest decades, the vast territory of the East-European sub-continent will experience the same climate-forming mechanism of atmospheric circulation as at present. These properties of self-similarity of the regional bioclimatic system allow the method of actualism to be used for the construction of both paleo-geographical and prognostic scenarios of landscape–ecological conditions based on modern hydrothermal relations.

A fairly clear picture of the distribution of zonal types and groups of plant formations in the hydrothermal field of the Volga River basin is given in Figure 6.3.

It is significant to evaluate the possible changes in the primary productivity of terrestrial ecosystems that are influenced by global warming. Reconstruction and prognosis

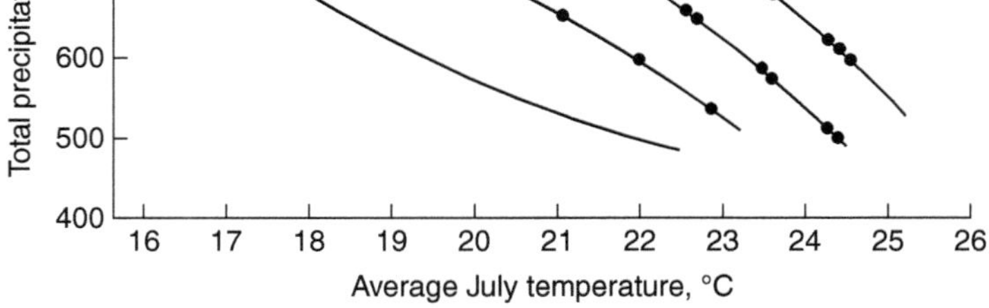


**FIGURE 6.2** Space interrelations between average July temperature and annual precipitation for different ages and dates (the correlation coefficient is shown in the brackets). The ages: 1 – contemporary conditions (0.80); the climatic prognosis by the GISS model for dates: 2 – 2050 (0.86); 3 – 2075 (0.87); 4 – 2100 (0.86).

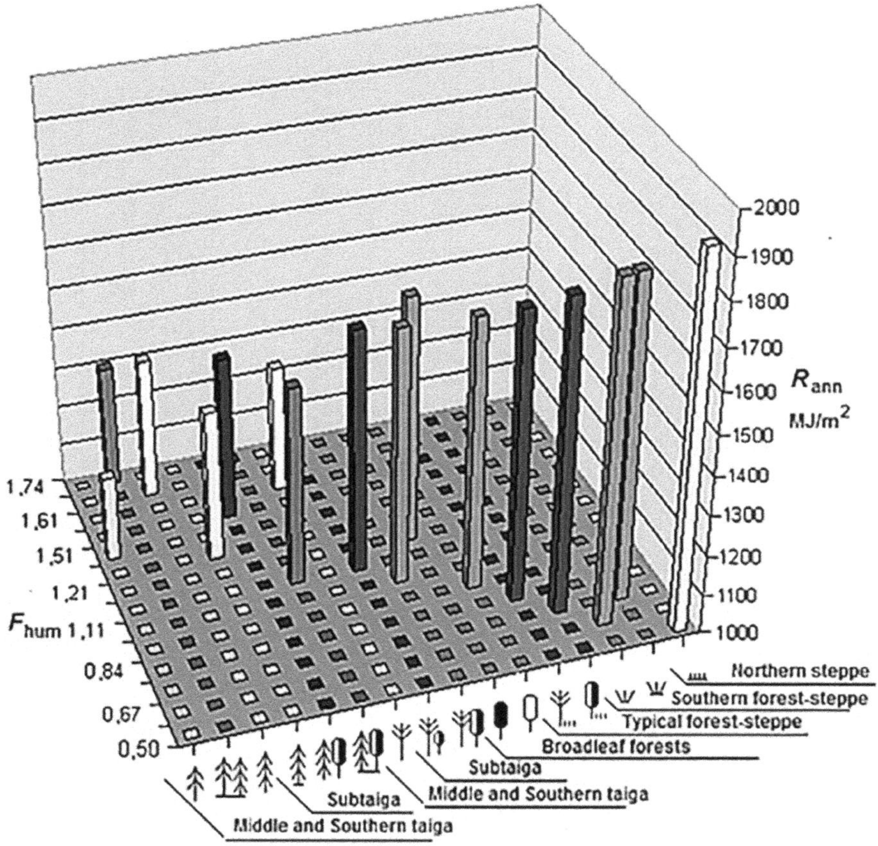


**FIGURE 6.3** The distribution of zonal types and groups of plant formations in the field of values of the annual radiation balance and the annual moisture coefficient. Designations of groups of formations are in Table 5.2.

of the primary productivity of natural ecosystems can be estimated by two methods. The first method is based on calculations of the past or future values of the annual radiation balance and the index of drought. For this purpose, we have obtained an empirical relation between the annual radiation balance (based on direct actinometric observations) and the average July temperature in the Volga River basin and adjacent regions of the Russian Plain: middle and southern taiga, Priladozhye and Prionezhye, as well as the Onega and Northern Dvina basins, the entire steppe and forest–steppe left bank of the Dnieper River, forest–steppe and northern steppe regions of the Central Russian Upland, and finally, dry steppe and semi-deserts of the Lower Don and Northern Caucasia.

As a result, an empirical relationship was obtained between the annual radiation balance and the mean July temperature. The connection takes the form of a parabola (Figure 6.4), and for most of the territory, it is straightforward:

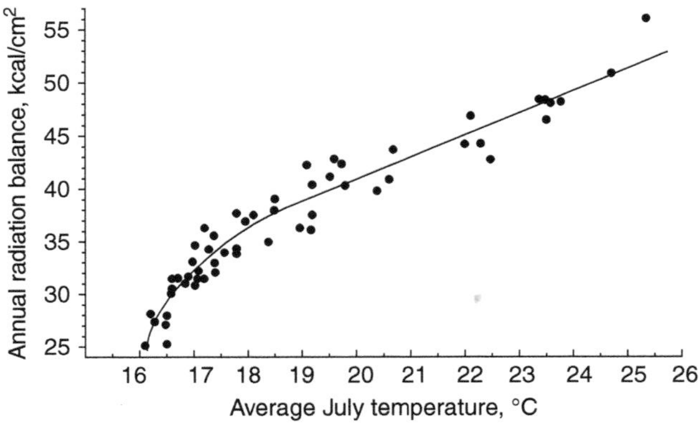


**FIGURE 6.4** Correlation between the annual radiation balance and the average July temperature for the territory of the Volga River basin and adjacent regions of the Russian Plain (data from actinometric and meteorological observations).

$$R_{ann} = 0.0214 \cdot t_{July} + 0.0044 \cdot t_{July}{}^{2} + 12.03; \; R = 0.88; \; R^2 = 0.77. \tag{6.13}$$

The map of contemporary primary productivity of natural ecosystems is built on the basis of previously established (Kolomyts et al. 1993) regional empirical relations of bioproductivity and annual radiation balance $R_{ann}$ and Budyko's radiation index of drought (aridity) $I_{Bud} \approx I_{dry}$ (Figure 6.5).

Utilizing this relationship, it is possible to derive the predicted values of annual radiation balance and, accordingly (with consideration of atmospheric humidification), the radiation index of aridity. Then the background values of the primary bioproductivity of landscapes are determined using an empirical nomogram (see Figure 6.5) for the given calculated epochs and periods.

The nomogram of relationships is based on selected regions of the Russian Plain and adjacent plain and mountain territories that, first, have known factual values of productivity for different zonal–regional plant formations from northern taiga to forest–steppe and northern steppes (Bazilevich 1993), and second, have available data on radiation balance and atmospheric precipitation. The nomogram utilized more than 80 points with factual values of the primary bioproductivity of landscapes, annual radiation balance, and annual precipitation total. To obtain a more leveled picture of bioproductivity, particular values of the latter in the fields of $R_{ann}$ and $I_{arid}$ were grouped into pleiads (groups), and the mean value of productivity was calculated for each of them. Thus, we managed to obtain a family of straight lines in the range of 3–14 ton/ha that characterized an unambiguous dependence of productivity on specified geophysical parameters for each pleiad.

Now the south taiga forests have the maximum annual $P_{prim}$ (10.5–11.5 tons/ha per year, while northern steppe formations have the minimum (6–7 tons/ha). According to prognostic calculations, the primary bioproductivity can increase everywhere by 1.5–2.0 tons/ha as early as the 2050s as a consequence of the rise in both temperature and precipitation; by 2075, even in dry steppe, it will reach 9.5–10.0 tons/ha. It can be

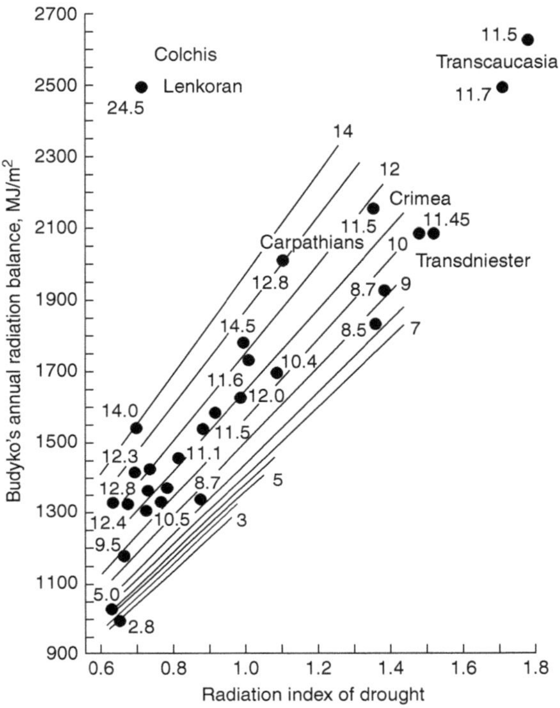


**FIGURE 6.5** Distribution of primary productivity of natural ecosystems (t/h/year) in the coordinate field of annual radiation balance and the drought index for the territory of the Russian Plain and its surroundings. Legend: 2.8, 5.0, 9.5, etc. – average productivity values for a given galaxy of points.

anticipated that in the 2100s hydrothermal fields will be far beyond their contemporary values for the Russian Plain territory (the same occurred in the optimum of the Mikulino interglacial period). At present, there are no combinations of high $R_{ann}$ values (more than 1,900–2,000 MJ/m$^2$) and very low values: $I_{Bud}$ (below 0.7–0.8), when the productivity of vegetation must exceed the value of 14 tons/ha per year. Such hydrothermal conditions are typical of contemporary humid forest formations in the extreme south of the sub-boreal belt and even the subtropical belt, where the annual productivity of forests exceeds 20 tons/ha. These data were used for prognostic maps of the primary productivity of natural ecosystems of the Volga River basin for the periods of 2075 and 2100.

Primary bioproductivity can also be calculated by the initial hydroclimatic characteristics: temperature and precipitation, which per se appear as climatic prognosis (Figure 6.6). Each nomogram has been created based on climatic standards from 120 meteorological stations of the Volga River basin and empirical data on primary bioproductivity. An important addition to the cited materials is the taxonomic (weighted average over the territory) norms of landscape–geophysical parameters for various groups of plant formations and landscape groups of the Volga River basin (see Kolomyts 2005). Tables 6.4 and 6.5 contain the norms for the basic parameters. Taxonomic norms

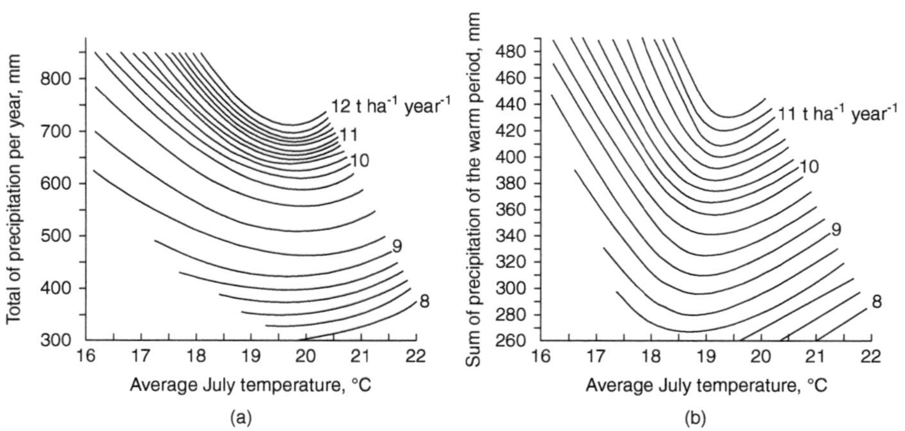


**FIGURE 6.6** Dependence of primary bioproductivity on average July temperature and total precipitation per year (a) and for the warm period (b).

**TABLE 6.4** Taxonomic norms of the main hydro-climatic parameters for various groups of plant formations of the Volga River basin

| *NATURE ZONES (SUB-ZONES)* | *GROUPS OF FORMATIONS* | *PARAMETERS* $R_{год}$ | $T_{JAN}$ | $T_{JULY}$ | $\Sigma T \geq 10°$ | $E_0$ | $R_{ANN}$ | $R_{WARM}$ | $I_{BUD}$ | $F_{HUM}$ |
|---|---|---|---|---|---|---|---|---|---|---|
| Middle and southern taiga |  | 1,212 | −12.2 | 16.0 | 1,517 | 415 | 768 | 477 | 0.77 | 1.80 |
| |  | 1,305 | −15.6 | 16.7 | 1,589 | 456 | 697 | 447 | 0.64 | 1.74 |
| |  | 1,352 | −11.6 | 17.2 | 1,722 | 430 | 759 | 466 | 0.70 | 1.43 |
| |  | 1,320 | −11.9 | 17.4 | 1,782 | 478 | 737 | 480 | 0.77 | 1.49 |
| Mixed forests |  | 1,411 | −10.2 | 17.3 | 1,948 | 461 | 763 | 456 | 0.80 | 1.55 |
| |  | 1,377 | −14.7 | 18.0 | 1,811 | 491 | 679 | 425 | 0.79 | 1.23 |
| |  | 1,496 | −13.9 | 18.6 | 2,011 | 531 | 630 | 406 | 0.91 | 1.08 |
| |  | 1,604 | −12.2 | 18.8 | 2,178 | 546 | 650 | 401 | 1.02 | 1.21 |
| |  | 1,635 | −12.0 | 19.2 | 2,300 | 568 | 652 | 402 | 1.03 | 1.12 |

*(Continued)*

**TABLE 6.4 (*Continued*)** Taxonomic norms of the main hydro-climatic parameters for various groups of plant formations of the Volga River basin

| *NATURE ZONES (SUB-ZONES)* | *GROUPS OF FORMATIONS* | *PARAMETERS* | | | | | | | | |
|---|---|---|---|---|---|---|---|---|---|---|
| | | $R_{год}$ | $T_{JAN}$ | $T_{JULY}$ | $\Sigma T \geq 10°$ | $E_0$ | $R_{ANN}$ | $R_{WARM}$ | $I_{BUD}$ | $F_{HUM}$ |
| Broadleaf forests |  | 1,617 | −11.0 | 18.7 | 2,199 | 559 | 655 | 405 | 1.02 | 1.07 |
| |  | 1,674 | −11.5 | 19.6 | 2,356 | 631 | 602 | 359 | 1.15 | 0.94 |
| Typical forest–steppe |  | 1,719 | −12.5 | 20.2 | 2,024 | 679 | 556 | 342 | 1.25 | 0.80 |
| |  | 1,775 | −11.1 | 20.5 | 2,686 | 691 | 536 | 318 | 1.13 | 0.90 |
| Southern forest–steppe |  | 1,842 | −12.3 | 21.6 | 2,580 | 792 | 507 | 301 | 1.44 | 0.72 |
| |  | 1,803 | −14.9 | 20.9 | 2,503 | 734 | 509 | 298 | 1.36 | 0.62 |
| Northern steppe |  | 1,918 | −14.3 | 22.2 | 2,608 | 830 | 438 | 282 | 1.69 | 0.57 |

*Note:* For the designation of groups of plant formations, see Table 5.2; the remaining notation is the same as in Table 5.2.

**TABLE 6.5** Taxonomic standards of any hydro-climatic parameters for various landscape groups of the Volga River basin (see in Figure 5.2)

| *NATURAL ZONES (SUB-ZONES)* | *TYPES OF MESO-SITES* | *LANDSCAPE GROUPS* | *HYDRO-CLIMATE PARAMETERS* | | | | | | | |
|---|---|---|---|---|---|---|---|---|---|---|
| | | | $R_{ANN}$ | $T_{JAN}$ | $T_{JULY}$ | $R_{YEAR}$ | $F_{HUM}(1)$ | W-20 | W-100 | $P_{PRIM}$ |
| Middle taiga | E | 1 | 1,260 | −13.0 | 17.1 | 725 | 1.89 | 34 | 162 | 9.4 |
| | TE | 2 | 1,305 | −15.6 | 16.7 | 697 | 1.68 | 27 | 134 | 11.0 |
| | T | 3 | 1,196 | −10.6 | 16.1 | 782 | 1.85 | 37 | 180 | 10.6 |
| | TA–A | 4 | 1,166 | −14.0 | 16.6 | 770 | 1.78 | 30 | 162 | 11.0 |
| South taiga | E | 5 | 1,366 | −12.5 | 17.4 | 727 | 1.72 | 32 | 172 | 11.2 |
| | TE | 6 | 1,348 | −15.1 | 18.0 | 676 | 1.44 | 26 | 130 | 10.3 |
| | T | 7 | 1,334 | −11.2 | 17.0 | 754 | 1.71 | 40 | 149 | 12.1 |
| | TA–A | 8 | 1,334 | −11.5 | 17.3 | 731 | 1.71 | 30 | 156 | 11.4 |
| Mixed forests | E | 9 | 1,429 | −12.0 | 17.6 | 714 | 1.36 | 23 | 99 | 11.5 |
| | TE | 10 | 1,498 | −11.2 | 18.1 | 698 | 1.17 | 22 | 118 | 10.9 |
| | T | 11 | 1,464 | −11.2 | 18.0 | 721 | 1.29 | 25 | 125 | 11.4 |
| | TA–A | 12 | 1,427 | −11.4 | 17.4 | 754 | 1.31 | 29 | 136 | 11.1 |

(*Continued*)

**TABLE 6.5 (*Continued*)** Taxonomic standards of any hydro-climatic parameters for various landscape groups of the Volga River basin (see in Figure 5.2)

| *NATURAL ZONES (SUB-ZONES)* | *TYPES OF MESO-SITES* | *LANDSCAPE GROUPS* | *HYDRO-CLIMATE PARAMETERS* $R_{ANN}$ | $T_{JAN}$ | $T_{JULY}$ | $R_{YEAR}$ | $F_{HUM}(1)$ | *W-20* | *W-100* | $P_{PRIM}$ |
|---|---|---|---|---|---|---|---|---|---|---|
| Broadleaf forests | E | 13 | 1,676 | −12.7 | 19.0 | 618 | 1.00 | 19 | 106 | 10.6 |
| | TE | 14 | 1,570 | −11.6 | 18.5 | 649 | 1.09 | 21 | 89 | 10.8 |
| | T | 15 | 1,684 | −11.4 | 19.6 | 640 | 1.09 | 16 | 79 | 10.6 |
| | TA–A | 16 | 1,590 | −11.2 | 19.0 | 688 | 1.12 | 17 | 86 | 11.9 |
| Forest–steppe and northern steppe | E | 17 | 1,774 | −13.8 | 21.1 | 523 | 0.73 | 11 | 56 | 8.4 |
| | TE | 18 | 1,786 | −10.8 | 21.4 | 512 | 0.78 | 10 | 45 | 9.0 |
| | T | 19 | 1,801 | −11.4 | 20.5 | 558 | 0.95 | 12 | 44 | 9.5 |
| | TA–A | 20 | 1,841 | −13.8 | 21.4 | 476 | 0.74 | 8 | 46 | 7.4 |

were calculated based on the results of an information statistical analysis of the connections between phenomena (phytocoenological, soil, or landscape units) with one or another factor – a geophysical attribute. For this purpose, a matrix was constructed for the distribution of all selected factors over the gradations of this phenomenon. To calculate the norm, we used the central values of all gradations of the geophysical parameter, as well as the matrix of normalized (columns) partial coupling coefficients (see A.D. Armand 1975; Kolomyts 1998), which were introduced into the calculations as "weighted" coefficients.

## 6.3 SPACE RELATIONS BETWEEN TEMPERATURE AND PRECIPITATION

For landscape–ecological paleo-reconstructions and prognoses, it is very important to assess the extent to which the contemporary character of the territorial conjugacy of temperature and precipitation was typical of the past and will persist in the future. Superposed graphs show (see above, Figure 5.4) that the Middle Belt of the Russian Plain, in the whole paleo-forecasting trend, is characterized by an inversely proportional statistical relationship between summer thermal conditions and annual atmospheric humidification: the warmer regions have always been and will be drier, and vice versa. An analogous pattern is observed in the many-year series of modern climatic variations. According to the data from the Nizhny Novgorod meteorological station (Myza) for a 44-year period (1929–1972), when the eastern and meridian types of circulation dominated over the territory of the Russian Plain and the western type was weak, there were weak negative correlations between $r_{ann}$ and rt with the mean temperature of the warm period ($R$=–0.59 and –0.52, respectively). As shown in Budyko (1971), an analogous situation should be anticipated in the middle latitudes under global anthropogenic warming.

One can suppose a certain general tendency towards decreasing atmospheric humidification as summer temperatures become higher, which is in agreement with contemporary spatial landscape–geophysical relations. There is actually no such correlation for winter warming ($R=0.34$). Thus, with respect to the warm year period, with its parameters, as has already been mentioned, determining the structure and function of natural ecosystems to the utmost, the properties of self-similarity of the paleo-prognostic system become quite apparent.

Hence, taking into account the very close correlation between $t_{July}$ and $r_{ann}$, one can consider that the late Pleistocene and Holocene had the same climate-forming mechanism of atmospheric circulation across the entire territory of the East-European sub-continent as it is at present, and this mechanism will obviously persist in the coming decades. This conclusion is of great methodological importance. It proves the validity of the method of actualism in the construction of prognostic scenarios of landscape–ecological conditions for a given territory.

The positive thermal anomalies of the warm half of the year were previously and will later be accompanied by a greater or lesser, but actually occurring throughout the entire territory, increase in atmospheric humidification. This is typical not only of the Mikulino (Eem) interglacial period and the predicted periods (by the GISS model) up to 2050 but also of the climatic optimum of the Holocene. While the first is in agreement with the generally accepted view, the second is rather unexpected. It is generally thought that the Atlantic period of the Holocene in the moderate climate of the Russian Plain was warmer but, for the most part, drier than at present (Girs 1971; Kolomyts 1998). However, this is not supported by the data obtained. The climatic optimum at the major watershed of the Volga River basin differed from the modern period apparently not only in higher temperature but also in higher annual (and therefore summer) rainfall.

Let us present some general conclusions. Prognostic constructions should proceed from the fact that the structure–functional reorganization of natural ecosystems in the southern half of the boreal belt and in the northern area of the sub-boreal belt of the Russian Plain, induced by global climatic variations, must be determined, first, by the alteration of hydrothermal conditions of the warm (but not cold) period of the year and, second, both by temperature increase or decrease and changes in precipitation (with the respective shifts in edaphic moistening), but not by temperature rise or decline. Meanwhile, the century and super-century variations of the climate must initiate the structural–functional shifts in vegetation and soils mainly via the changes in summer productive moisture in the soil.

## 6.4 REGIONAL SCENARIOS OF CHANGES IN PRIMARY BIOPRODUCTIVITY IN THE SYSTEM OF BACKGROUND CLIMATE FLUCTUATIONS

To implement such a regional scenario, we turn to the nomograms of dependences of background bioproductivity on the main climatic characteristics: the mean July temperature *t*July, annual precipitation $r_{ann}$, and precipitation of the warm period $r_{warm}$ (see Figure 6.6). The nomograms show quite a dramatic change in the sign of the bonds visible in the range $t_{July} = 19°–20°$; for annual precipitation, the critical value is $t_{Jul} \approx 19.5°$. The increase in summer temperature is accompanied by a decrease in precipitation at $t_{July} < 19.5°$ and by an increase at $t_{July} > 19.5°$. By crossing the critical temperature threshold, the hydrothermal structure of the annual humidity factor per se changes (see Table 6.2). It turned out that the isolation of the critical radius takes place on the East-European sub-continent in the transition zone from the boreal to the sub-boreal belt (Bazilevich et al. 1986; Kolomyts 2005), and in many places, it approaches the main landscape boundary of the Russian Plain, according to Mil'kov (1981).

In this relatively narrow transitional zone, drastic changes occur in the natural ecosystems of the zonal type due to the most important climatic boundary: the $F_{hum}$ crossing 1. The boreal belt differs from the northern (forest–steppe and steppe) part of the sub-boreal belt in terms of the migration of substances (mainly abiotic in the first case and biogenic in the second case), the type of background soil formation process (podzolic and soddy, respectively), and, finally, the reserves of organic matter and primary bioproductivity (Bazilevich et al. 1986).

Based on empirical nomograms (see Figure 6.6) and using the ergodic properties of the climate system (Rauner 1981), it was possible to obtain model ideas about the changes in productivity with the increase in the mean July temperature $t$ with a step ($t_0 \rightarrow t_1$) and the simultaneous change in precipitation $r$ (Figure 6.7). It was taken into account that warming can be accompanied by both an increase ($r_0 \rightarrow r_1 \rightarrow r_2 \rightarrow$...) and a decrease ($r_0 \rightarrow r_1 \rightarrow r_2 \rightarrow$...) in precipitation. Let us consider the model situations of the ratios of considered parameters ($\Delta t$, $\Delta r$, and $\Delta$PC) separately for each bioclimatic belt. We emphasize that these situations relate mainly to the zonal (and sub-zonal) types

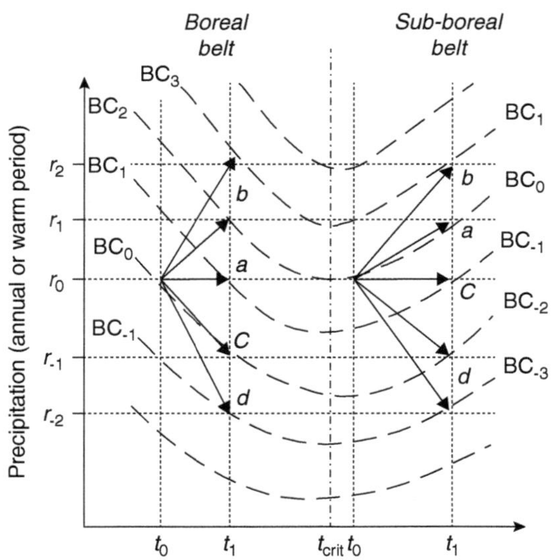


**FIGURE 6.7** Model representations of the dynamics of primary bioproductivity under various scenarios of changes in average July temperature and total precipitation (annual or warm period). Explanations are in the text.

(and sub-types) of plant cover, its local representatives being biogeocenoses of the eluvial (placental) series.

In the *boreal belt*, with an increase in the mean July temperature ($+\Delta t = t_1 - t_0$), the following scenarios are possible:

a. If precipitation remains unchanged ($\Delta r=0$), the primary bio productivity increases, although it is relatively low ($+\Delta(PC)=PC_0 \rightarrow PC_1$); for example, in the range $r_{ann}$=700–800 mm, an increase in $t_{July}$ by 1° (17° → 18°) gives an increase in productivity by 600–900 kg/ha; with annual precipitation of 570–630 mm, this increase does not exceed 200–300 kg/ha;
b. A much greater increase in the productivity of boreal forests will occur during the thermo-humid trend when summer warming is accompanied by a progressive increase in precipitation ($+ \Delta r=r_0 \rightarrow r_1 \rightarrow r_2; +\Delta(PC)=PC_0 \rightarrow PC_2 \rightarrow PC_3$);
c. Bioproductivity can remain at the initial level ($\Delta(BC)=0$) if precipitation decreases with increasing temperature ($-\Delta r=r_0 \rightarrow r_{-1}$), and very significantly; the same increase in $t_{July}$ by 1° should be compensated by a decrease in annual precipitation from 40–50 to 100–110 mm and precipitation in the warm period to 80 mm;
d. Under conditions of explicit thermal aridization when warming is accompanied by a noticeable decrease in atmospheric moisture content ($-\Delta r=r_0 \rightarrow r_{-1}$), productivity will drop significantly ($-\Delta(PC)=PC_0 \rightarrow PC_{-1}$).

The following scenarios are characteristic of the *sub-boreal belt*:

a. A relatively minor increase in precipitation ($+ \Delta r=r_0 \rightarrow r_1$) does not cause any noticeable changes in primary bioproductivity ($\Delta(PC)\approx 0$), despite the increase in summer temperature. $E_0$, at $\Delta t_{July}$=1° (20° → 21°); to preserve the initial productivity, it is necessary to increase $r_{ann}$ by 25–50 mm and $r_{warm}$ by 10–20 mm;
b. Productivity can increase ($+ \Delta(PC)=PC_0 \rightarrow PC_1$) only with a more significant increase in precipitation ($+\Delta r=r_0 \rightarrow r_2$), i.e., under conditions of a significant thermo-humid trend;
c. The increase in summer temperatures not accompanied by an increase in atmospheric moisture content ($\Delta r\approx 0$) initiates the development of the thermo-arid trend, which causes a decrease in primary bioproductivity ($-\Delta(PC)=PC_0 \rightarrow PC_{-1}$); at a constant annual precipitation, on average, for every 1° increase in $t_{July}$, there is a decrease in productivity by 200–800 kg/ha;
d. Further development of the thermo-arid trend caused by a progressive decrease in precipitation ($-\Delta r=r_0 \rightarrow r_{-1} \rightarrow r_{-2}$) against the background of rising temperatures leads to an even greater decrease in the bioproductivity of sub-boreal forests ($-\Delta(PC)=PC_0 \rightarrow PC_{-2} \rightarrow PC_{-3}$).

Thus, boreal and sub-boreal zones, even near the border that separates them, dramatically differ from each other in the main functional parameter of forest ecosystems (primary productivity) in response to background climatic fluctuations. The patterns revealed earlier in the general qualitative analysis (Bazilevich et al. 1986; Kolomyts 2005) are given a quantitative interpretation, disclosing the ecological relations.

## 6.5 MAIN REFERENCES

It has been established that equifinal links in the system of regional connections of hydrothermal parameters are hydro-edaphic characteristics: summer reserves of productive moisture in soil and soil humidity, which have a direct effect on the distribution of phytocoenotic and soil objects. The controlling action of the global climatic system on the condition of not only regional but apparently also local natural ecosystems is realized just through hydro-edaphotops. The conception offers quite new ideas of the regional bioclimatic system, which characterizes climate-genic exo-dynamic properties of the soil–phytocoenotic "core" of natural complexes and is a scientific–methodical basis for landscape–ecological prognoses. A system of statistical parameters characterizing spatial relations between hydrothermal attributes has been established. These relations are focused on vegetation and soils, as well as on primary bioproductivity.

# Empirical Simulation of Climate-Genic Changes in Biological Cycle

# 7

## 7.1 FUNCTIONAL ORDINATION OF FOREST ECOSYSTEMS ACCORDING TO ARID CLIMATIC TREND

### 7.1.1 Introductory Remarks

The functioning of natural complexes is an object of the physical and ecological direction in landscape science and, in the broadest sense, is a set of all processes of material and energy exchange of inert and living matter in geo(eco)systems (D.L. Armand 1975; Beruchashvili 1990). All these processes are united by the general concept of metabolism, with a focus on living matter. As a result of a great debate about the content of the concepts of "function" and "metabolism", it was proposed to combine them into a single concept of "physiology of ecosystems" (Global Change 1990, Chapter 6). In the metabolic approach, the priority is to study the parameters of energy and mass transfer in the ecosystem and between it and the abiotic environment, as well as the analysis of the corresponding information interactions that form the structure of the ecosystem. The fundamental nature of this approach is confirmed by the fact that the parameters of matter–energy metabolism in biogeosystems are their most important classification features (Timofeev-Resovsky and Tyuryukanov 1966).

Ecosystem metabolism, as a process of updating and maintaining the mass of its living matter, is carried out through the interaction of two opposite processes (Kerzhentsev

DOI: 10.1201/9781003733119-7

1999): *anabolism* (creation of organic matter by the assimilation of mineral substances with the help of photosynthesis) and *catabolism* (decomposition of complex organic compounds into simple ones). Initial anabolism, or *autotrophic biogenesis*, according to Glazovskaya (1981), is carried out by phytocenosis, which creates primary biological production, while the function of catabolism is performed by the soil as a community of heterotrophic organisms. In addition, another process is involved in the functioning of the ecosystem – *necrobolism*, which consists of the death (necrosis) and the return (translocation) of resources useful for offspring (assimilators) to reserve or reproductive organs and tissues. The structure of a functioning ecosystem can be represented as the ratio of three phase components of the integral "eco-mass": living biomass, necromass, and mineral mass. The successive transformation of these partial masses into each other is the process of ecosystem functioning in the most general form (Kerzhentsev 2006).

## 7.1.2 Ergodic System of Nature Complexes

We have used the method of binary ordination not only for revealing the mechanisms of inter-component connections but also for factor analysis of the state of integral ecosystem objects: landscapes, biogeocoenoses, forest types, and the types and kinds of soils. Ecological factors of the zonal–regional level are general hydro-climatic conditions (air temperature and atmospheric precipitation). The direct ecological factor of the local level is hydro-edaphotops as the most important intermediate in the transmission of external hydro-climatic signals to the functional state of topo-ecosystems.

A kind of *empirical simulation modeling* of local representations of the global–regional changes of natural environments is based on the schemes of binary ordinations, resulting in a certain advance toward elucidating the local mechanisms of these changes. The essence of modeling is as follows.

As is known, the landscape mosaic of any territory is composed of the areals of local geo(eco)systems, combining the features of both zonal–regional physical–geographical backgrounds and local factors (natural and/or anthropogenic). In the "purest" form, the zonal–regional background is represented by *flat-interfluve* landscape formations (see Section 5.3.4). The latter, according to G.N. Vysotskiy (1909, 1960), are formed at moderately drained, gently sloping watersheds, with a significant depth of groundwater-table occurrence, in the absence of washout or deposition of soil particles. The soils and plants developing at flat-interfluves are of zonal types. Hence, it is considered that flat-interfluve natural complexes represent the zonal–regional bioclimatic "norm" of a given territory. The *primary landscape facies* (*biogeocoenoses*) formed at flat-interfluves are a certain equifinal dynamic state of elementary geosystems corresponding to the harmonic (optimal) combination of their components (Sochava 1978), i.e., flat-interfluve conditions.

In terms of soil geochemistry, a flat-interfluve is an eluvial, or autonomous, "elementary landscape" (Polynov 1956; Glazovskaya 1964; Perelman 1975). It is characterized by the independence of soil formation processes from groundwater, the absence of matter influx by fluid or solid lateral flow, and the presence of matter consumption through overland flow and infiltration. All of these soil–hydrological and biotic peculiarities of flat-interfluves provide them with the role of "zonal representative" in a given community of natural–territorial complexes.

Local geomorphological and edaphic factors alter the climatic background represented in flat-interfluves and create the respective factorial–dynamic rows of local geo(eco)systems (Krauklis 1974). These rows consist of a set of so-called *extra-zonal natural complexes* (Sprygin 1986), which are local representatives of other zonal–regional ecosystems, often rather distant from the given region. Extra-zonal complexes cause the "fuzzification" of zonal ecological–geographical boundaries (Mil'kov 1986), i.e., they create a subregional (local) ecotonization of the territory. Such ecotonization is often so large scale that it reaches the regional level.

Thus, within a territory with certain boundary conditions of zonal–regional bioclimatic background, we can construct a series of factorial–dynamic rows of elementary natural complexes (landscape fascia and their groups) by positioning the members of each row in the order of intensification (or weakening) of the hypertrophic effect of the respective local factor. From the viewpoint of climate-genic changes in natural environments, we are interested in hydrothermal rows of extra-zonal topo-ecosystems (together with zonal representatives), as well as their separate components, in the direction of their increasing aridity or, on the contrary, humidity. At the same time, each of these rows may be presented as a certain analog of the time row of the -e trend (Table 7.1).

In consequence, we obtain an *ergodic operation system,* which is of extremely important methodological significance for geographical research. The ergodic approach in itself is nothing but a modern modification of the traditional comparative geographical method initiated, as is known, as early as the beginning of the 19th century by Alexander Humboldt. The idea of ergodicity lingers also in the fundamental chronological conception of A. Gettner (1930). He showed that the temporal stages of spatial development of geographical events were filled with their spatial equivalents and became a synthetic scientific product, chorologic in essence (Zamyatin 1999). The principle of ergodicity is quite widely used in the study of hydrometeorological processes and allows for the performance of the most important procedure of interchanging spatial and temporal coordinates (Rauner 1981). The arrangement of spatial states of an object in accordance with their temporal sequence forms a *phase space* of this object (Puzachenko 1998). Based on the properties of self-similarity of the ergodic operation system, we substitute temporal sequences for spatial ones and, by analyzing the vector row of flat-interfluves and extra-zonal topo-ecosystems, make a kind of *simulation of local representation of the background climatic trend*, which becomes the basis for revealing the real spectrum of possible directions of functional–structural transitions between ecosystems.

In this case, the *vector rows of aridization of flat-interfluve and extra-zonal topo-ecosystems* and their separate components for each regional ecosystem were constructed in conformity with the supposed global and regional climatic changes. It concerned not only the quantitative parameters of their state but also qualitative characters per se. On the binary ordination graphs, all states of the objects under study were arranged from left to right by the *X*-axis (for phenomenon *A*) and from top to bottom by the *Y*-axis (for factor *B*) in such a way that the change of states was of vector character and would be adequate to the general tendency of aridization of local landscape–ecological conditions. The latter may be, e.g., transitions from the waterlogged

**TABLE 7.1** Taxonomic (weight-average) standards of basic values of summer reserves of productive moisture (mm) in different soil layers for local ecosystems of various ecoregions of the Middle Volga region

**Zhiguli low-mountain range**

| *RESERVES OF MOISTURE IN SOIL LAYERS OF* | *GROUPS OF BIOGEOCOENOSES* | | | | | |
|---|---|---|---|---|---|---|
| | 1 | 2 | 3 | 4 | 5 | 6 |
| 0–20 cm | 48 | 42 | 52 | 91 | 57 | 73 |
| 0–50 cm | 74 | 88 | 91 | 166 | 117 | 153 |
| 0–100 cm | 115 | 171 | 121 | 355 | 233 | 325 |

**Denudation-outwash high plain (Green Town)**

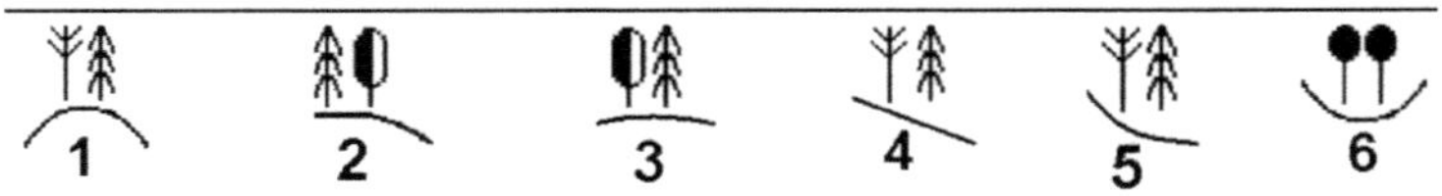

| *RESERVES OF MOISTURE IN SOIL LAYERS OF* | *GROUPS OF BIOGEOCOENOSES* | | | | | |
|---|---|---|---|---|---|---|
| | 1 | 2 | 3 | 4 | 5 | 6 |
| 0–20 cm | 15 | 40 | 42 | 35 | 49 | 112 |
| 0–50 cm | 38 | 72 | 97 | 96 | 110 | 266 |
| 0-100 cm | 61 | 155 | 212 | 239 | 185 | 688 |

**Zavozhsko-Kerzhensky outwash forested lowland**

| *RESERVES OF MOISTURE IN SOIL LAYERS OF* | *GROUPS OF BIOGEOCOENOSES* | | | | | |
|---|---|---|---|---|---|---|
| | 1 | 2 | 3 | 4 | 5 | 6 |
| 0–20 cm | 15 | 23 | 54 | 58 | 58 | 80 |
| 0–50 cm | 68 | 55 | 132 | 142 | 166 | 222 |
| 0–100 cm | 107 | 117 | 254 | 220 | 327 | 431 |

*Note:* Conventional meaning of biogeocoenosis groups can be seen Table 5.3.

bottoms of small river valleys to the steep, well-drained slopes of watersheds, from hydromorphic to xeromorphic states of the edaphotop, and from sedge–sphagnum coniferous or mixed forests to mesophytic grass oak forests, and further to dry pine forests (belomoshniks), etc.

The changes in quantitative characters of the functioning of biogeocoenoses, adequate to the arid climatic trend, may be unambiguous: in accordance with the position of this group of objects (biogeocoenoses, forest, or soil types) in the flat-interfluve and extra-zonal row, i.e., depending on how they correlate with the local ecological optimum of the given territory. Thus, the hierarchically ordered row of vector ordinations of geo(eco)systems and their separate components simulates the order of hydrothermal signal transmission by the channels of inter-component connections, as well as the respective changes in the states of systems in the direction determined by the background climatic prediction.

The method of empirical simulation of the bioclimatic trend proposed by us as the initial procedure for predictive landscape–ecological constructions is a certain implementation of one of the declared methods of geographical forecasting – the method of spatio-temporal analogies (Zeidis et al. 2001). This method is based on the ergodicity properties of the considered operating system mentioned above. Vector placor-extra-zonal schemes of binary ordination of topo-ecosystems are very convenient for predictive analysis and, as a tool for ecological forecasting itself, are quite universal. According to them, it is possible to "play" (based on the principle of actualism) the real directions of functional and structural transformations of phytocenotic, soil, and landscape complexes under the influence of natural or anthropogenic climate changes, as well as large-scale hydro-melioration.

### 7.1.3 Hydrothermal Ordination of Biological Cycle Parameters

The mechanisms of biotic regulation of the carbon cycle under global warming may be described by climate-genic dynamics of the indices of internal turnover and vegetable matter balance in ecosystems. For this purpose, it is necessary to carry out empirical simulation modeling of the functional parameters of topo-ecosystems. It was necessary to establish the local mechanisms of biotic regulation of the carbon cycle and the regional patterns of this regulation based on predicted changes in discrete parameters of the small biological cycle in forest biogeocoenoses at the specified variants of climate prediction for a particular period. The potential of biotic regulation of the carbon cycle was assessed by the *hydrothermal ordination of discrete metabolic parameters of forest ecosystems* under different zonal–regional and local conditions of the Volga River basin.

Next, *eco-mass parameters* (tons/ha) have been used for the calculation of carbon balance: (1) skeletal tree–shrub phytomass, BS; (2) root mass, BR; (3) total verdure mass, BV; (4) forest litter mass, ML; (5) debris – dead skeletal mass (brushwood and dead wood), WD; (6) humus mass in organic-mineral layers of the soil, HU. Live phytomass was calculated by the general and regional tables of biological productivity of fully stocked (normal) stands (Shvidenko et al. 2008) using the average age and the quality of locality of each species – the initial parameters obtained by forest taxation on test plots. The WD, ML, and HU parameters were obtained empirically. The transition from eco-masses to their carbon content was performed using the known carbon coefficients (Kobak 1988; Isaev and Korovin 1999; Dobrovolsky and Nikitin 2006).

The ordination, i.e., gradient analysis, was carried out. Methods of ordination analysis, developed at one time in geo-botany, have found wide application in landscape ecology (Ramenskiy 1971; Sochava 1974; Whittaker 1975). "Ordination is the ordering of objects (species or communities) along some axes that determine the variation of vegetation" (Rozenberg 1984, p. 47). We used direct gradient analysis, in which the identification of leading environmental factors is based on the study of the statistical variation of certain objects (biogeocenoses or their individual features) in the space of values of these factors. Functional ordination made it possible to carry out a kind of empirical simulation of the expected climate-genic changes in the biological cycle. This simulation is based on the ergodicity (self-similarity) properties of our operating system, according to Simonov (1966) and Rauner (1981). Within the territory with the boundary conditions of the zonal–regional bioclimatic background, a series of spatial factorial–dynamic series of the functional characteristics of topo-ecosystems was constructed. The members of each series were arranged in order of strengthening or weakening of the partial impact of the considered hydrothermal factor in such a way as to obtain some analog of the time series of topo-ecosystems corresponding to the vector of background climatic changes (trend) in the given region.

The thermo- and hydro-edaphic ordination of metabolic characteristics of topo-ecosystems was made by two geophysical parameters: the temperature of soil 50 cm deep ($t_{50}$) and summer productive moisture reserves in the 0–50 cm soil layer ($W$–50). This parameter is most closely connected with atmospheric humidification. The functional characteristics of forest ecosystems also show the highest correlation with these parameters. As one can see (Table 7.2), correlation connections are not always high, though quite significant (Pearson's test of significance, $P \ll 0.05$). In cases of weak connections, the latter can be interpreted only as a general tendency of changes in the given metabolic parameter under the influence of geophysical trends against the background of the significant "noise" effect of other local factors.

As is known (Montgomery and Peck 1982), *violations of the basic principles of statistical analysis are inevitable when describing complex multicomponent biological systems*. In particular, the principle of linear independence of predictors is not observed, and the correlation and determination coefficients cannot be high, since the regression equation most often describes the relationship of the object under consideration with environmental conditions, while the biological system and its components exist according to their own internal laws. Dissymmetrization and fragmentation of hydrothermal niches, with a predominantly Poissonian distribution of biogeocoenotic units, indicate a complex process of their climate-genic transformations differentiated in space (Kolomyts 2008).

In landscape–ecological systems, the futility of searching for binary relationships "factor-phenomenon" with high, almost functional coefficients of correlation $R$ and determination $R^2$ is quite obvious. Not only in foreign but also in domestic literature has this been long abandoned, and values of $R^2 \geq 0.30$–$0.35$ are considered quite acceptable. In many publications, the values of $R$ and $R^2$ are not given at all in the constraint equations, but the Pearson significance indicator is usually indicated (in our works, this indicator is also used). Here are two foreign articles as examples (Buccolini et al. 2012; Calogero Schillaci et al. 2017). In the first of these works Figures 7–9 (binary relationship plots) show the following $R^2$ values: 0.194, 0.314, 0.332, and 0.207. In the second work, the correlations in Figures 5 and 6 are presented without showing $R$ or $R^2$ at all, and in Table 7.2, $R^2$ values equal to 0.470, 0.034, and 0.127 are shown.

For this reason, correlation and determination coefficients are sometimes not given at all in ordination analysis (Martin-Benito and Pederson 2015; and others). The main task of ordination research is to separate the signal under study from "noise" (and not

**TABLE 7.2** Examples of the calculated models describing the changes in the biological cycle parameters of biogeocoenoses in different regional ecosystems of the Middle Volga Region under the trends of soil temperature and soil moisture content during the vegetation period

| *MODEL KIND (CALCULATE FORMULA)* | *REGIONAL ECOSYSTEMS (SEE FIGURE 5.4)* | *PARAMETERS (SEE IN THE TEXT)* | *COEFFICIENTS* | | | *STATISTIC CHARACTERISTICS* | | |
|---|---|---|---|---|---|---|---|---|
| | | | $B_0$ | $B_1$ | $B_2$ | $R^2$ | *1. P* | *2.* $S_Y$ |
| $y=b_0+b_1x_1+b_2x_2$ | By-Sura | BV | 21.93 | –0.7415 | **–0.7135** | 0.441 | 0 | 3.09 |
| | | HU | 5.99 | 0 | **55.75** | 0.484 | 0 | 20.29 |
| | Green Town | BR | 8.712 | 5.838 | **–0.0678** | 0.339 | $0.1 \cdot 10^{-3}$ | 17.92 |
| | | HU | 16.34 | –0.2402 | **0.07426** | 0.508 | 0 | |
| | Shchelokovsky Farmstead | ML | 27.18 | –2.333 | **0.04896** | 0.514 | 0 | 6.51 |
| $y=b_0+b_1x_1+b_2x_1^2$ | Shchelokovsky Farmstead | WD | 642.1 | –114.7 | 5.142 | 0.876 | 0 | 2.21 |
| $y=b_0+b_1x_2+b_2x_2^2$ | Zhiguli | ML | 44.08 | –0.3611 | 0.00091 | 0.489 | 0 | 17.75 |
| | Shchelokovsky Farmstead | BS | –34.93 | 0.9472 | 0.00403 | 0.591 | 0 | 90.49 |
| | | HU | 12.5 | 0.1712 | 0 | 0.421 | 0 | 9.31 |
| | Kerzhenets | WD | 22.85 | –0.2588 | 0.00149 | 0.518 | 0 | 12.78 |
| | | HU | 8.84 | 0.0091 | 0.0006 | 0.724 | 0 | 8.03 |
| | By-Oka-Terrace reserve | BS | 244.6 | 0.3956 | –0.00312 | 0.310 | $0.7 \cdot 10^{-3}$ | 71.4 |
| | | HU | 25.31 | 0.0159 | 0.00042 | 0.333 | $0.4 \cdot 10^{-3}$ | 16.32 |
| $y=\exp(b_0+b_1/x_1)$ | Zhyguli | BS | 3.706 | 15.42 | | 0.373 | 0 | 39.3 |
| | | BR | 1.768 | 26.26 | | 0.377 | 0 | 12.25 |
| | Green Town | ML | 0.0845 | 35.08 | | 0.256 | 0 | 16.0 |
| | | WD | –4.389 | 64.63 | | 0.412 | 0 | 7.0 |
| | Vyksa | ML | 0.3274 | 46.81 | | 0.637 | 0 | 0.34 |

*Note:* In calculating formulas: $x_1$ – soil temperature at a depth of 50 cm ($t_{50}$); $x_2$ – stored soil moisture in the layer 0–50 cm (*W*–50). Statistical characteristics: $R^2$ – coefficient of determination; *P* – significance level (Pearson's criterion); $S_y$ – 0073tandard deviation. Coefficients for significant arguments in equations of multiple regression are shown in bold print.

to achieve unambiguous connections) and to quantify the magnitude of this signal. This signal often turns out to be much less powerful than many "noise" influences, but the important thing is that it is constantly acting and unidirectional, and this is its effectiveness.

The futility of searching in complex landscape–ecological systems for binary "factor–phenomenon" relations with high, almost functional correlation coefficients $R$ and determination $R^2$ is quite obvious. This has long been abandoned not only in Russian but also in the international literature, and the values of $R^2 \geq 0.30–0.35$ are considered quite acceptable. Let us refer again to the very authoritative authors Montgomery and Peck (1982). In particular, they write that in the case of complex biological systems, A violation of the basic principle of analysis of "linear independence of predictors" is allowed when there is no possibility of its implementation, and the problem requires a solution.

The main task of ordination landscape–ecological research is to separate the signal under study from "noise" (and not to achieve high unambiguous connections) and to quantify the magnitude of this signal. This signal can be much less powerful than many "noise" influences. In our studies, the $R^2$ values ranged from 65%–75% to 25%–35% (Kolomyts 2008). However, this signal must be constantly acting and unidirectional, and this is its effectiveness. At the same time, the components of information "noise" are usually multidirectional and therefore can significantly cancel each other out.

The range of the greatest scattering of the mean summer soil (and air) temperatures between the extreme facies groups in their topological order is about 4°. Such an interval almost completely corresponds to the maximal deviations of July temperature according to GISS for the end of the 21st century, but for HadCM3, it is adequate to deviations only for the period until 2050. Environmental prediction by this model for the second half of the century is entirely of an extrapolation character.

The ordination, i.e., gradient analysis, was used to establish the parametric and spatial correlations of predicted changes in the two branches of a small biological cycle, production and detritus, in broadleaf, mixed, and light-coniferous forests of the Middle Volga Region. The most effective way of assessment and prediction of the functional states of the ecosystem is the analysis of separate links of the biological cycle and their ratios. We give a brief overview of empirical simulation of climate-genic changes of the biological cycle.

On the territory of southern and typical forest–steppe, there is an unambiguous and quite clear decline in the production potential of forest and forest–steppe communities at higher summer soil temperatures and particularly at lower soil moisture content (Figure 7.1a). Obviously, in southern marginal forests, global warming should result in a lower amount of $CO_2$ captured by phytocoenoses from the atmosphere during photosynthesis. The ascending branch of the biological cycle will cause an additional intensification of the atmospheric greenhouse effect and thus aggravate global warming.

Carbon deposits in skeletal phytomass will reduce most quickly in flat-interfluve mesophilic elm–lime and oak forests of erosion–loamy uplands and in forest communities of deep small valleys. Their role as a sink for atmospheric carbon will diminish. In the aqueous–glacial plains of the nemoral–forest sub-zone, the most abrupt diminution of productivity is expected to occur in the excessively moist birch–spruce–pine forests. These groups of facies can markedly reduce the rates of carbon conservation in skeletal phytomass. At the same time, xeromorphic biogeocoenoses, particularly

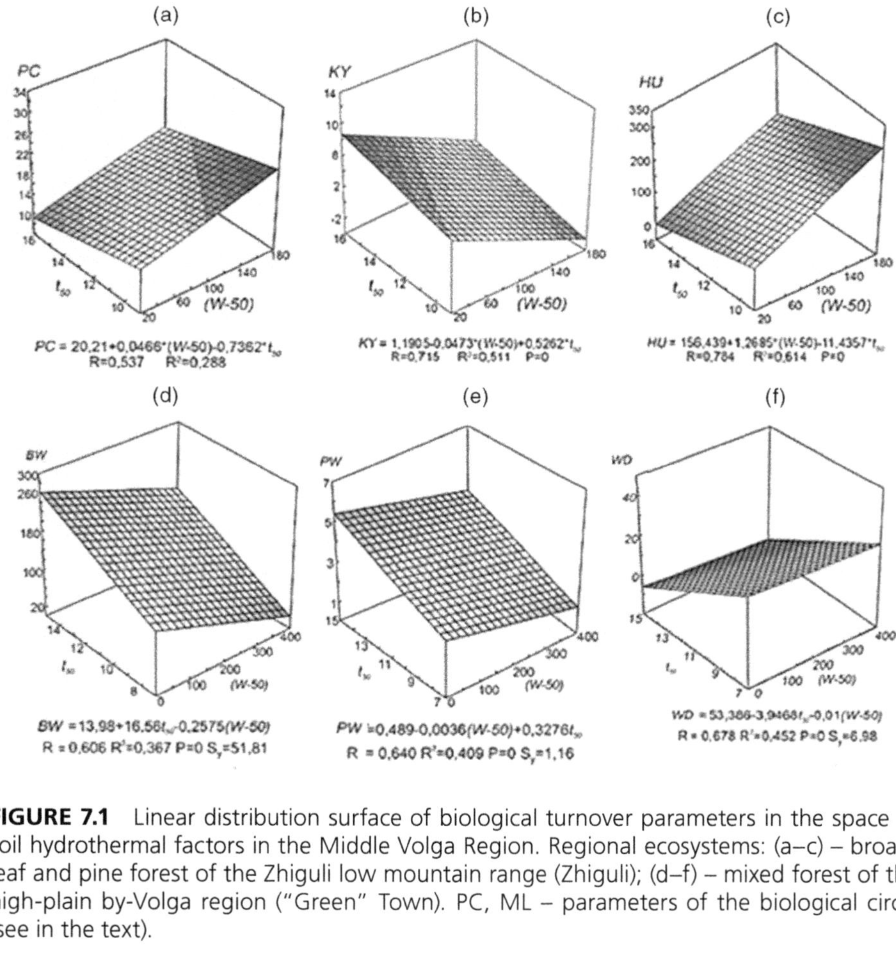


**FIGURE 7.1** Linear distribution surface of biological turnover parameters in the space of soil hydrothermal factors in the Middle Volga Region. Regional ecosystems: (a–c) – broadleaf and pine forest of the Zhiguli low mountain range (Zhiguli); (d–f) – mixed forest of the high-plain by-Volga region ("Green" Town). PC, ML – parameters of the biological circle (see in the text).

forest–meadow–steppe communities on steep sunny slopes and pine steppified forests of well-drained watersheds, will have almost no effect on the base rates of production, so that the contribution of their autotrophic biogenesis to $CO_2$ concentration in the atmosphere will remain at the previous level.

In the northern regions of the forest–steppe zone, under slightly excessive atmospheric humidification, broadleaf and mixed forest biogeocoenoses will increase (and quite significantly) both their primary productivity and reserves of living organic matter alongside a simultaneous rise in temperature and reduction of soil moisture content (Figure 7.1d and e). Accordingly, the conservation of atmospheric carbon in perennial skeletal phytomass of these topo-ecosystems will increase, and as a consequence, the significance of this phytocoenotic pool as a carbon sink will also increase.

The moisture deficit on the zonal forest–steppe ecotone will increase under global warming and impede the decomposition and mineralization of dead organic matter. The growth of forest litter mass *ML* (see Table 7.2) results in a greater share of the dead fraction in the total above-ground phytomass. Simultaneously, the amount of dead mass utilized by saprotrophs and used in subsequent trophic chains becomes less. Thus, in the marginal forest communities, a general reduction of flow capacity of the detritus branch of metabolism will occur, inducing a residual accumulation of dead mass and, as a consequence, an inevitable slowdown of the whole biological cycle. The conservation of dead organic matter in forests will increase, followed by a certain weakening of $CO_2$ flows to the atmosphere.

However, the resultant deficit of atmospheric carbon dioxide will be surpassed with interest by the much greater quantity that will remain in the atmosphere due to intensive humus mineralization (Figure 7.1c) and the abrupt drop in the annual productivity of forests (see Figure 7.1a).

With the transition from southern marginal forests to the northern band of the forest–steppe zone and further to the zone of mixed forests, the thermo-arid trend increasingly intensifies the processes of decomposition of dead organic matter and correspondingly reduces all dead mass fractions (Figure 7.1f), including forest litter and humus. As a consequence, the litter-fall index will be much lower, and the degree of dead phytomass utilization will increase. Thus, the "work" of the detritus branch of the above forest communities will lead directly to the additional release of $CO_2$ from soil–plant cover to the atmosphere, with the inevitable enhancement of the greenhouse effect.

On outwash forested lowlands of the sub-taiga zone, the limiting role of soil moisture content for autotrophic biogenesis will increase under global warming. The thermo-arid trend will induce a slight growth in timber stand productivity but simultaneously a decrease in total annual forest production, which must result in a lower amount of $CO_2$ bound in autotrophic biogenesis. Much more unambiguous and intensive changes must occur in the detritus branch of the biological cycle of outwash forested lowlands. Soil temperature increases and desertification abruptly activate the degradation of dead organic matter, followed by a decrease in forest litter mass and total dead mass, with a higher degree of its utilization and additional release of $CO_2$ into the atmosphere.

In the presented system of chain reactions, detritogenesis plays the key role as a functional unit that primarily responds to hydrothermal signals (this may be judged, in particular, by parameters $R$ and $R^2$ in Table 7.1). Being the most narrow transmission channel of connections in the biological turnover and determining, in the long run, its intensity and equilibrium (Odum 1971), the detritus branch is the main limiting factor that influences the stability of forest ecosystems and, at the same time, the main "conductor" of their structural shifts under the corresponding external impacts.

### 7.1.4 Empirical Simulation of Climate-Genic Changes in the Biological Cycle

#### *7.1.4.1 Autotrophic Biogenesis on Zonal Ecotones*

1. On the southern outpost of forest-to-steppe transitions, in the low mountain range of Zhiguli, the overwhelming majority of metabolic parameters show

a linear, parabolic, or exponential dependence on summer moisture content in soil (see Table 6.1), which acts here as the main transfer member of the effect of climatic signals on biogeocoenoses. The southern and typical forest–steppe, with $F_{hum}(1) \leq 1$, demonstrates an unambiguous and rather distinct decline in the production potential of forest and forest–steppe communities both at increasing summer soil temperature and, even to a greater extent, at decreasing soil moisture content (Figure 7.1a). Apparently, *in southern marginal forests,* global warming is expected to reduce the amount of $CO_2$ captured by phytocoenoses from the atmosphere for the creation of net production. Consequently, the ascending branch of biological turnover will induce an additional increase in the atmospheric greenhouse effect and thus intensify global warming.

At the same time, the response of different facial groups of the Zhiguli mountain range to the thermo-arid signal will be significantly differential. The maximum decrease in total annual production, from 18–23 to 13–17 t/ha on average, occurs with a reduction of ($W$–100) from 200–300 to 130–140 mm, which is in the range of base moisture content values for the most hydromorphic biogeocoenoses: flat-interfluve elm and lime forests, lime oak forest of shady and neutral slopes, and maple–birch communities of the bottoms of small valleys. The role of these forest communities as an atmospheric carbon sink will significantly decline under global warming.

In the remaining interval of soil aridization, with a reduction of ($W$–100) to 60–40 mm, forest productivity shows no definite tendencies of change, remaining at a generally low level with fluctuations of 8–10 to 16–18 t/ha per year. This means that the thermo-arid trend induced by global warming will have little effect on production in the forest–meadow–steppe facies.

The thermo-edaphic factor affects production much more significantly than the hydro-edaphic factor. Overall, there is a tendency for the productivity of marginal forests to decrease from 18–20 to 12–15 t/ha as $t_{50}$ increases from 8°–9° to 13°–14°.° Accordingly, wood reserves decline as well. As can be seen, both factors interfere in the same direction, which provides a rather high efficiency of the thermo-arid trend.

2. On the *sub-taiga–(forest–steppe) ecotone* of the nemoral–forest sub-zone (the regional systems of By-Sura and Green Town), the most abrupt diminution of productivity is expected to occur in the most hydromorphic biogeocoenoses: trans-accumulative and eluvial-accumulative birch–spruce–pine forests with bilberry, sphagnum, and green moss. Here, at a $t_{50}$ increase from 7°–8° to 12°–13°, the mean growth values of both wood (PW) and roots (PR) are reduced by more than half: from 3.0–3.5 to 1.4–1.7 t/ha per year on average. Such trends in PW and PR are favored by concurrent soil aridization. These facies groups are capable of markedly reducing the rates of carbon conservation in skeletal phytomass. However, a drastic increase in wood reserves is likely in moderately humidified topo-ecosystems: transit and trans-accumulative coniferous and mixed forests, with base values ($W$–50) ≤ 90–100 mm. With soil aridization to the level of ($W$–50) = 30–40 mm, the mean biomass weight (BW) value in these forests will increase from 80–110 to 200–240 t/

ha, significantly approaching those in the most productive flat-interfluve lime and oak forests. At the same time, xeromorphic biogeocoenoses, particularly forest–meadow–steppe communities of steep sunny slopes and pine steppified forests of well-drained watersheds, will have almost no effect on the base rates of production, so the contribution of their autotrophic biogenesis to $CO_2$ concentration in the atmosphere will remain at the previous level.

Hydromorphic and meso-hydromorphic forest communities (high-grass small-leaved and spruce–pine forests of the lower catena links, with their coldest and wettest soils) will be maximally sensitive to the thermo-arid signal. In the range of temperature rise $t_{50}$=8°–9° →→ 11°–12°, their total productivity (PC) is expected to increase from 6–8 to 13–15 t/ha per year on average (in some cases, up to 17–19 t/ha), the annual growth of the skeletal portion from 2.2–3.0 to 4.5–6.0 t/ha, and green mass reproduction from 3–4 to 6–8 t/ha per year. At the same time, wood reserves will increase from 40–80 to 160–200 t/ha, i.e., 3–4-fold. The similar wood increment will also be determined by the reduction of moisture reserves from base values (*W*–50) ≥ 250–300 mm to 150–120 mm. Concurrently, the productivity of most other ecosystems (mesomorphic and meso-xeromorphic nemoral–grassy, as well as xeromorphic cowberries–bluejoint coniferous and broadleaf forests (from eluvial to transit)) will be affected only by temperature increase, and even to a much lesser extent. Here, the _$t_{50}$ growth from 10.5°–11.5° to 14°–15° will cause PC to increase by 0.5–1.5 t/ha per year and BW to increase from 120–160 to 220–260 t/ha on average.

#### *7.1.4.2 Detritus-genesis*

On the zonal ecotone of forest and steppe, the moisture deficit increasing under global warming will impede degradation and mineralization of brushwood and forest litter. As the whole 1-m soil layer is being dried out, ML increases with rather high correlation coefficients (Figure 6.1b). The accumulation of litter mass will imply additional conservation of the carbon assimilated from the atmosphere during photosynthesis. As the litter mass increases, the litter-fall index CY and the share of the whole dead fraction BD in total above-ground phytomass also become higher. Concurrently, the amount of detritus utilized by saprotrophs and used in subsequent trophic chains is reduced, and the time of ID required for complete degradation of this amount of dead mass significantly increases.

The thermo-arid trend must also induce a general intensification of deadwood (WD) degradation, with additional $CO_2$ emissions to the atmosphere. It will be particularly marked in hygro- and mesomorphic forest communities with their low temperatures. However, in xeromorphic biogeocoenoses with base values (*W*–50) ≥ 11°–11.5°, the temperature rise will have almost no effect on WD value. Thus, the marginal forest ecosystems of the forest–steppe zone under global warming must show a total reduction of the flow capacity of the detritus branch of metabolism, which will cause residual accumulation of dead mass and, as a result, an inevitable slowdown of the entire biological turnover. Conservation of dead organic matter in forests will increase, and accordingly, the $CO_2$ flows to the atmosphere will weaken. However, the resulting deficiency of atmospheric carbon dioxide will be outweighed by its much larger amount accumulating in the atmosphere due to intensive mineralization of humus (see Figure 6.1c) and a drastic decrease in the annual productivity of forest ecosystems.

With the transition from southern marginal forests to the northern band of the forest–steppe zone and further to the zone of mixed forests, the thermo-arid trend increasingly intensifies the processes of decomposition of dead organic matter and correspondingly reduces all dead fractions (including forest litter) and humus (Figure 7.1c and f). As a consequence, the litter-fall index will be much lower, and the degree of detritus utilization will increase (see Table 6.1). The time required for complete degradation of this amount of dead mass will also increase. Thus, the "work" of the detritus branch in the functioning of the above forest communities will be directed toward the additional release of $CO_2$ from soil–plant cover to the atmosphere, with an inevitable enhancement of the greenhouse effect.

The maximal response to the thermo-arid trend should be expected in topo-ecosystems of the lower link of landscape couplings, with their most humid and cold soils. One should particularly note the multifold decrease in the litter-fall index, which points to an increase in the flow capacity of the detritus branch in the biological cycle of excessively humid forests. The pool of BW, ML, and HU in these forests will be the main source of additional carbon release into the atmosphere under global warming. At the same time, forest communities in moderately humid and dry habitats will almost not change their characteristics of dead mass. Here, the negative effect of changes in detritogenesis on the biotic regulation of carbon content in the atmosphere will be apparently insignificant.

#### *7.1.4.3 The driving forces of climate-genic successions*

In the forest ecosystems of the boreal belt and the extreme north of the sub-boreal belt, global warming should cause a very significant "expansion" of the carrying capacity of the descending branch of the biological cycle, which, generally speaking, should contribute to the overall acceleration of metabolism. However, this will take place against the background of a decrease in the efficiency of the production process and a decrease in the annual turnover of living phytomass in forest ecosystems. The thermo-arid trend causes similar negative trends in autotrophic biogenesis not only in the conditions of loamy sandy elevated plains but also in lowland outlands, and against the background of both an increase and a decrease in forest productivity.

The emerging imbalance of the main components of the biological cycle indicates a violation of the sustainable functioning of the forest community. Such a situation cannot exist for an arbitrarily long time and should serve as an impetus for the structural restructuring of ecosystems, aimed primarily at increasing the efficiency of living phytomass. This provides a new level of balance between the production and decomposition of organic matter, which is the main condition for the existence of an ecosystem, by definition (Odum 1971; Gerasimov 1985; Glazovskaya 1981; Glazovskaya 1992). As N.V. Timofeev-Resovsky emphasized in his extended interpretation of the Darwinian law of natural selection (Tyuryukanov 2001), when external conditions change, the selection of populations and communities proceeds along the line of development of those that ensure the continuity and closure of biogeochemical cycles in the new energy environment level.

Obviously, with a thermo-arid trend, a new level of balance can only be achieved by replacing the forest community first with a forest–meadow–steppe community, and then with a steppe community, as demonstrated in the digraphs of predicted landscape–ecological transitions for the Zhiguli massif (see Chapter 12). With a pronounced hydrothermal signal, which, in particular, the HadCM3 model predicts already for the forecast period of 2050, direct transitions from forest to steppe are possible. Such dynamic trends are assumed, for example, for upland mesomorphic oak forests and pine–broadleaf forests of the By-Sura region, as well as trans-eluvial xero-mesomorphic pine forests and sub-boreal forests of the Green Town region.

The results of gradient analysis have been used to formulate a general model scheme that describes the fundamental mechanism of transition from functional changes in ecosystems to their structural reorganizations under the influence of global warming. This mechanism must become most apparent in the zonal ecotones of forest and steppe. Under sustained development of the thermo-arid climatic signal, the scheme of transformation of the forest ecosystem into a steppe one is as follows:

Acceleration of the "work" of the detritus branch of metabolism, with a concurrent decrease in the coefficient of efficiency of phyto-production

↓

Disturbance of biological cycle equilibrium (closure) as the most important condition for the stable functioning of an ecosystem

↓

Degradation and decomposition of the forest community, its replacement first by arid open wood (typical forest–steppe) and then by grassy–steppe formation

↓

Increase in the coefficient of efficiency of autotrophic biogenesis

↓

Reaching the equilibrium of biological turnover at a new ("meadow–steppe" or "dry steppe") level of functioning of an ecosystem, which brings the latter into a new stable state.

Obviously, the same mechanism of chain reactions must work at landscape–ecological transitions between forest formations themselves: from more to less hydromorphic. In the presented system of chain reactions, detritus-genesis plays the key role as a functional unit that primarily responds to the hydrothermal signal (this may be judged, in particular, by parameter $R^2$ in Table 6.2). Being the most narrow transmission channel of connections in the biological turnover and determining, in the long run, its intensity and equilibrium (Odum 1971), the detritus branch is the main limiting factor that influences the stability of forest ecosystems and, at the same time, the main "conductor" of their structural shifts under the corresponding external impacts.

# 7.2 REGIONAL AND LOCAL HUMIDITY FACTORS AND THEIR SIGNIFICANCE FOR ECOLOGICAL PROGNOSES

## 7.2.1 Issues of the Problem and Basic Materials

The ratio between radiation heat and atmospheric humidity, as is known (Köppen 1938; Vysotskiy 1960; Grigor'ev 1966; D.L. Armand 1975; Isachenko 1985), is the most important landscape-forming factor responsible for the physiognomic traits of natural–territorial habitats (geo(eco)systems) of almost all structural levels of the biosphere – from global to local (topological). Along with that, this factor is the main link between these complexes and the biological cycle, which characterizes the functional aspect of landscape organization.

The most comprehensive view of the hydrothermal regime of natural complexes is given by the atmospheric humidity factor as a landscape–geophysical parameter that reflects primarily the background climatic conditions of the territory. Quite a few such factors are known: the annual and seasonal coefficients of the heat-to-humidity ratio proposed by Köppen, Vysotsky–Ivanov, Budyko, Bazylevych, Thornthwaite, Marthonn, Ryabchikov, Mirkin, Selyaninov, and Richter. In the present work, we have used the Vysotsky–Ivanov annual atmospheric humidity factor $F_{hum}(1)$, which is a ratio of annual precipitation $r_{ann}$ to evaporativity $E_0$ (see Chapter 5). The values of the $E_0$ parameter are calculated for each month:

$$E_0(\text{month}) = 0.0018(25+t)2(100-w), \tag{7.1}$$

where $E_0$(month), $t$, and $w$ are the monthly values of evaporative, temperature, and relative air humidity, respectively. The Vysotsky–Ivanov annual atmospheric humidity factor is widely used in Russian hydrometeorology and physical geography. Firstly, this parameter allows the landscape–ecological analysis of territory based on the known information about the connections of global and regional geo(eco)systems and their components with the heat and humidity ratio. Secondly, it is the most suitable for calculations because it is based directly on the data of long-term observations at meteorological stations.

According to our estimations for the territory of the Russian (East European) Plain, parameters $E_0$ and $F_{hum}(1)$ are formed mainly by the mean July temperature ($t_{July}$) with high coefficients of correlation ($R$) and determination ($R^2$):

$$E_0 = 1384 - 161.6\, t_{July} + 6.245\, t_{July}^2, \quad R = 0.93, \quad R^2 = 0.87; \tag{7.2}$$

$$F_{hum}(1) = 12.09 - 0.9095\, t_{July} + 0.01744\, t_{July}^2, \ R = 0.94, \ R^2 = 0.88. \tag{7.3}$$

The global climate prediction models are widely used for both planetary-and regional-level ecological forecasts, as they generally describe the hydrothermal trends by two initial parameters: temperatures and precipitation, which are predictors for the calculation of $F_{hum}(1)$. In these forecasts, the annual atmospheric humidity factor is usually compared with the zonal types/subtypes of plant formations and their longitudinal–sectoral variants (Sochava 1978; Puzachenko and Skulkin 1981; Isachenko 1985). The tightness of these connections is rather high. For instance, for the zonal spectrum of the central part of the Russian Plain (from the middle taiga to the southern steppe), we have obtained an explicit deterministic series of distribution of the groups of primary plant formations by the gradient of the annual atmospheric humidity factor (Table 7.3). The whole territory of the basin has about 170 meteorological stations, 120 agro-meteorological stations, and more than 300 observation points. Climatic niches of vegetation form a single continuous series by this factor, without any marked leaps and with continuously changing taxonomic norms of $F_{hum}(1)$. Such series points to a sequence of anticipated phytocoenological transitions at this or that climatic trend. For example, at a $F_{hum}(1)$ decrease from 1.85–1.65 to 1.23–1.12, the fir woods of middle and south taiga of the upper Volga must transform into broadleaf dark coniferous and/or pine formations, and then the latter into the more continental dark coniferous sub-taiga of the lower Kama River and into broadleaf and pine forests.

Statistical analysis revealed highly significant correlations between the positions of zonal (sub-zonal) boundaries and $F_{hum}(1)$. Covariance is $\leq$4–6, and only for the boundaries of the forest–steppe zone, it is up to 10%–11%. These correlations were used for making a prognostic map of natural zones (sub-zones) in the region on the basis of the respective prognostic $F_{hum}(1)$ maps (Kolomyts et al. 2012).

The annual atmospheric humidity factor corresponds to the hydrothermal conditions of flat-interfluve low-dimensional natural complexes – landscape facies (biogeocoenoses) as local representatives of the zonal–regional bioclimatic background of the given territory (Vysotskiy 1960; Sochava 1978). On this level, however, there is also a broad range of extra-zonal topo-ecosystems, which are able to simulate other zonal types of geographic environment, not only neighboring but also rather distant. Such ecological deviations from flat-interfluve natural complexes are formed under the influence of local geomorphological and edaphic factors refracting the given climatic background. Primarily, there are localized changes in "expense items" of the water balance: total evaporation and runoff, which create the local structural and functional diversity of topo-ecosystems. That is why the connections between local geo(eco)systems and the $t_{July}$, $r_{ann}$, and $F_{hum}(1)$ parameters are very weak and often statistically unreliable.

Thereby, it is necessary to find a factor of the heat and humidity ratio that would adequately show the topological differentiation of the natural environment and could be used as the initial hydrothermal predictor for local landscape–ecological forecasts. Such forecasts are especially relevant in light of the modern and upcoming global climate changes. As is known (Timofeev-Resovsky and Tyuryukanov 1966), the origins of global environmental changes lie at the level of elementary structural units of the biosphere. In the present work, we have presented some results of the scientific search in this particular field.

**TABLE 7.3** Ecological optimums (×) and "blurred" parts of ecological niches (●) of plant formation groups in the space of the values of the Vysotsky–Ivanov's humidity factor $F_{hum}(1)$ on the territory of East-European Plain

**FORMATION GROUP ASSOCIATIONS (SEE *TABLE 5.2*)**

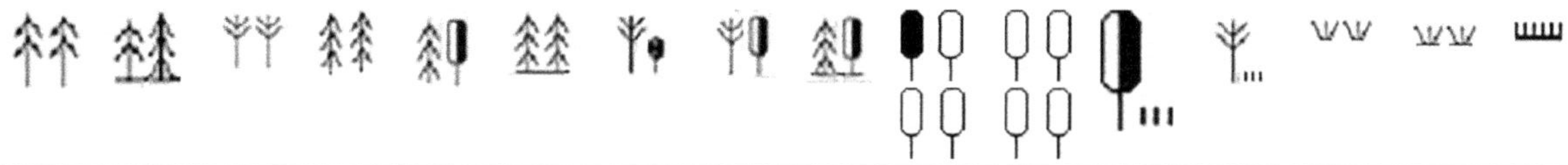

| *GRADATION OF HUMIDITY FACTOR* | | | | | | | | | | | | | | | | |
|---|---|---|---|---|---|---|---|---|---|---|---|---|---|---|---|---|
| 0.50–0.61 | | | | | | | | | | | | | | ● | × | × |
| 0.61–0.73 | | | | | | | | | | | | | ● | × | × | ● |
| 0.73–0.82 | | | | | | | | | | | | ● | × | × | ● | |
| 0.82–0.96 | | | | | | | | | | ● | × | × | ● | ● | | |
| 0.96–1.04 | | | | | | | | | × | ● | × | ● | | | | |
| 1.04–1.12 | | | | | | ● | ● | × | × | × | | | | | | |
| 1.12–1.23 | | | | | | × | × | ● | ● | ● | | | | | | |
| 1.23–1.33 | | | ● | ● | ● | ● | × | | | ● | | | | | | |
| 1.33–1.45 | | | × | × | × | | ● | | | | | | | | | |
| 1.45–1.55 | | ● | ● | × | | | | | | | | | | | | |
| 1.55–1.65 | ● | ● | × | ● | ● | | | | | | | | | | | |
| 1.65–1.75 | ● | × | ● | ● | | | | | | | | | | | | |
| 1.75–1.85 | ● | × | | | × | | | | | | | | | | | |
| 1.85–2 | × | ● | | | ● | | | | | | | | | | | |
| Taxonomic norm of $F_{hum}$ | 1.80 | 1.73 | 1.47 | 1.42 | 1.54 | 1.23 | 1.21 | 1.12 | 1.07 | 1.07 | 0.94 | 0.90 | 0.80 | 0.72 | 0.62 | 0.57 |

*Note:* The sign "×" denotes the optimum of dominance of the equivalence classes; "" is the "blurred" part of ecological niche.

## 7.2.2 The Space Diversity of Annual Atmospheric Humidity Factor

The local response to the global and regional climatic signals is formed by the transfer functions in landscape connections, which transform these signals while passing through the coupled ensembles of natural complexes and their geo-components. It was established empirically that the main channel of connections between local ecosystems and the background climatic system passes through the *summer soil moisture content*, which is a rather reliable geophysical indicator of the state of ecosystems, on the one hand, and is the most powerful ecological factor predetermining their functional parameters, on the other hand (Kolomyts 2005). It has also been revealed that the main factor of climate-genic transformation of natural ecosystems in the region is the hydrothermal conditions of the vegetation period. These are the initial ideological positions of functional ordination analysis.

The empirical simulation modeling of the functional parameters of topo-ecosystems is based on the biosphere approach proposed previously by Timofeev-Resovsky (1970), namely, the principle of the spatial organization by living substance of the matter and energy turnover in biogeocoenoses as discrete elementary structural units of the biosphere. Spatial diversity of the matter and energy turnover in the biosphere is created by the structural heterogeneity of biogeocoenoses. In the Volga River basin, the July reserves of productive moisture in soil layers of 0–20 cm (*W*–20), 0–50 cm (*W*–50), and 0–100 cm (*W*–100) deep were found to be closely connected with $F_{\text{hum}}(1)$, the most important landscape–geophysical index reflecting zonal conditions. These connections for forest ecosystems of the Russian Plain are as follows:

$$(W-20)=\exp\left[2.281+0.8068F_{\text{hum}}(1)\right];\ R=0.83;\ R^2=0.69; \tag{7.4}$$

$$(W-50)=\exp\left[2.622+1.071F_{\text{hum}}(1)\right];\ R=0.89;\ R^2=0.79; \tag{7.5}$$

$$(W-100)=\exp\left[3.225+1.134F_{\text{hum}}(1)\right];\ R=0.88;\ R^2=0.77. \tag{7.6}$$

On the other hand, there is a rather strong dependence of forest topo-ecosystems distribution on summer soil moisture capacity. Within a local space, the spatial background soil moisture content is transformed by the type of site on the catena, slope exposure and steepness, and the composition of the soil-forming rock. As a result, there is a particular set of soil moisture content values at the same $F_{\text{hum}}(1)$, with the respective diversity of facial groups. According to our measurements and calculations, informational interrelationship coefficients *C*(*A*;*B*) of biogeocoenotic groups with parameters (*W*–20), (*W*–50), and (*W*–100) in each forest ecoregion of the Middle Volga River basin are equal: from 0.128–0.154 to 0.173–0.200. This significantly increases the critical threshold of significance of connections (0.070).

The Kud'ma and By-Sura ecoregions, with their weakly broken relief, can be described by one value of the $F_{\text{hum}}(1)$ parameter each: 1.29 and 1.07, respectively. For the By-Sura, such a heat-to-humidity ratio is natural for the zonal border between mixed and broadleaf forests in the southern sector of the Volga River basin (Kolomyts 2008), which is represented in the landscape structure of this region. However, the situation with the Zhiguli ecoregion appears to be much more complicated, where the solar exposure factor plays a key role in the local differentiation of geosystems. The placor elm–linden grooves of the area near the watershed of the Zhiguli plateau have $F_{hum}(1)=0.95$, which is compliant with the transition from broadleaf forest to typical forest steppe. To determine the humidity factors on differently oriented slopes, we used the results of action-metric observations conducted in various mountainous areas similar to the Zhiguli in climatic conditions (Shcherbakov 1970; Vygodskaya 1981). Based on the experimental data (in the annual averaging and under the actual conditions of cloudiness), the $k(E_0)$ correction indices were found for evaporation. They made it possible to determine the deviations from total radiation, radiation balance, and, thereby, the $E_0$ value under the sub-horizontal (upland) conditions (see Table 6.3). The average weighted norms of $k(E_0)$ for all six groups of Zhiguli biogeocoenoses (see Table 5.3) are as follows: 1 – 1.17; 2 – 1.3; 3 – 0.95; 4 – 1.00; 5 – 0.90; 6 – 1.05.

As we can see, the difference in the annual evaporativity between geosystems of steep sun-exposed and shadowy slopes is 25%–30%. At the same time, on the slopes of the western and eastern quarters of the horizon and on the northern low-gradient slopes (in the facie groups 2 and 3), the deviations of $E_0$ from its values on uplands do not exceed 3%–5%. The bottoms of small valleys are additionally heated by approximately the same value. In the calculation of $F_{hum}(1)$, the annual precipitation for all topogeosystems was taken to be equal. As is known (Gartsman 1977), at the topological level, i.e., at the scale of the small water catchment area, the water balance of the territory varied mostly due to its expenditure part (evaporation and runoff), while the incoming part (precipitation) varies insignificantly.

The calculations resulted in obtaining the local atmospheric humidity factors for topo-ecosystems of the Zhiguli Mountains. The comparison of these factors with their zonal–regional (background) analogs made it possible to establish zonal–geographical conditions for the classification of the hydrothermal regimes of particular extra-zonal groups of low-mountain biogeocoenoses. Thus, we performed the quantitative simulation of landscape–zonal systems by hydro-edaphotopes of the local placor and extra-zonal series. The local $F_{hum}(1)$ can be viewed as a kind of simulation measure allowing for a more precise determination of the entire range of local deviations from background zonal conditions of a given territory. In the Zhiguli ecoregion, this variety appeared to be rather wide. It matches the zonal range of atmospheric humidification from the northern steppes to the mixed forests.

### 7.2.3 Edaphic Humidity Factor and the Method of Its Calculation

The atmospheric humidity factor characterizes the hydrothermal conditions of a territory only at a first approximation and is rather generalized (D.L. Armand 1975; Isachenko 1985). It is even believed that such ratios as a single value are irrelevant for comprehending the topological variety of hydrothermal conditions of geo(eco)systems (Sochava et al. 1970).

The data on $F_{hum}(1)$ for the plain conditions of the By-Sura is its obvious confirmation. Here, local variations of this factor are actually absent. The background atmospheric humidity factor is an input parameter for geo(eco)systems of the topological level. It does not take into account the local diversity of the processes of heat and moisture cycles, which gives a wide range of states of hydro-edaphotopes at equal ratios of incoming solar energy and precipitation. However, in the case of highly dissected relief, the local $F_{hum}(1)$ makes certain adjustments through the evaporative processes related to the solar exposure of slopes (as has been shown for the Zhiguli). Nonetheless, the topological contrasts of expenditure parts of the water balance (evaporation and runoff), which determine soil moisture content with respect to the incoming amount of moisture, are not taken into account.

It is necessary to apply directly to the output parameter of the state of hydro-edaphotopes: the stored soil productive moisture. Hydro-edaphotopes refract the heat and moisture ratios in accordance with the local peculiarities of landscape coupling. This results in a certain topological diversity of the coefficients of humidification, not atmospheric but edaphic: $F_{hum}(2)$.

Through the edaphic humidity factor, it is possible to perform the quantitative simulation of zonal hydrothermal conditions by geo(eco)systems of the local level for regions with slightly dissected relief, where the local diversity $F_{hum}(1)$ is absent. The representativeness of such a simulation is provided by the rather strong correlation relationships between the summer-stored soil moisture and the atmospheric humidity factor (see above).

The parameter $F_{hum}(2)$ was established by the relationship between $F_{hum}(1)$ and the mean July stored soil productive moisture, predetermined on the basis of data from 120 agro-meteorological stations and five water balance stations of the Volga River basin. These relationships for the forest sites were as follows ($R=0.89$–$0.90$; $R^2=0.79$–$0.81$):

$$F_{hum}(1) = 0.242 + 0.06284(W-20) - 0.00062(W-20)^2; \tag{7.7}$$

$$F_{hum}(1) = 0.1226 + 0.025(W-50) - 0.00008267(W-50)^2; \tag{7.8}$$

$$F_{hum}(1) = 0.1596 + 0.01236(W-100) - 0.00002178(W-100)^2. \tag{7.9}$$

Substituting the average multi-year values of the stored soil moisture into these formulas for each group of forest biogeocoenoses within the given ecoregion, we obtain the respective set of local edaphic humidity factors $F_{hum}(2)$. Each of these factors characterizes the local heat and moisture ratio corresponding to the fixed norm of the summer soil moisture content in the topo-geosystem; that is why these factors are referred to as edaphic.

## 7.2.4 Local Coefficient of Common Humidification

One of the parameters often used in hydro-climatology is common moistening $W_{com}$ as the moisture resource of the evaporation process (Mezentsev et al. 1974). The $W_{com}$ parameter, in addition to atmospheric precipitation recorded by precipitation gauges, takes into account the moisture coming to the Earth's surface due to the condensation of near-ground water vapor (dew, hoarfrost) and moisture exchange with the underlying soil layers. The general humidity was calculated by the annual values of evapotranspiration, river discharge, and evaporativity, yielding the following multiple regression equations ($R=0.88$; $R^2=0.77$):

$$W_{com} = 1,211 - 31.66\, t_{July} + 0.1329\, r_{ann} + 0.4108(W - 50); \tag{7.10}$$

$$W_{com} = 1200 - 31.81\, t_{July} + 0.119\, r_{ann} + 0.2259(W - 100) \tag{7.11}$$

Based on the given dependencies, we estimated the $W_{com}$ values and then the local general (atmospheric-soil) humidity factors $F_{hum}(3)$ as a $W_{com}$ to $E_0$ ratio. For the By-Sura and Zhiguli placor topo-ecosystems, the $F_{hum}(3)$ parameter proved to be 1.25 and 1.07, respectively. This means that the former and latter regional ecosystems, with respect to general humidity, should be classified as a mixed forest zone and a borderland between mixed and broadleaf forests, respectively (Kolomyts 2003). According to $F_{hum}(3)$, these ecoregions should be shifted northwards by a whole sub-zone compared to the data provided by $F_{hum}(1)$.

Judging by the values of local edaphic and general humidity factors, the natural potential of basic zonal norms of humidity is rather high not only in the plain By-Sura but also in the Zhiguli Mountains, providing steady growth of forest communities on upland habitats in the past centuries. At the zonal forest and steppe ecotones, forest communities can exist even under conditions of evident deficiency of atmospheric humidification.

Thus, we have succeeded in substantially extending the methodological basis of the estimates of heat and moisture ratio, which allowed us to use this criterion much more efficiently for revealing the local diversity of hydrothermal structures of geo(eco)systems both in their initial state and at all stages of the forecast period. For the By-Sura and Zhiguli biogeocoenoses, the basic humidity factors: $F_{hum}(1)$, $F_{hum}(2)$, and $F_{hum}(3)$ – and their forecasted values for different facial groups are presented in Tables 7.4 and 7.5. The prediction calculations of the summer stored moisture content were made as described in the works of Kolomyts (2008) and Kolomyts et al. (2012).

**TABLE 7.4** Distribution of biogeocoenosis groups of the By-Sura ecosystem in the field of their levels of functioning and local coefficients of atmospheric $F_{hum}$ (1), edaphic $F_{hum}$ (2) and total $F_{hum}$ (3) humidification, which are determined by the taxonomic norm of productive moisture reserves in the soil

| LEVELS OF FUNCTIONING | | | | |
|---|---|---|---|---|
| I Macro-metabolic | |  | | |
| II Sub-macro-metabolic |  | | | |
| III Meso-metabolic | | |  | |
| IV Sub-micro-metabolic | | |  | |
| V Micro-metabolic | 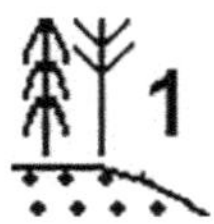 | | |  |
| $F_{hum}(1)$ | 1.07 | | | |
| $F_{hum}(2)$ | 4 – 1.40<br>1 – 0.88 | 3 – 2.05 | 2 – 1.43<br>5 – 1.75 | 6 – 2.69 |
| $F_{hum}(3)$ | 4 – 1.22<br>1 – 1.19 | 3 – 1.25 | 2 – 1.20<br>5 – 1.22 | 6 – 1.25 |

*Note:* Meanings of biogeocoenosis groups can be seen in Table 5.3.

# 7.3 HUMIDITY FACTORS AND LEVELS OF FUNCTIONING OF GEO(ECO)SYSTEMS

Twenty-three parameters of the phytocoenotic functional unit of forest biogeocoenoses (see Kolomyts 2008) were used to estimate their measures of inclusion relations according to Andreev (1980) and to determine the *ecological groups (ecotypes) of facies* corresponding to different levels of functioning: both production and decomposition of phytomass. The maximum values of live phytomass and productivity, as well as the highest rate of decomposition of dead organic matter, are typical of the dominant, i.e., macro-metabolic topo-geosystems. Sub-macro-metabolic and meso-metabolic systems

**TABLE 7.5** Distribution of biogeocoenosis groups of the Zhiguli ecosystem in the field of their levels of functioning and local coefficients of atmospheric $F_{hum}$ (1), edaphic $F_{hum}$ (2), and total $F_{hum}$ (3) humidification, which are caused by the solar exposure contrasts of the lowlands

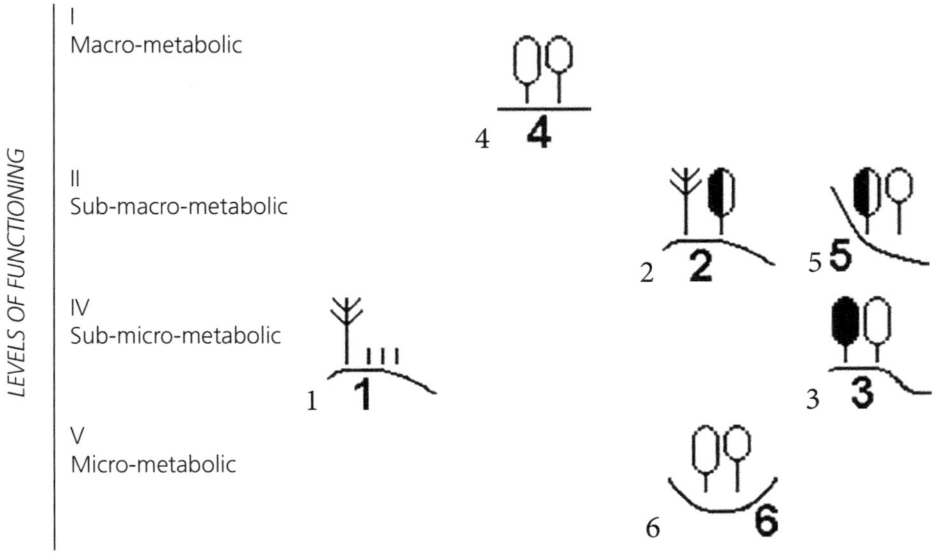


| | | | | |
|---|---|---|---|---|
| $F_{hum}(1)$ | 1 – 0.82 | 4 – 0.95 | 2 – 0.93<br>6 – 0.91 | 5 – 1.06<br>3 – 1.05 |
| Zones/sub-zones – analogs, eastern sector of the Russian Plain (Zonal humidify factor) | Typical and southern forest–steppe (0.76–0.95) | Broadleaf forests (0.95–1.05) | Typical forest–steppe (0.86–0.95) | Mixed forests (1.05–1.35) |
| $F_{hum}(2)$ | 1 – 1.29 | 4 – 1.80 | 2 – 1.64<br>6 – 1.88 | 5 – 1.86<br>3 – 1.64 |
| $F_{hum}(3)$ | 1 – 0.75 | 4 – 1.12 | 2 – 1.00<br>6 – 1.01 | 5 – 1.30<br>3 – 1.16 |

*Note:* Meanings of biogeocoenosis groups can be seen in Table 5.3.

have a lower level of functioning. Finally, reduced geosystems (sub-microbolic and microbolic) are characterized by the minimum reserves of live phytomass, the lowest productivity, and a highly impeded biological cycle. Thus, there is a rather high contrast between the functional states of forest biogeocoenoses in each ecoregion and ecosystem, which should create the respective diversity of their responses to the same background hydrothermal signals.

Under the conditions of the slightly dissected elevated plain (the ecoregion of By-Sura), the local differentiation of natural territorial complexes is determined almost solely by the edaphic humidity factor (see Table 6.5). The upland nemoral–forest complexes (facies group 3) occupy the median position in the hydro-edaphic series, being characterized by the highest level of functioning and, therefore, the maximum productivity. They are distinguished by a relatively high degree of soil–ground moisture content ($F_{hum}(2)>2$). Only the eluvial-accumulative and super-aqual coniferous and mixed forest geosystems (group 6) are more humid.

The taxonomic basic norm of the summer stored productive moisture in the 1-m layer of upland soil is about 250 mm in By-Sura and more than 350 mm in the Zhiguli. It exceeds the maximum of stored moisture with the minimum moisture-holding capacity of sod-mesopodzol and gray forest loamy soils (180–200 mm). As is commonly known, the summer stored moisture in the soil at the moisture-holding capacity level provides optimal growth of plants (Rode 1965; Verigo and Razumova 1973), which perfectly corresponds to the eurytopic functional category of placor oak–linden grooves.

The entire extra-zonal pleiad of the By-Sura topo-ecosystems is located on the lower steps of functioning relative to uplands. At the same time, almost all of them, except for the sixth facies group, are under conditions of lower edaphic humidity compared to uplands. Although their taxonomic norm is ($W$–100) = 175–185 mm, i.e., close to the optimum, these forest communities are not characterized by high productivity. The transition to the pine–broadleaf forests of the moraine plains of By-Sura and further to the spruce–pine forests and pine forests (facies groups 2 and 1) is accompanied by a further decrease in the stored soil moisture content to 120–90 mm, which corresponds to the level of slower plant development (Rode 1965). There is also a simultaneous decrease in $F_{hum}(2)$ to 0.90. The limiting factor of productivity of these mesometabolic and micro-metabolic forest ecosystems is the soil–ground moisture deficit, even though their atmospheric humidity is quite sufficient ($F_{hum}(1)>1$).

The local hydrothermal structures under conditions of deeply dissected low-mountain relief (Zhiguli ecoregion) are much more varied and contrasting. Here, not only the edaphic humidity factor but also atmospheric factors $F_{hum}(1)$ and $F_{hum}(3)$ are territorially differentiated, which is caused almost solely by the solar exposure contrasts of mountain relief elements.

Concerning atmospheric humidity, the placor linden groves of the Zhiguli Mountain Range are characterized by enhanced xero-morphism: their $F_{hum}(1)$ matches the heat and moisture ratio of the most southern borderland of the broadleaf forest sub-zone. At the same time, these forest communities are well-developed at highly hydromorphic edaphotopes ($F_{hum}(2) \rightarrow 1.8$), which relates them to the placor oak–lime forests of By-Sura.

As has been mentioned above, there is a rather wide range of extra-zonal natural territorial complex in Zhiguli. In their semiarid pleiad, the minimum edaphic humidity is typical of the 1st facies group of forest–meadow–steppe zonal habitus. These sparse steppificated pine forests belong to the typical steppe–forest and are at the lower (sub-micro-metabolic) level of functioning. On the other hand, the semi-humid pleiad is formed, in addition to upland elm–linden groves, by sub-macro-metabolic pine–broadleaf forests and sub-micro-metabolic oak groves of shady and neutral slopes (facies groups 2 and 3) gravitating towards the mixed forest zone. The increased hydromorphism of soils is also typical of small valley bottoms and the adjacent slopes (groups 6 and 5).

Thus, the local aridization of hydro-edaphotopes (except for super-aqual sites) while ascending along the catena is accompanied by a decrease in the total level of functioning of forest communities. This consistency is notable for the fact that it simulates the pattern of transition of the functional states of forest ecosystems under the forecasted thermo-arid climatic trend.

## 7.4 THE PREDICTIVE ESTIMATES OF EDAPHIC HUMIDITY CHANGES AND THEIR ECOLOGICAL EFFECTS

Let us consider, by the example of the nemoral–forest sub-zone of the Volga Upland, the possible changes in the state of broadleaf and mixed forests in the context of expected thermo-aridization of soils, according to the HadCM3 climate prediction scenario (see above Section 4.8). At present, the forest cover of this territory is not more than 15%–25% (Stupishin 1964). The remaining upland oak–linden groves and their small-leaved derivatives on the clay–loam substrate (Kud'ma ecoregion, facies groups 1 and 3) initially have summer stored soil moisture hardly providing continuous forest cover (see Table 7.5 and Figure 7.2a). At the end of the 21st century, the average values of (*W-20*), (*W-50*), and (*W*-100) here will be 29–30, 45–55, and 145–155 mm, respectively, which will lead to a tendency to transition of these forests into light forests.

On sands with the prevalence of pine forests and pine–linden–oak groves (group 2), the initially stored soil moisture was already close to the third critical level, almost precluding the existence of forests. Beginning from 2030 to 2050, the indicated values of stored soil moisture will decrease here to 20, 45, and 95 mm, leading to the replacement of these forests by northern (meadow) steppe with separate spots of aspen and birch groves. Sound forest communities will be preserved only in the currently over-wetted valleys of small rivers. In the typical forest–steppe (By-Sura ecoregion), the productive summer stored soil moisture is expected to decrease by 2050 to 65% to 70% of the base norm (see Figure 7.2b), which will threaten the sustainable functioning of most forest phytocoenoses. By the end of the 21st century, the summer stored soil moisture in this area will possibly decrease 1.5 to 1.7 times. The watershed pine groves and pine–broadleaf forests of the fluvio-glacial plain will also be affected.

In the oak–linden groves of clay–loamy plateaus (group 3), the decrease in soil moisture is expected to be relatively low: to 30, 60, and 175 mm in the soil layers of 0–20, 0–50, and 0–100 cm, respectively. Nonetheless, already in 2030, these forests will start losing their appearance as zonal representatives of typical forest–steppe, slowly turning into outliers and open tree formations.

In spruce–lime oak forests of gentle-concave slopes (group 5) and in paludal coniferous forests of outwash depressions (group 6), the decrease in (*W*–100) will be 170–140 mm (the decrease of stored soil moisture below the second critical level). This should cause the beginning of the fragmentation of contiguous forest cover into prairified outliers.

The soil grounds of the Zhiguli Mountain Range will undergo the strongest aridization, especially the forest communities of the semi-arid pleiad (facies groups 1 and

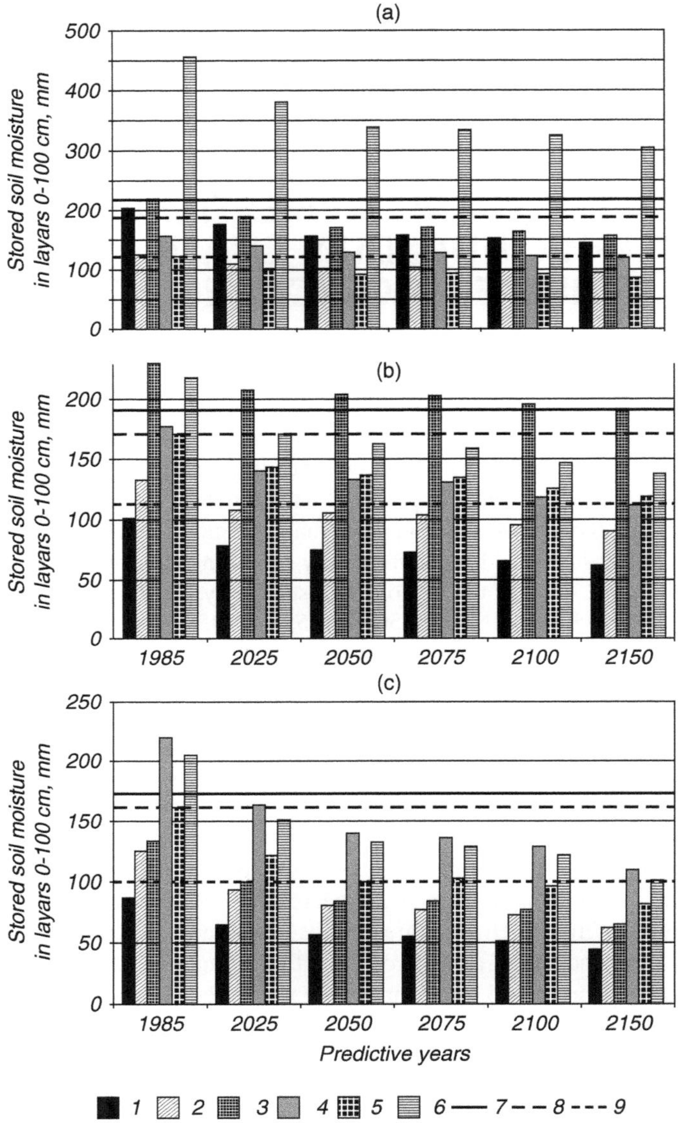


**FIGURE 7.2** The predictive changes of the stored soil productive moisture layer 1 m in the forest ecosystems of the forest–steppe in the Middle Volga Region, according to the HadCM3 model. Regional ecosystems: (a)– Kud'ma (sub-zone of broadleaf forests); (b) – By-Sura (sub-zone of typical forest steppe); (c) – Zhiguli (sub-zone of south forest–steppe, on the boundary with the north steppe). *1–6* – groups of landscape facies, or biogeocoenoses (see Table 5.3); *7, 8* and *9* – correspondingly first, second, and third critical levels of July stored soil productive moisture in layer 1 m (see further Table 9.1).

2 – see Figure 6.2c). By the middle of the 21st century, the steppified open pine forests of sunny slopes and pine–broadleaf forests of neutral and shady slopes will preserve their stored soil moisture at only 60% to 65% of the standard norm; by 2100, the productive moisture in this area will decrease almost twofold. With regard to the summer stored moisture, the pine–broadleaf forests of Zhiguli will cross the hydro-edaphic boundaries of their existence, even as open tree formations, and forest–meadow–steppe communities will begin to transform into the southern dry steppe and then desert steppe.

The maple–elm–lime forests of the Zhiguli plateau (facies group 4), as well as their upland analogs in the By-Sura, will be most stable due to increased initial stored soil moisture. Nonetheless, they can also undergo catastrophic changes because of global warming. The standard norm of (*W*–100) of about 220 mm in these forests will decrease to 140 mm by 2050 and twofold by the end of the 21st century, which should lead to the complete collapse and extinction of upland broadleaf forests in the southern forest–steppe.

Thus, the predicted dynamics of summer stored soil moisture indicate the inevitable worsening of forest growth conditions in the vast forest–steppe transitional area of the Great Russian Plain under global warming. As is commonly known (Morozov 1949), the main factor of stability and productivity of forest communities is the reforestation of forest-forming tree and shrub species. Special studies have shown that an extremely dry climate leads to a rapid decrease in the intensity of spontaneous seeding and development of tree and shrub undergrowth, higher activity of insect pests, and enhanced zoogenic pressure on spontaneous seeding and undergrowth (Matveev and Novikova 1988). All these processes lead to the inevitable extinction of forests and considerably complicate artificial forest cultivation under the conditions of a steppe climate.

## 7.5 SOME INFERENCES

This chapter concludes, so to speak, the pre-forecast stage of local and regional geo-ecological monitoring. The conducted experience of empirical–statistical modeling of structural characteristics and functional states of forest ecosystems of the main catchment area of the Volga River basin not only confirmed the already well-known zonal–regional patterns of organization of the biogeocenotic cover in the domestic literature but also revealed new layers of knowledge about this organization, based on multidimensional cross-analysis of inter-component and inter-complex connections in local-level ecosystems. This stage allows us to obtain the empirical base necessary for adjusting and interpreting the results of numerical landscape–ecological forecasting.

# Peculiarity of Landscape–Ecological Forecasting

# 8

## 8.1 THE WAYS OF DEVELOPMENT OF REGIONAL PREDICTION CONCEPT

Fundamental problems of ecology and geography include, as already mentioned, the problem of global changes, which is the core of the "International Geosphere-Biosphere Program". The program is designed for a long-term outlook and envisages the development of scenarios for the near future of the biosphere in terms of physical models describing basic processes and events. As is known, global geosystem monitoring is most up-to-date and actually realizable on the scale of individual ecological regions. However, natural processes and events at the regional hierarchical level are characterized by the greatest diversity and high discreteness; therefore, the regional response to global climatic changes inevitably takes the form of multiple reactions of vegetation, soils, and landscapes as a whole to background climatic signals. So far, there is no distinct notion of this multiplicity because the measure of sensitivity of soil–biotic components to climatic changes in different zonal–climatic and geomorphological conditions has not yet been estimated.

Forecasting is, as is well known, the ultimate goal and the main feature of natural scientific theories. Expressed in the form of the requirement for reproducibility of the published result in an experiment under certain controlled conditions, forecasting "is directly related to such goals of scientific research as 'explanation' or 'disclosure of the essence of a phenomenon'" (Rosenberg et al. 2008, p. 172). The scientific nature of any forecast implies its provability, derivability, reproducibility, and the presence of limitations (Puzachenko 1998).

The conception of regional manifestations of global environmental changes is undoubtedly of fundamental importance for not only regional but also global ecology


DOI: 10.1201/9781003733119-8

because global biosphere processes and events are perceived most deeply through the level of ecological regions. This perception is aided by the high diversity of their nature–territory structures, as well as the complexity of functioning and evolutionary transformations of regional ecosystems. The ecological region of the main drainage area of the Volga River basin is quite a favorable natural model for prognostic analysis because it is entirely included in the East-European boreal ecotone – a system of close zonal boundaries that is very sensitive to global climatic changes. Here, zonal boundaries are characterized by high dynamism, which is indicated by the entire Holocene history of aboriginal flora in the Middle Belt of the Russian Plain. The main attention should be paid to the mechanisms of shifts in the patterns of vegetation, soils, and landscapes on the model territory under given scenarios of disturbing impacts of the climatic system, which are envisaged in the foreseeable future.

At present, there are actually no experiments on the determination of the most probable directions and rates of distribution of hydrothermal signals from global to topological (local) levels. The response of ecological niches of local ecosystems to the behavior of the background climatic system is still not clear either. The regional hydrothermal trends are generally described by two initial parameters: temperatures and precipitation, the connection of which with landscape facies (biogeocoenoses) is weak and often statistically unreliable. It is necessary to establish the causal mechanisms of local response to global and regional signals by revealing transfer functions in landscape connections.

To use these results in further scientific and practical research in the field of environmental protection, it is necessary to evaluate the common prognostic landscape–ecological situation for each continental region of the Earth. However, the direct use of the global prognostic parameters for regional levels is impractical and can be erroneous. Global prognoses sketch the broad outlines of possible changes in the environment and are often accompanied by schematic maps. Unfortunately, the scales of such maps are too small (no more than 1:30–40,000,000), and only shifts of boundaries of nature zones are shown on them (Leemans 1992; Houghton et al. 1996; Kobak et al. 2002; etc.). Ecological predictions themselves are based for the most part on paleogeographical reconstruction (paleoanalogies) by a very scarce network of points with palynological (spores and pollen) data (Velichko 2010), so they do not describe the regional and local diversity of ecological situations.

The greater impact appears to be in the adoption of the theoretical (calculated) estimation of global climatic changes derived from the models of general atmospheric circulation – AOGCMs (Budyko et al. 1992; Houghton et al. 1996). However, special systematic difficulties emerge here. The main problem is the transformation of global climatic signals into regional and local ecological ones on a rather large scale of natural landscape types, including sub-regional and local plant societies. Up to now, regional tests have not been carried out adequately to determine the most probable directions and relative rates of spreading of external signals both on the system of connection of the nature component and on the lines of landscape chains in space. Therefore, the response of ecological (climatic) niches of these phytocoenoses to global climatic changes is not clear.

This book expounds the main statements of an original geo-ecological concept of prediction: "Global Changes on the Regional Levels", as a basis for regional bio-ecological and geosystem monitoring under global anthropogenic climatic changes. The present work is one of the experiments of regional landscape–ecological prognosis at a much larger scale (1:5–6,000,000).

The objects of our research are forest and forest–steppe landscape–zonal systems of the headwater of the Volga River basin. The ecological region of the main drainage area of the Volga River basin is quite a favorable natural model for prognostic analysis because it is entirely included in the East-European boreal ecotone – a system of close zonal boundaries, which is very sensitive to global climatic changes.

## 8.2 LANDSCAPE–ECOLOGICAL SYSTEM AS AN OBJECT OF PREDICTIVE MODELING

The landscape–ecological approach to modeling the functional parameters of local ecosystems is based on Timofeev-Ressovsky's bio-chorological concept (Timofeev-Ressovsky and Tyuryukanov 1966) concerning the spatial organization of matter and energy cycles by living matter in biogeocoenoses as discrete, elementary structural units of the biosphere. According to theoretical developments in this field, the biosphere is regarded as a statistical ensemble of biogeocoenoses as bio-chorological units that weakly interact with one another but have a highly ordered internal organization (due to stabilizing selection). Each ecosystem at the level of facies or their groups is assigned specific functional characteristics related to the structure of local matter and energy fluxes. This approach permits the prediction of the behavioral patterns of forest communities under different geomorphological and edaphic conditions on the basis of comprehensive empirical data collected in the course of field studies, while retaining the statistical methods of their analysis.

We have developed the numerical methods of regional and local landscape–ecological forecasting based on discrete empirical–statistical models of natural ecosystems. The regional and topological forecast analyses of climate-induced transformations of natural complexes are based on the *numerical landscape–ecological forecasting methodology developed by the author*. The calculations use the methods of information theory, descriptive ("fuzzy") sets, and Markov chains. The ways and efficiency of using these principles in the assessment of the past and future states of natural ecosystems are demonstrated as well.

Procedures were carried out with the ecological niches of landscape–zonal and local ecosystems based on their inclusion relations as descriptive vectors, giving the probabilities of stability (stabilization) of each object and its transitions into other objects according to the given climatic trend. Original methods were used to construct the matrices and orientational graphs of probabilities of landscape and facies transitions, describing the general picture of ecosystems' response to climatic changes.

The calculation technique has been brought to the prescribed level and can be included in the scientific and methodological arsenal of geographical ecology. It organically combines a fairly rigorous formalized approach to solving forecasting problems with procedures for collecting, processing, and analyzing empirical material accessible to a wide range of researchers. In contrast to the well-known domestic and foreign approaches, this method accounts for the multiple nature of the transformation of natural complexes for a fixed climate trend.

A significant feature of the methodology of ecological and geographical forecasting is its *experimental character*. Computational models reproduce the variation in predicted processes by *empirically simulating* the spatially distributed parameters of the basic ecological niches of the studied objects. The forecast itself is essentially *a controlled experiment*, as defined by Helbert (2008a,b). The researcher sets the input variables and, at the output, obtains the pattern of predicted structural and functional states of the studied objects in this statistical sample, with the identification of new objects outside the sample. In particular, the following were used:

1. local empirical imitation of a regional bioclimatic trend using models of binary hydro- and thermo-edaphic ordination of forest topo-ecosystems;
2. the use of the poly-zonality property of local ecosystems as a way of their reaction to global climate changes;
3. empirical imitation of climate-induced changes in the biological cycle based on hydrothermal ordination analysis of its parameters;
4. empirical imitation of changes in productivity and the carbon sink in forest ecosystems during climate fluctuations.

All predictive ecological models proceed somehow from the principle of actualism; however, they are unambiguously determined in some cases and probabilistic in others. The determined models do not take into account the diversity of states of sub-regional and local geo(eco)systems under the same background climatic conditions, which is especially typical of mountain territories. Ecological–geographical prediction is always more reliable and practically significant if it is characterized by certain stochasticity and diversity, i.e., if it is probabilistic. We proceed from the premise that "the numerical predictions of prognostic models ... should be considered ... rather as information for thinking about the likely future trajectories of ecosystems ... taking into account the significant uncertainty of forecasts" (Shvidenko 2012, p. 55). The predictive modeling undertaken by us is fundamentally probabilistic.

The forecast strategy initially identifies phytocoenotic, soil, or landscape units for certain values of the basic climate conditions and then quantitatively assesses the most probable transformations of these objects following anticipated climate change for a specified period. Since the probabilistic ecological forecast is conducted through operations with modern and assumed hydrothermal niches of geo(eco)systems, each such niche is represented as a "fuzzy" (descriptive) set of system states in the form of a row (or column) vector. The vector components are the normalized partial coupling coefficients. According to the proposed principle of landscape–ecological forecasting, the climatically determined functional transformation of one geo(eco)system into another is more significant the smaller the degree to which their climate niches intersect in the initial state.

Predictions of ecosystem transformations include two stages of analysis: (1) evaluation of the probabilities of changes in the functional states of ecosystems and (2) calculations of the rates of ecosystem transformations. Quantitative predictive analysis is carried out through operations with hydrothermal niches of regional or local ecosystems. The algorithms of predictive calculations are described in detail for both stages of analysis. In conclusion, the methods of calculation and construction of average weighted (by territory) matrices and directed graphs of landscape–ecological transitions (for prognosis) and deviations (for paleoreconstructions) have been considered.

The procedure for predicting the state of forest ecosystems, based on empirical–statistical simulation, has been carried out. As already mentioned, the prediction strategy is based on the actualism method. It consists of the original identification of plant units with specific parameters of basic climatic conditions and subsequent quantitative estimation of the most probable transformations of these objects corresponding to the expected climatic changes within a specified time. This approach assumes multifold, probabilistic transformations of ecosystems at the same value of the hydrothermal trend, where a new state may have characteristics of several states existing at the initial moment, rather than a single state. Note that functional relaxation, as the primary response of ecosystems to an external factor, is a priority object of landscape ecological prediction. The main prediction based on ecological estimation of the forthcoming global climate warming is expected to be functional rather than structural–morphological.

## 8.3 IDEOLOGICAL FOUNDATIONS AND STRATEGY OF PREDICTIVE RESEARCH

The previous generalization by the authors of field and office studies made it possible to develop a substantiated and internally consistent *ideology of local and regional landscape ecological predictions*. This ideology, also relying on authorities in its basic statements, can be summarized as follows:

1. The stable functioning of geo(eco)systems is provided, first of all, by the soil–biotic mechanisms of their monosystem organization and self-regulation (Sochava 1974; Sochava 1978). These mechanisms are based on *inter-component interactions* – the vertical landscape connections realizing the stabilizing selection, which maintains the modal phenotype of this system, according to Pianka (1978), and provides it with certain *geotropic areal stability* (Kolomyts 2024).
2. The impetus for geo(eco)system transformation at a new climatic signal is given by the aggravation of *inter-complex interactions* (Zlotin 1987; Tishkov 1988), i.e., horizontal (lateral) connections and relations in the *system of landscape couplings* (Glazovskaya 1988), or catenas (Gerrard 1981). The multivalued transformation of this signal by local geomorphological and hydro-edaphic

factors creates and aggravates a heterogeneous environment. This aggravation is determined by the multivalued transformation of the background signal in the elements of micro- and meso-catenas under the influence of local geomorphological and hydro-edaphic factors (Vysotsky 1960).

3. The developing disruptive selection promotes the formation of two or more modal phenotypes (Pianka 1978). There is inevitable *spatial diversity of local and regional* functional and structural *rearrangements* of ecosystems at the same background signal (Kolomytz 2022). This system-forming process is aimed at a new *evolutionary climax* of phytobiota (Krištofovič 1946) under the given zonal–regional conditions.
4. In the case of disturbed catenary organization of ecosystems, *phytobiota* as the most dynamic geo-component plays the key transforming role and determines the "fate" of the entire ecoregion (Semenov-Tyan-Shansky 1928; Simonov 1982). The main driving force of ecosystem rearrangements is the competitive relationship between the ecological niches of plant communities (Svirezhev and Logofet 1978) in the space of new hydrothermal fields. Due to competition, there is ecological diversification, i.e., the division of niches that favors the development of more complex biotic formations (Pianka 1978). Evolution is implemented via migration processes in the plant cover (Vasil'ev 1946) and, therefore, ecosystems change into the *new type of areal stability – migration*, which is accompanied by the enhanced efficiency of utilization of matter and energy resources, always more or less limited.

So, the driving force behind the functional–structural transformation of this geo(eco) system is the aggravation of horizontal (inter-ecosystem) connections, which occurs as a result of the destructive effect of external signals (both natural and anthropogenic), leading individual ecosystems or even the entire ecosystem cover away from a state of former stability. In this case, the biota, which was a stabilizing factor for the geosystem, turns into a dynamically transforming factor, due to its inherent ability for lateral movement – the tendency of life to spread, according to the famous expression of Vernadsky (1994). This is especially evident in the transformation of the vegetation cover, which, having the highest rate of turnover of its mass, has a powerful interactive effect on the material and energy processes and other geo-components, thereby rebuilding the entire natural–territorial mosaic.

It is the aggravation of lateral (inter-complex) natural interactions during periods of disturbance of the previously achieved (to one degree or another) stability of natural ecosystems under the influence of new climatic signals that is the root cause and driving force behind the functional and structural rearrangements of geo(eco)systems. Such is the initial ideological premise laid by us as the basis for the development of methods for regional and local ecological and geographical forecasts. The aggravated lateral relationships become the subject of an independent predictive study, and the parameters of these relationships should be directly included in the corresponding calculation models. The proposed ideological concept is a working tool for applying the landscape approach to ecological prediction. On this basis, we have developed specific ways of constructing *predictive models within the framework of geographical ecology.*

The strategy of landscape–ecological prognosis may be summarized in the following algorithm: first, to carry out an identification of selected ecosystem objects (zonal and sub-zonal phytocoenological systems) in relation to certain values of contemporary climatic conditions, and then to estimate the most probable transformation of the revealed ecological (hydrothermal) niches of the given objects according to expected climatic changes for the given prognostic date. Furthermore, it is suggested that there may be multiple transformations of ecosystems at a fixed value of the hydrothermal trend, where a new state can encompass several pathways of not one but several states that are known at present moments in time.

The general principle of the prognosis is expressed as follows: the value of the climatic-caused transformation of one ecosystem into another is greater the lesser the degree of intersection of their climatic niches in the initial states; that is, the greater the contemporary contrasts of their functional states, the larger the range of the overlap of niches after the rapprochement of ecosystems due to the given climatic factor. By the maximum values of transformation, the most probable directions and the conversion degree of the ecosystem are estimated. At the same time, an unambiguous character of the transformation of regional ecosystems is proposed at a fixed value of the geophysical trend, when the new state may have features of not one but several states existing at a given moment.

The basic *principle of landscape–environmental forecasting* that we propose states that the climatically induced functional transformation of one ecosystem into another is more significant the less the degree of overlapping (a measure of mutual inclusion, see below) of their climatic niches in the initial state; i.e., the more pronounced the initial contrast of their states, and the greater the value of niche overlap after the systems converge based on this geophysical attribute. This principle corresponds to one of the tenets of the ecology of communities, which concerns the relationships between competing populations. Developing the well-known position on the multidimensional nature of the concept of an ecological niche (Whittaker 1975), it can be argued that the described ecological rule is applicable to the phytocoenotic community at any hierarchical level of landscape organization: from biogeocoenosis to regional or zonal ecosystems.

In the predictive assessments going forward a few decades in terms of the characteristic times of objects, the primary focus should not be on structural changes in ecosystems but on directional shifts in their functioning (Kolomyts 2018). This refers, above all, to changes in the rate of the small biological cycle and the transition of the ecosystem to a new level of balance between the productive and detrital branches of metabolism. Thus, the priority object of landscape–environmental forecasting is the functional relaxation of ecosystems as their primary response to an external impact.

The results of the local prognostic analysis are presented in both graphic–analytical models and large-scale maps. The transition from the local to the regional level of prognostic modeling is carried out using inductive hierarchical extrapolation, a method that we developed based on the empirically established phenomenon of the polyzonal nature of local ecosystems as a response to global climate changes.

The author has developed the regional and local landscape–ecological forecast based on discrete empirical–statistical models of natural ecosystems, which permit the operation with a relatively small number of the most informative characteristics and provide results that may not be quite exact but have a higher (geographical) level of generalization than known simulation modeling.

It is also postulated that the primary object of landscape–ecological prognosis within a century must be not the structural evolution of ecosystems but the directed alteration of their functioning, i.e., shifts in the small-scale biological cycle (phytomass production and destruction), which take several years in the taiga zone and a year in the subzone of broadleaf forests. These typical times of functional relaxation approximately correspond to the duration of the carbon cycle in forest (live and dead) phytomasses and mobile soil humus.

We have established empirically that the main channel of connections of regional and local ecosystems with the background climatic system passes through the summer soil moisture content which, on the one hand, serves as a rather reliable geophysical indicator of the state of ecosystems and, on the other hand, is the most powerful local ecological factor (see Chapter 7). It has also been decided that the object of priority in landscape–ecological prediction within a century must be not the structural evolution of ecosystems but the directed change of their functioning, i.e., the shifts in the small biological cycle (phytomass production and degradation).

## 8.4 FUNCTIONAL CHARACTER OF LANDSCAPE–ECOLOGICAL PROGNOSIS

### 8.4.1 Preliminary Remarks

The duration of structural transformations in a natural ecosystem is usually many times longer than the period for which a prognosis is made, but this period may be regarded as commensurate with the characteristic (typical) time of ecosystem functioning. A vital role in the ecosystem belongs to its minor biological cycle, which has two opposite segments: the ascending bio-production segment (autotrophic biogenesis, or anabolism) and the descending detrital segment (catabolism) (Odum 1971; Kerzhentsev 1999). Thus, from the standpoint of object relaxation times, specialists making an ecological prognosis for the nearest several decades must concentrate on the irreversible change of ecosystem functioning rather than on the structural evolution of ecosystems.

Our method of landscape–ecological prognosis (Kolomyts 2008; Kolomyts 2020) is aimed precisely at assessing the functional changes in ecosystems, such as the readjustment of physiological processes in plants and modification of soil–memory parameters (Targulyan and Sokolov 1978). These events occur within several months to several years, which is commensurate with the time of the general shift in landscape–ecological conditions. Such a change of ecosystem states may be referred to as *functional relaxation*. Functional changes in the system, in turn, determine the rate and period of its *structural relaxation*, or its morphological adaptation to new conditions.

In a first approximation, the biological cycle may be roughly characterized by two discrete integrated parameters: (1) for the ascending segment, this is the index of net primary production ($P_{prim}$, see Table 6.1), which characterizes the efficiency of resource utilization by a phytocenosis, on the one hand, and energy expenditures for biological processes in the soil, on the other (Glazovskaya 1981); (2) for the descending segment, this is the coefficient of annual phytomass destruction (CY, see Section 5.4), which reflects the ratio between the rate of annual influx of fallen-off green mass and the rate of decomposition of forest litter mass (in herbaceous ecosystems, dead plants).

Taken together, $P_{prym}$ and CY characterize the ecological efficiency of the soil–phytocoenotic nucleus of an ecosystem (Odum 1971). The ratio of these parameters reflects the degree of balance in metabolic processes; hence, they characterize the performance of the plant component in maintaining ecosystem functioning. Specialists consider that the maintenance of balance depends mainly on the detrital branch, which is the narrowest channel in the system of connections and, hence, plays a crucial role in the cycle of matter (Bazilevich and Rodin 1971; Odum 1971). This opinion is based mainly on data from studies on forest ecosystems and applies primarily to boreal forests.

Using the example of the Volga River basin, we set ourselves the task of revealing the system of relationships between the ecological niches of regional vegetation–soil units and the discrete parameters of the minor biological cycle and, on this basis, estimating the probability of functional shifts in natural ecosystems in the context of different scenarios of global climate change.

### 8.4.2 Functional Isomorphism of Ecosystems and Its Quantitative Estimate

Natural ecosystems are regarded as relatively weakly integrated formations that change their structure under external influences in order to maintain their functional level (first of all, primary production) as an invariant property ensuring the stability of a system (Sochava 1978; Kerzhentsev 1999). On the other hand, functional parameters, compared to structural ones, must respond to external signals much earlier because their characteristic times are smaller (Odum 1971; Zeidis et al. 2001).

This contradiction may be resolved by assuming that the preferable pathway of any structural rearrangement in a given ecosystem is directed toward ecosystems that are closely similar to it with respect to the taxonomic norm of $P_{prim}$ or CY and the structure of corresponding ecological niches, or, in other words, are functionally isomorphic. Isomorphism of two systems implies that their components are interchangeable, which provides for their structural and functional similarity (Paffengolts 1973). We consider the priority of *functional isomorphism*, which allows the ecosystem exposed to some influence to remain more or less invariant in the functional aspect, maintaining the initial rates of autotrophic biogenesis and organic matter decomposition, or the rate of the minor biological cycle. This must facilitate the transition of the system to a new state with a qualitatively new structure. Conversely, structural rearrangements directed toward systems markedly differing (less isomorphic) in the functional aspect, i.e., in the norm of $P_{prim}$ or CY and the structure of corresponding ecological niches, will be impeded. In other words, an ecosystem disturbed by external influence "chooses" the pathway of structural rearrangements that provides for the smallest change in its functioning and the shortest period of recovering stability in a new state.

The proposed *principle of the functional isomorphism of ecosystems* agrees with the hypothesis of ecological duplication (Rozenberg 1991), which is well known in autecology. The existence of species similarity with respect to ecological and chronological gradients allows the ecosystem to avoid catastrophic changes (Tishkov 1991). The systems undergo transformation along the preferable pathway, with minimal manifestations of nonlinearity (Zeidis et al. 2001), which allows their behavior to remain more or less linear. Functional isomorphism is the factor that makes it possible to identify the most preferable pathways of phytocoenological and soil transitions among the variety emerging in the course of prognostic ecological calculations.

The generalization measure of functional isomorphism $N(A,B)$ for two objects $A$ and $B$ may be calculated as a coefficient of their similarity (see Section 4.5) by taxonomic norms of parameters $P_{\text{prim}}$ and CY:

$$N(A,B) = g_1 \frac{2\left[P_{\text{prim}}(A) \cap P_{\text{Prim}}(B)\right]}{P_{\text{Prim}}(A) + P_{\text{Prim}}(B)} + g_2 \frac{2\left[\text{CY}(A) \cap \text{CY}(B)\right]}{\text{CY}(A) + \text{CY}(B)}. \tag{8.1}$$

Here, the symbol "$\cap$" means the intersection of the values of objects as descriptive ("fuzzy") vectors; $\_g_1$ and $\_g_2$ – "weight" coefficients. Normalized (reduced to 1) coefficients of information (*C*(*A*/*B*), see Equation (4.8) in Chapter 4) have been used in the capacity of these "weight" coefficients. The _g_1 and _g_2 proved to be 0.353 and 0.647, respectively, for the groups of plant formations and 0.285 and 0.715, respectively, for soil groups. As we can see, spatial changes in the soil and plant cover are much more tightly coupled with variations in the destructive branch of the biological cycle than with fluctuations in its productive branch. This regularity, which we have noted previously for local geo(eco)systems (Kolomyts 2020), is equally distinct at the regional level as well, i.e., it may be considered a certain ecological phenomenon.

The above formula was used for the calculation of the measures of functional isomorphism of the considered phytocoenological and soil units (Table 8.1), making it possible to reveal an interesting pattern of metabolic closeness/remoteness of regional ecosystems in the entire zonal spectrum of the boreal ecotone. First of all, it turned out that plant formations (and, as a whole, soil–plant complexes presented in the Table) of the subboreal belt are characterized by a higher degree of similarity in the processes of their functioning compared to formations of the boreal belt. Consequently, the key component of the hydrothermal signal (temperature in the boreal belt and pluvial component in the subboreal belt) must cause more effective functional (and then structural) mutual transitions in the forest–steppe and steppe ecosystems compared to the row of taiga and mixed forests. In other words, under the conditions of a climatic trend, when both temperatures and precipitation change simultaneously, the primary functional and structural shifts should be expected in the northern part of the subboreal belt but not in the south or middle of the boreal belt. The results of our analysis of phyto-ecological consequences of climate changes in the last 100 years confirm the validity of this hypothesis.

It is typical for shady broadleaf forests to be rather close in functioning not only to formations of typical forest steppe but also to southern forest–steppe phytocoenological

**TABLE 8.1** Average weighted measures of functional isomorphism between the plant formation groups and dominating for them soil groups on territory of the Volga River basin and its encirclement

| *SOIL–PHYTOCOENO-LOGICAL SYSTEMS* | $PP^{ILL}_{IRON}$, $PS^{GL}_{POD}$ | $PS^{PEAT}_{POD}$, $PS^{HMD}_{POD}$ | $SPS^{GL}_{SUR}$, $SPS^{HM}_{POD}$ | $PS^{PEAT}_{POD}$, $SPS^{GL}_{SUR}$ | $SPS^{HM}_{POD}$, $SPS^{GL}_{SUR}$ | $SPS^{H}_{POD}$, $GS_{LIGHT}$ | $SPS^{BLEA}$, $GS_{LIGHT}$ | $SPS^{ILL}_{IRON}$, $SPS^{CARB}$ | $SPS^{D}_{POD}$, $SPS_{BLEA}$ | $GS_{PROP}$ | $GS_{DARK}$ $GS_{LIGHT}$ | $CHER_{POD}$, $GS_{DARK}$ | $CHER_{POD}$, $CHER_{MED}$ |
|---|---|---|---|---|---|---|---|---|---|---|---|---|---|
| $PS^{peat}_{pod}$, $PS^{hmd}_{pod}$ | **0.80** | – | | | | | | | | | | | |
| $SPS^{gl}_{sur}$, $SPS^{hm}_{pod}$ | **0.92** | 0.77 | – | | | | | | | | | | |
| $PS^{peat}_{pod}$, $SPS^{gl}_{sur}$ | **0.96** | 0.78 | **0.96** | – | | | | | | | | | |
| $SPS^{hm}_{pod}$, $SPS^{gl}_{sur}$ | 0.68 | 0.58 | 0.76 | 0.72 | – | | | | | | | | |
| $SPS^{h}_{pod}$, $GS_{light}$ | 0.79 | 0.64 | **0.82** | 0.75 | **0.88** | – | | | | | | | |
| $SPS^{blea}$, $GS_{light}$ | **0.89** | 0.72 | **0.92** | **0.88** | 0.78 | **0.89** | – | | | | | | |
| $SPS^{ill}_{iron}$, $SPS^{carb}$ | 0.62 | 0.54 | 0.67 | 0.64 | **0.88** | 0.80 | 0.70 | – | | | | | |
| $SPS^{d}_{pod}$, $SPS_{blea}$ | **0.45** | **0.42** | *0.48* | *0.47* | 0.60 | 0.54 | *0.49* | 0.68 | – | | | | |

*(Continued)*

**TABLE 8.1 (*Continued*)** Average weighted measures of functional isomorphism between the plant formation groups and dominating for them soil groups on territory of the Volga River basin and its encirclement

| *SOIL–PHYTOCOENO-LOGICAL SYSTEMS* | $PP^{ILL}_{IRON}$, $PS^{GL}_{POD}$ | $PS^{PEAT}_{POD}$, $PS^{HMD}_{POD}$ | $SPS^{GL}_{SUR}$, $SPS^{HM}_{POD}$ | $PS^{PEAT}_{POD}$, $SPS^{GL}_{SUR}$ | $SPS^{HM}_{POD}$, $SPS^{GL}_{SUR}$ | $SPS^{H}_{POD}$, $GS_{LIGHT}$ | $SPS^{BLEA}$, $GS_{LIGHT}$ | $SPS^{ILL}_{IRON}$, $SPS^{CARB}$ | $SPS^{D}_{POD}$, $SPS_{BLEA}$ | $GS_{PROP}$ | $GS_{DARK}$, $GS_{LIGHT}$ | $CHER_{POD}$, $GS_{DARK}$ | $CHER_{POD}$, $CHER_{MED}$ |
|---|---|---|---|---|---|---|---|---|---|---|---|---|---|
| $GS_{prop}$ | 0.69 | 0.59 | 0.76 | 0.72 | **0.99** | **0.89** | 0.79 | **0.89** | 0.61 | – | | | |
| $GS_{dark}$, $GS_{light}$ | **0.49** | **0.41** | *0.48* | *0.47* | 0.62 | 0.59 | 0.53 | 0.72 | **0.90** | 0.63 | – | | |
| $Cher_{pod}$, $GS_{dark}$ | **0.49** | **0.41** | *0.48* | *0.46* | 0.62 | 0.58 | 0.53 | 0.71 | **0.92** | 0.62 | **0.98** | – | |
| $Cher_{pod}$, $Cher_{med}$ | 0.44 | **0.38** | 0.54 | *0.42* | *0.49* | 0.51 | *0.47* | 0.61 | **0.90** | 0.54 | **0.86** | **0.87** | – |
| $Cher_{leach}$, $Cher_{sou}$ | **0.45** | **0.38** | 0.59 | *0.43* | 0.58 | 0.54 | *0.49* | 0.67 | **0.90** | 0.58 | **0.94** | **0.95** | **0.86** |

*Note:* For conventional meanings of plant formation groups, see Table 5.2. Groups of soil, by Liverovsky (1974): $PP^{ill}_{iron}$, – illuvial-iron and illuvial-humus podzols, as well as peat and podzolic-gley soils; $PS^{gl}_{pod}$ – proper podzolic soils, finely, shallowly, and deeply podzolic; $PS^{peat}_{pod}$, and $PS^{hmd}_{pod}$ – sod-podzolic finely and shallowly podzolized; $SPS^{gl}_{sur}$, $SPS^{hm}_{pod}$ – sod-podzolic illuvial-iron and deeply podzolized; $PS^{peat}_{pod}$, $SPS^{gl}_{sur}$ – sod-podzolic, with a second humus horizon, clarified and residually carbonate; SLs – light gray forest; SL – gray forest; SLt – dark gray forest; Chop – podzolized chernozems; Chl – meadow–chernozem soils and leached chernozems; Cht – typical and non-carbonate chernozems; $Cher_{pod}$ – ordinary chernozems; $Cher_{med}$ – southern chernozems.
High degree of isomorphism (more 80%) is showed by half-rich print, the highest (more 90%) – with underline; abnormally high (more 70%) – simple underline; minimal degree (less 50%) – italics print.

systems. Their functional approach is promoted by the predominance of the same soil types in all three cases: podzolized and leached chernozems (see Table 8.1). Being relict formations in lime forests as a heritage of typical and southern forest steppe from the xerothermal period, these soils seem to impart a general forest–steppe appearance to mesophilic nemorose–herbal formations, which transgressed here in the later sub-Atlantic period. The described group of subboreal ecosystems obviously characterizes the most preferable phytocoenotic and soil transitions within this belt under a thermo-arid or, conversely, cold–humid climatic trend.

On the contrary, more xerophilic lime–oak forests gravitate toward sub-taiga coniferous–broadleaf forests and complex nemorose–herbal fir–spruce and pine forests. Obviously, oak forests may act as primary prototypes of new functional states of these boreal ecosystems under climate aridization. Such transitions have taken place in the recent past, as demonstrated by the wide spread of light gray forest soils typical of oak forests under the mixed forests (see Table 8.1).

It would be interesting to note the obvious functional isolation of pine–broadleaf forests on peaty-podzolic soils from other formations of the sub-taiga zone and their relatively high isomorphism with the broad phytocoenological spectrum of the entire forest steppe zone: from shady broadleaf forests with their grey forest soils to rich herb–sheep's fescue–feather grass steppes on typical and southern chernozems. Such functional unity seems to be nonrandom. It demonstrates deep genetic relations between pine–broadleaf forests and forest–steppe formations, their frequent interchanges during climatic fluctuations in the past, and the most probable phytocoenotic transitions in the future. It may be supposed that under global warming leading to climate aridization, the process of steppification of boreal forests will begin with their transition into more xerophilic pine–broadleaf forests (via the stage of nemorose–herbal pine forests with oak in the undergrowth, see Table 8.1), which will then be transformed at an even higher rate into typical and southern steppe. Such functional and structural rearrangements should be expected primarily in the eastern, more continental sector of the Volga River basin, which is predisposed to climatic aridization.

### 8.4.3 Influence of Functional Isomorphism of Ecosystems on Their Structural Transformation

The ratio of ecological $P_{prim}$ and CY gradients determines the type of functional isomorphism that prevails under given zonal–ecological conditions and, therefore, indicates the most probable directions of shifts in ecosystems at a given climatic trend among the multitude of their calculated variants. In the boreal zone, the change of plant formations and soil varieties is mainly determined by the rate of organic matter decomposition in the forest litter. Parameter CY varies from 13 to 30–36 (Figure 8.1); i.e., its ecological gradient is fairly extensive. Primary production, however, varies within a narrow range, with its values remaining high: 9–10 to 11.5–12 t/ha per year. Therefore, the effect of the functioning of boreal ecosystems on their spatial organization is determined mainly by the catabolic segment of their biological cycle, being only weakly dependent on autotrophic biogenesis. This fact confirms the well-known concept that the forest functions mainly as a detritus ecosystem (Odum 1971).

Middle and southern taiga fir–spruce and pine forests on podzols proper and on illuvial humic and peat-gley podzolic soils are characterized by the lowest rates of organic matter decomposition. In these forests, CY > 17–20 years; i.e., it falls into the stagnant category according to Bazilevich and Rodin (1971), showing that the metabolic cycle as a whole is retarded. The main factor responsible for this retardation is a high value of the annual humidity factor $F_{hum}(1)$, whose taxonomic norm (average weighted mean) reaches 1.75–1.85. Nevertheless, the heat-to-moisture ratios in the southern taiga subzone provide for a sufficiently high level of primary production, averaging 11–11.5 t/ha per year (Table 8.2), and the metabolic level is low because of high CY values. Farther south, in conifer–broadleaf forests with soddy podzolic soils, the rate of the biological cycle becomes markedly higher, as follows from CY decreasing to 11–13 years (the retarded category).

Conventional meanings are seen in Table 8.1.

Formations of the mixed forest subzone, along with oak forests and nemoral pine forests on gray and dark gray forest soils, grow under conditions of zonal ecological optimum. Its characteristic features include the highest values of primary production

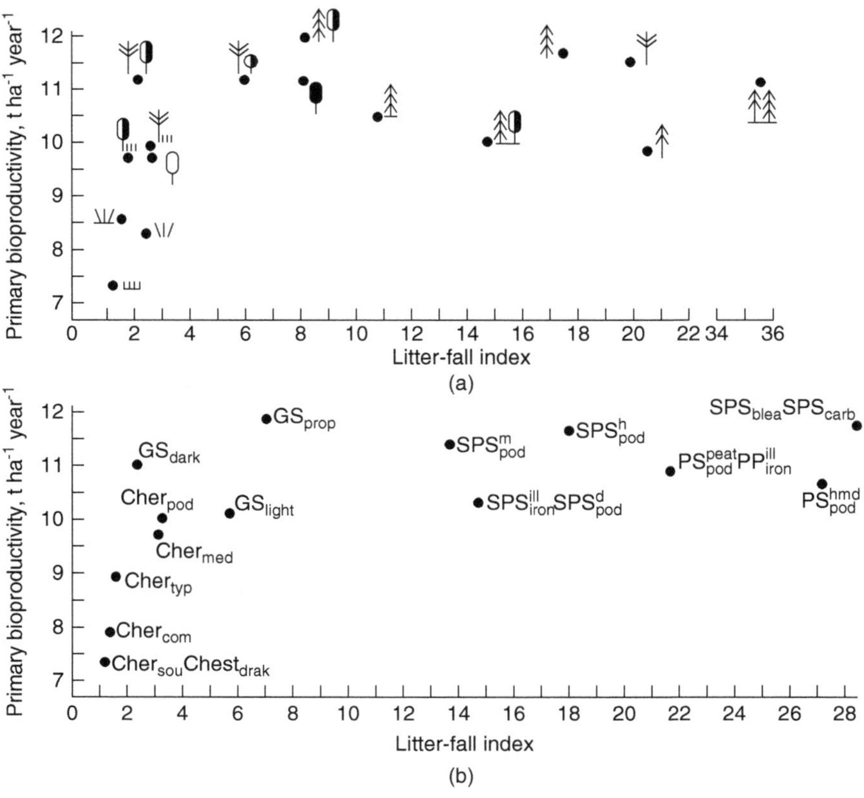


**FIGURE 8.1** Distribution of plant formation groups (a) and soil groups (b) in the field of taxonomic (average weighted) norms of primary bio-productivity ($P_{prim}$) and litter-fall index (CY).

**TABLE 8.2** Ecological niches of classes and groups of plant formation in field of values of primary bio-productivity for the Volga River basin

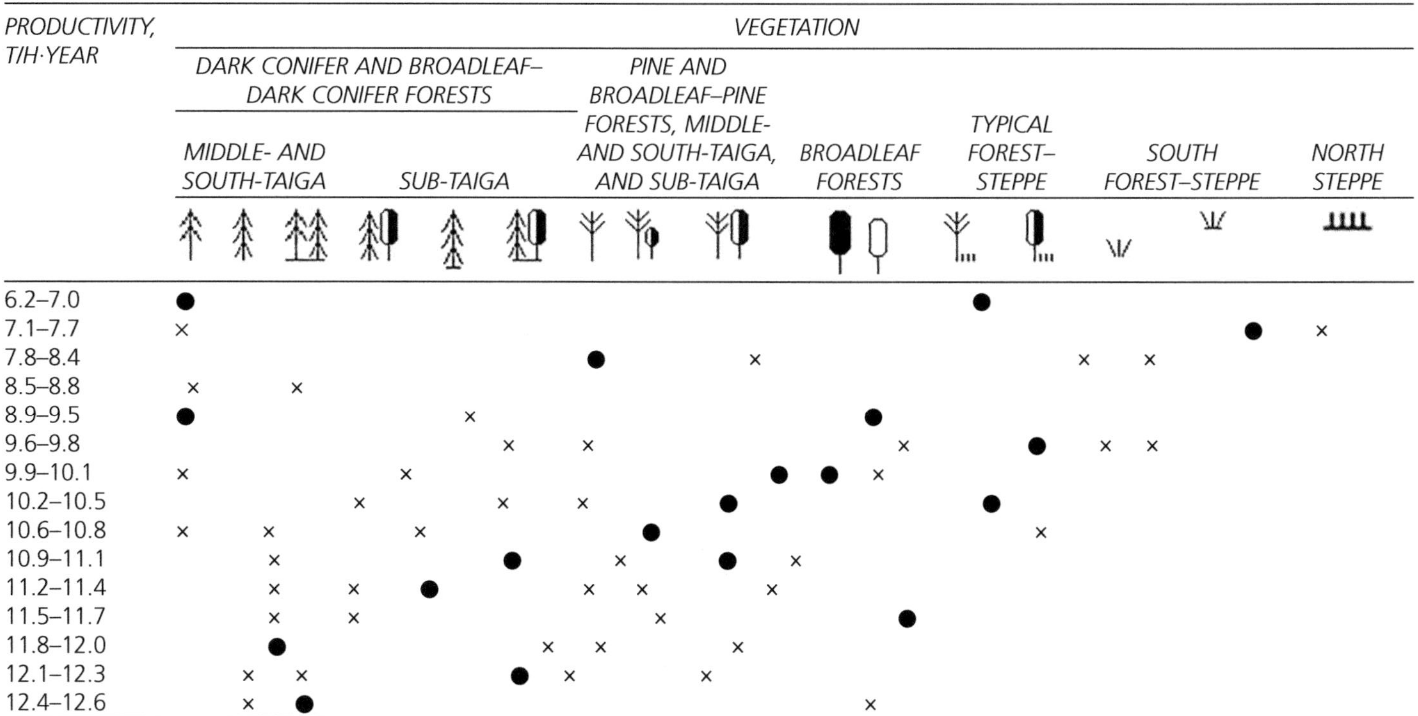


*Note:* For conventional meanings, see Table 5.2 on p. 56.

(no less than 11–12 t/ha per year) and moderate or even increased (in broadleaf–pine forests) rates of phytomass catabolism (CY=2–8 years, the accelerated category). At this optimum, the zonal ratio of the two segments of the biological cycle changes radically, because the ecological gradient of the first (ascending) metabolic segment markedly increases, and that of the second (descending) segment becomes sharply reduced.

Upon transition to the sub-boreal zone, where the temperatures characteristic of the warm period are markedly higher, the rate of organic matter decomposition reaches maximum values (CY < 1.5–2 years). On the other hand, primary production becomes lower because of a sharp decrease in the amount of precipitation and a corresponding reduction in the atmospheric humidity factor (to 0.9–0.8 in the typical forest–steppe and below 0.6 in the northern steppe). This parameter does not exceed 10 t/ha per year in the typical forest–steppe zone with leached and podzolized chernozems and only 7–8 t/ha in the steppe formations on common and southern chernozems (see Table 8.2).

Thus, in the boreal zone, both components of the metabolic cycle are sufficiently effective, although the catabolic component plays a leading role; in the boreal zone, the bio-production component gains priority. This means that the functional machineries of boreal and sub-boreal ecosystems respond differently to the same background hydrothermal signal, and the resulting structural effects are also different. In the former case, the decomposition of organic matter accumulated in the forest litter is the most responsive process; in the latter, this is autotrophic biogenesis accounting for live phytomass production.

If a boreal ecosystem remains within its bioclimatic zone, its initial changes will be directed toward the ecosystems with the closest CY values. For example, if the climate has a thermo-arid trend (temperature rises while moisture supply decreases), middle taiga fir–spruce forests will be first replaced by pine forests and, only later, by southern taiga dark conifer forests (see Figure 8.1a). Sub-taiga nemoral herbaceous fir–spruce forests will be converted not only into relatively humid mixed forest formations but also directly into linden–oak forests.

At the same time, inter-conversions of sub-boreal ecosystems within their bioclimatic zone must depend almost exclusively on the production process because the ecological amplitude of the detritus metabolic component is small. Conversions accompanied by the smallest changes in the annual amount of plant production are preferable. Under conditions of a weak thermo-arid trend, this may apply to the replacement of shaded broadleaf (oak–linden) forest by a typical forest–steppe (with oak or pine stands), with light gray forest soils tending to transform into podzolized and leached chernozems (see Figure 8.1b). If the thermo-arid trend is strong, nemoral forest formations may be converted directly into the southern forest–steppe (rich herb–sheep's fescue–feather grass steppe). Soil-forming processes will lead to the development of typical and carbonate-free chernozems (this is evident when Figures 8.1a and b are compared).

A more complex picture should be expected upon prospective transitions of ecosystems from the boreal to sub-boreal zone or vice versa. In such cases, both segments of the biological cycle have equal effects on the pathway of rearrangements in ecosystems, although the roles of each type of isomorphism do not remain constant. Consider two possible variants of such interzonal transitions corresponding to the alternative bioclimatic trends: cold humid and thermo-arid. In the first variant (a decrease in temperature with a simultaneous increase in atmospheric humidity), sub-boreal plant formations transform into boreal formations; the regulatory role of autotrophic biogenesis decreases, whereas the processes of organic matter decomposition become more significant. In other words, the destruction isomorphism gains system-forming significance at the expense of production isomorphism.

In the second variant, with a strong and consistent thermo-arid signal, a reverse transformation of boreal ecosystems into sub-boreal is inevitable, with the destruction isomorphism giving way to the production isomorphism. In boreal ecosystems with high $P_{prim}$ and CY values, such as fir–spruce or pine forests of the southern taiga subzone, initial rearrangements (until the stage of mixed forests) will follow the pathway providing for the smallest loss of above-ground dead organic matter, under the regulatory influence of destruction isomorphism. The subsequent transformation of the resulting mixed forest ecosystem will be increasingly dependent on the minimal change (in most cases, decrease) in the annual phytomass increment, indicating the increasing role of the production isomorphism.

### 8.4.4 Ecological Forecast in Models of Bio-Production Isomorphism

A wide range of similarity measures of production vectors for both vegetation and soils of the Volga River basin (from 0.10–0.20 to 0.60–0.70 and more) indicates very contrasting fields of functioning of these partial ecosystems, which predetermines an initially high selectivity of their most preferable mutual transitions with a particular hydrothermal signal.

Such selectivity can be revealed using the well-known method of correlation pleiades, which allows us to identify the potentially most real directions of their functional, and then structural transformations. Obviously, the foci of the transition network are the most isomorphic binary bundles – correlation pleiades of the highest level of similarity of production niches; they form the primary chains of phytocoenological and soil transformations. As the isomorphism of niches decreases, the significance of the transitions of objects should decrease.

In the models of bio-production isomorphism (Figure 8.2), correlation constellations of phytocoenological units are revealed by setting two levels of homogeneity ($N$):

1. at $N>0.60$ (constellation of the first level of similarity);
2. at $N>0.40$ (constellation of the second level). At the second level of homogeneity, the main background of spatial differentiation of similarity measures and corresponding networks of functional transitions is created, while the first level forms a kind of foci of such transitions.

In our example, at the second level of similarity, two large constellations are formed. One of them (constellation I) covers the western, less continental sector of the Volga basin, and the other (constellation II) covers the eastern, more continental sector. The first clearly expressed constellation includes the Upper Volga southern taiga fir–spruce forests, middle and southern taiga pine forests, and broadleaf forests; they are also adjacent to the dark coniferous sub-taiga.

The listed formations, possessing maximum functional isomorphism, are distinguished by the highest taxonomic productivity rate (11.2–11.8 t/ha-year), as well as very

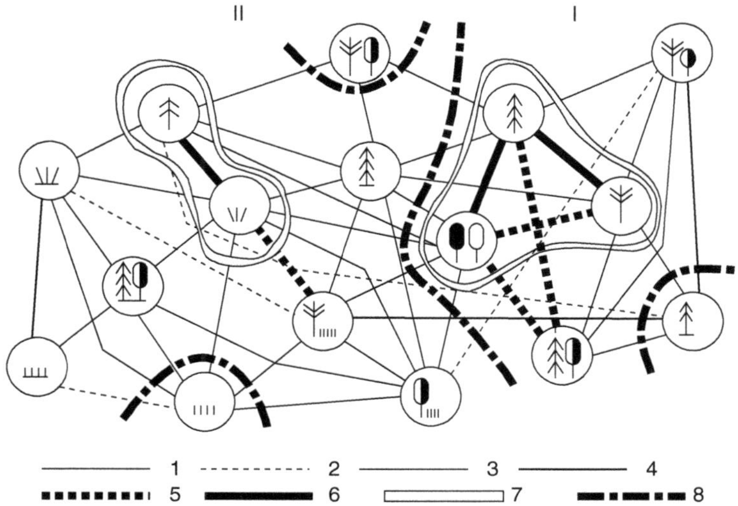


**FIGURE 8.2** Correlation constellations of plant formation classes of the Volga River basin by measures of similarity of vectors of their partial coefficients of connection with primary bio-productivity. Similarity measures: 1 – 0.11–0.20; 2 – 0.21–0.30; 3 – 0.31–0.40; 4 – 0.41–0.50; 5 – 0.51–0.60; 6 – 0.61–0.70. Boundaries of homogeneity levels: 7 – first (>0.60), 8 – second (>0.40). Pleiades I covers the less continental western sector of the Volga River basin, and Pleiades II covers the eastern sector.

extended (up to seven positions of the vector-column) and mutually intersecting production niches (see Table 8.2). They function in a very wide range of landscape–geophysical conditions and, therefore, have a fairly high stability.

It can be expected that with an increase in the temperature of the vegetation period and a general increase in atmospheric precipitation, the southern taiga fir–spruce forests of the Upper Volga and Oka River basins can be relatively quickly replaced by broadleaf–dark coniferous forests and even directly by oak–linden forests; with the aridization of the climate, they may be replaced by pine forests. At the same time, with an increase in climate humidity, the transformation of pine and broadleaf–pine formations into dark coniferous–broadleaf forests will be significantly hindered; replacing them with pure oak–linden forests will be much preferable.

Correlation Pleiades II covers almost the entire zonal spectrum of the eastern sector of the Volga basin, from the Kama-Pechora southern taiga to the East Kazakhstan steppes. These groups of formations are distinguished by a reduced taxonomic norm of phyto-production – from 10.2 t/ha·year in the southern taiga to 7.1 t/ha·year in the dry steppe. In addition, the phytocoenological units here have much more compressed vectors of production niches, i.e., a comparatively narrow range of their inherent hydrothermal space, which indicates their reduced stability. Apparently, they are capable of transforming earlier and/or faster with the same parameters of the geophysical signal.

The western and eastern sectors of the Volga basin also differ significantly in meridional changes in the optimum functioning of the boreal phytobiota (see Table 6.4). In the direction from the middle taiga to the mixed forests in the Upper Volga region, the bio-production optimum sharply decreases, while in the Kama region it increases just as sharply. In the first case, the maximum primary production is created by the middle taiga fir–spruce and pine forests, and in the second by the sub-taiga broadleaf–dark coniferous forests. The factor of increased continentality of the climate thus shifts the optimum functioning of the Kama-Ural taiga-forest formations to the southern boundary of the boreal belt. Only in the northern strip of the sub-boreal belt, in the direction from the broadleaf forests – mainly in the western sector – to the Trans-Volga steppes, can a system of unambiguous decline in productivity be traced.

Judging by the structure of Pleiades II, it can be assumed that in the most continental sector of the basin, with climate changes from cold and humid to warm and dry, the most preferable processes for the direct steppe formation of southern taiga forest formations in the past and in the future will involve the replacement of fir–spruce forests by forb–grass pine forests, bypassing the sub-taiga and nemoral forest stages.

The process of meadow and steppe formation of boreal pine forests, as well as the middle dark coniferous taiga in the western sector, may be even more distinct. With the opposite climatic trend, an active transgression of boreal forests into the forest–steppe and even into the northern steppe should take place.

## 8.5 SOME INFERENCES

Our concept of the functional isomorphism of ecosystems as a factor correcting their climatogenic transformation agrees with the well-known ideas of N.V. Timofeeff-Ressovsky, who extrapolated to biogeocoenoses the Darwinian theory of natural selection (cited from Tyuryukanov 2001). In particular, he considered that bio-chorological systems are selected by the criterion of their contribution to the rate and continuity of the total cycle of matter. Thus, there is reason to use the concept of the functional isomorphism of ecosystems as a working hypothesis for making ecological predictions. Based on the measures of functional isomorphism of ecosystems, we can arrange the probable variants of transitions between them in time, identifying the most probable and secondary variants, and gain deeper knowledge of the prospective pathways of ecosystem transformations.

# 9 Empirical-Simulation Predictive Modeling of Forest Ecosystems

## 9.1 THE VALUE OF EMPIRICAL SIMULATION IN LANDSCAPE–ECOLOGICAL FORECASTS

Numerical forecasts of biospheric processes and phenomena are usually made with the help of *simulation models*. The latter consist of one replacement or another of the real "system complexity" by some abstract structure, decomposing and simplifying the regularities of its structure and functioning under conditions of a priori uncertainty (Rozenberg 1984). The greatest progress has been made in the simulation modeling of global biosphere processes, in the creation of conceptual balance models of different types of geographical environments and individual types of plant communities, and in the construction of models of ecosystem behavior affected by technogenic pollution of natural habitats (Chertov et al. 2003; Tarko 2005). The whole system of material and energy balances is broken down into separate blocks, which allows us to model the response of various ecosystem components more adequately to external impacts based on the known methods of studying individual biospheric processes.

At the same time, while offering sufficiently thoroughly developed forecast scenarios for individual geo-components, simulation models cannot provide us with a holistic view of the future shape of the geo(eco)system itself or the degree of its stability under different combinations of environment-forming and functional parameters. Predictive

DOI: 10.1201/9781003733119-9

constructions based purely on mathematical models fail to reflect the complex nature of changes happening in the system. In addition, such models do not capture the territorial diversity of succession shifts under the same background impact; i.e., they may be "inadequate in space" (Lischke et al. 1998). Overall, the universality and simplicity of simulation models, with a sufficiently deep elaboration of the relationships they consider and the relatively high accuracy of forecasting, face their own imminent limitations due to the very narrow scope of environmental processes and phenomena that these models describe. Obviously, geographic ecology cannot be limited to this type of modeling, since it deals with spatially differentiated and multicomponent geo(eco)systems at the landscape level.

The landscape–environmental analysis of the spatial organization of natural complexes and their climate-genic dynamics (Kolomyts 2005; Kolomyts 2022) developed by us is based on the construction of *discrete empirico-statistical models* of their structure and functioning, as defined by Rozenberg (1984). For the construction of such models, we have proposed and implemented a *method of empirical simulation* of the forecasted geo-ecological situations (Kolomyts 2022). In it, the materials of field studies are used not as a verification of certain calculated data (which is the practice when creating the known simulation models) but as a factual basis for the modeling itself. The possibility of applying widely in the forecast analysis the results of field landscape studies, as well as various cartographic materials, makes it possible to obtain results with a high degree of spatial resolution and to justify broad geographical generalizations empirically.

An integral multicomponent elementary (geo-topic) unit, biogeocoenosis, in its fundamental definition (Sukachev 1972), acts as the initial object for empirical and statistical modeling. The sphere of geo-topes represents the most complex and active part of the natural environment, its functional "core" (Sochava 1978). "The material and energy cycles of biogeocoenoses are interconnected and form a giant cycle of the Earth's biosphere" (Timofeev-Resovsky 1970, p. 90). The territorial diversity of material and energy cycles in each ecoregion results from the structural heterogeneity of the biogeocoenoses.

Empirical–statistical models, being static models, have their drawbacks, the main one of which is the uncertainty of the results in time (Lischke et al. 1998). They give an idea of the final equilibrium states to which these natural complexes will strive in their functional and structural transformations (in line with their relaxation times). This shortcoming can be largely eliminated if we assume, according to Puzachenko (1998), that the trajectory of the predicted functional changes of the geo(eco)system at each discrete prediction step is a stationary nonequilibrium process, which is reversible. Then each new step of the predicted trajectory represents a jump in the functional change of the system.

The principle of the *functional isomorphism of ecosystems* (Kolomyts 2022), proposed by us, is used for the correction of the net of functional–structure transitions. The models of functional isomorphism bring significant restrictions in the probability pattern of predicted trajectories of changes in geo(eco)systems, and this pattern becomes more ambiguous and ordered. The foci of the transition network are maximally isomorphic binary links – correlation pleyads of the highest level of similarity of bio-productive niches; they form priority chains of soil–phyto-coenological transformations.

## 9.2 STATISTICAL ANALYSIS OF FUNCTIONAL BIOGEOCOENOTIC RELATIONSHIPS

Landscape–environmental formations are multicomponent systems, with a huge gap in the relaxation times of their constituent geo-components and a regularized system of direct and return environmental links (Sochava 1978), exhibiting non-linear behavior and the dominance of fundamentally probabilistic relationships (Zeydis et al. 2001). These objects belong to a vast group of natural systems with relative independence in the behavior of their components, characterized by nonequilibrium wave dynamics, self-organization, and hierarchy (Puzachenko 1998). They are relatively loosely connected natural formations, which is one of the main conditions for the manifestation of *polymorphism properties* in them, i.e., multiple patterns of formation of stable biogeosystems under different abiotic factors (Kolomyts 2022).

The description of such systems may not follow the basic principles of statistical analysis. In particular, the principle of linear independence of predictors is violated. When describing complex multicomponent biogeosystems, correlation coefficients $R$ and determination coefficients $R^2$ cannot be high (see Table 7.2), since regression equations most often describe the relationship of the object under consideration with the environmental conditions, whereas the biological system and its components exist according to their own internal laws (Montgomery and Peck 1982). The values of $R^2 \geq 0.20–0.35$ are considered quite acceptable, and the level of significance of the relationships is determined by Pearson's test. Dissymmetrization and fragmentation of hydrothermal niches, with a predominantly Poisson-type distribution of biogeocoenotic units, point to a complex, spatially differentiated process of their links with the climate (Kolomyts 2022).

In the discrete empirical–statistical modeling, we have used two types of well-known models: (1) information theoretic models (Puzachenko and Skulkin 1981), which reveal the closeness of inter-component coupling, the system of ecological niches of forest communities in the space of abiotic factors, as well as the channels of chain reactions to external perturbations; (2) set theory models (of the class of fuzzy sets models) describing the structure of matter–energy fields and material objects by means of similarity–contrast and inclusion measures (Andreev 1980). The former model made it possible to calculate the normalized coefficients of contingency of features $C(A, B)$. The latter model was used to distinguish the *functional ecotypes* of forest biogeocoenoses (see Section 5.3): from dominants (eurytopic) through mesotopes (temporal) to reductants (pessimal), which are quite clearly differentiated by the degree of development of discrete parameters of the biological cycle (productivity–detritogenesis). The ecotype system can be used as one way of introducing measures into the well-known but very abstract concept of *biodiversity of the natural environment.*

# 9.3 ORDINATION AS A BASIS FOR EMPIRICAL–STATISTICAL MODELING

Methods of ordination analysis, once developed in geobotany, have been widely adopted in landscape ecology (Whittaker 1975). We used a direct gradient analysis, in which the identification of the leading environmental factors is based on the study of statistical variation of biogeocoenoses and their individual features in the space of values of these factors. Within the territory with the boundary conditions of the zonal and regional bioclimatic background, a series of spatial factor-time series of the functional characteristics of biogeosystems was constructed. Among the local factors that refract this zonal and regional physico-geographical background, of prime importance are, as already mentioned, lithomorphization and hydromorphization, which replace each other in the catena when the type of location changes (see Table 5.3). This constitutes the leading role of edaphic factors in the formation of the intra-landscape structure of the territory. The functional ordination allowed for *empirical simulation of projected* climate-genic changes in the parameters of the biological cycle (Kolomyts 2018; Kolomyts 2020). It has been established empirically that the main throughput channel of connections of regional and local ecosystems with the background climatic system is through the *summer soil moisture content* (Kolomyts 2005). The most representative are the average July stocks of productive moisture in the 0- to 50-cm soil layer $(W\text{–}50)_{\text{VII}}$. They are related to the known Vysotskii–Ivanov atmospheric moistening coefficient $F_{\text{hum}}(1)$ (Vysotskiy 1960), and also partially to the mean temperatures of January ($t_{\text{Jan}}$) and July ($t_{\text{July}}$) and with the annual precipitation ($r_{\text{year}}$) by the following regression equation:

$$(W-50)_{VII} = \exp\left[2.622 + 1.071\, F_{\text{hum}}(1)\right];\ R = 0.89;\ R^2 = 0.79; \tag{9.1}$$

$$(W-50)_{\text{VII}} = 0.589t_{\text{Jan}} - 11.16t_{\text{July}} + 0.028r_{\text{Year}};\ R = 0.84;\ R^2 = 0.70). \tag{9.2}$$

On the other hand, there is a rather close dependence of the distribution of forest topo-ecosystems on the summer soil moisture content. Thus, nemoral forest biogeocoenoses of the Middle Volga region have information coefficients of contingency $C(A, B)$ with the ($W$–50) parameter in the range of 0.128–0.162, which significantly exceeds the critical threshold of significance of relationships –0.070.

Table 7.2 presents the results of ordination analysis of relationships between the metabolic properties of forest ecosystems and soil hydrothermal parameters in different ecoregions of the Volga River basin. An abiotic ordination of topo-ecosystems is carried out according to information models (Figure 9.1). First, it is carried out in the "lithomorphism–hydromorphism" series (Kolesnikov 1956), i.e., according to the leading complex factor of differentiation of the local ecological space. The ecological niche system of soil–plant complexes in the parameter space of the abiotic environment is established on this basis. This makes it possible to find their initial geo-topically stable states (Kolomyts 2022). Each factorial–dynamic series represents a regional system of

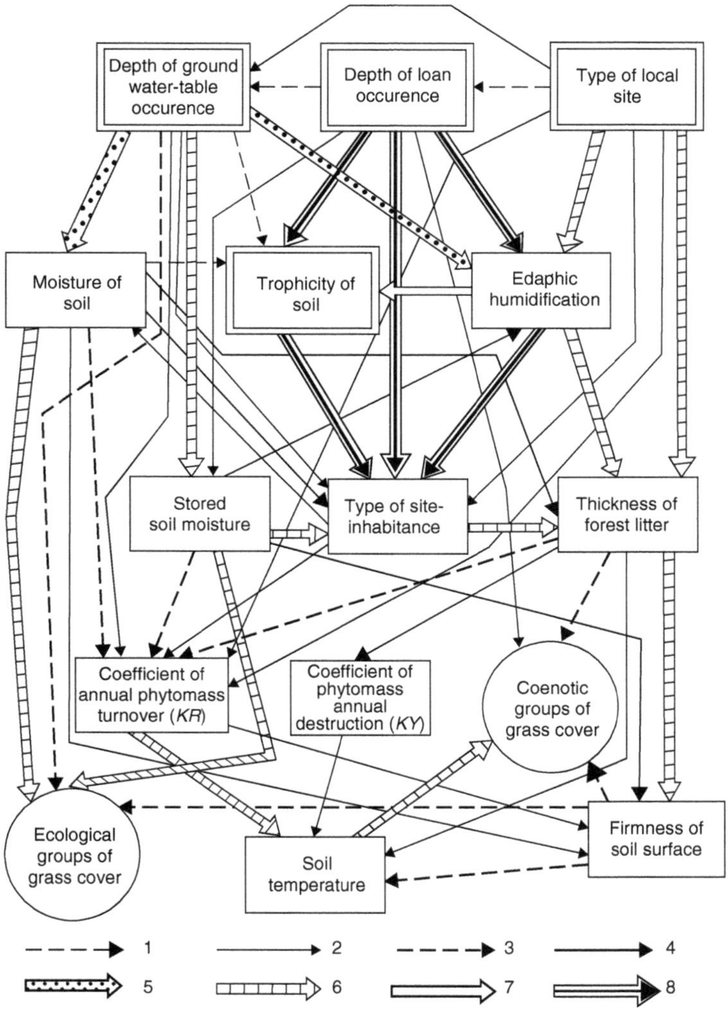


**FIGURE 9.1** General information statistical model of local organization of coniferous-forest ecosystems in the test site Vyksa (see Figure 4.1). Normalized coefficients of interrelation C(X;Y) between the geo-component attributes: 1 – 0.07–0.09; 2 – 0.10–0.12; 3 – 0.13–0.15; 4 – 0.16–0.19; 5 – 0.20–0.23; 6 – 0.24–0.27; 7 – 0.28–0.31; 8 – more than 0.31.

localized natural zoning. It serves as a topological model of the ecological–geographical space, which is able to describe the mechanisms of the reaction of local ecosystems to regional climate changes.

The hypothesis that we have adopted is that the hydrological regime of soils and grounds serves as the main input channel for external geophysical impacts on both the structural and functional parameters of the state of topo-ecosystems. This hypothesis is empirically confirmed. The most important information signals arrive through this channel, which are then distributed along the main functional parameters – in particular, through the characteristics of the processes of decomposition and accumulation of dead organic matter to the thermal regime of the soil. It is important to note that the soil thermals, initially determined by hydrological conditions, are also regulated by the detrital branch of the long-term cycle of phytomass. Here, the reverse influence of the biogeochemical block of the functioning of topo-ecosystems on the hydrothermal block is manifested.

Hydro-thermoedaphic ordination of the parameters of the biological cycle helped to make a probabilistic assessment of their changes under the climatic trend. We have obtained a step-by-step schematic model describing the *driving forces of climatogenic successions*, i.e., transformations of forest ecosystems into forest–steppe and steppe ecosystems under the influence of global warming. These are the following stages: (1) activation of the letritic branch of metabolism with a simultaneous decrease in the efficiency of the production process; (2) disruption of the biological cycle; (3) degradation and decay of the forest community, which is replaced by arid open woodland and then by a grass–steppe formation; (4) increased efficiency of autotrophic biogenesis; (5) reaching a new (forest–steppe and then steppe) level of closure of the biological cycle.

## 9.4 LOCAL EMPIRICAL SIMULATION OF THE REGIONAL CLIMATIC TREND

Massive empirical material from ground-based research makes it possible to establish certain initial "norms" of variables describing the states of biogeocoenoses and to estimate the allowable limits of change. The framework concept presented in this book was developed as a result of a generalized analysis of large-scale (1:25,000–50,000) landscape surveys conducted by the author between 1987 and 1998 in the forest belt of the main catchment area of the Volga River basin. Verification of the predictive ecological scenarios was based on empirically established, statistically reliable local and regional landscape relationships, which were viewed as mechanisms of the metabolic response of forest communities to current global warming.

Within the territory with the boundary conditions of the zonal and regional bioclimatic background, a series of spatial factor-time series of functional characteristics of biogeosystems is constructed. The members of each series are arranged in the order of strengthening or weakening of the partial impact of the considered hydrothermal factor in such a way as to obtain some equivalent of the time series of topo-ecosystems, which would correspond to the vector of background climatic changes (a trend) in a given region. Thus, we obtain an *ergodic operational system*, which is known to be of exceptional methodological importance in geographical research. The arrangement of the spatial states of an object in accordance with their temporal sequence forms the *phase space* of this object (Pope et al. 2000). Based on the self-similarity properties of the ergodic operational system, we replace spatial sequences with temporal ones and, analyzing the vector series of topo-ecosystems, conduct a kind of *simulation of the local mapping of the background climatic trend*, on the basis of which we identify

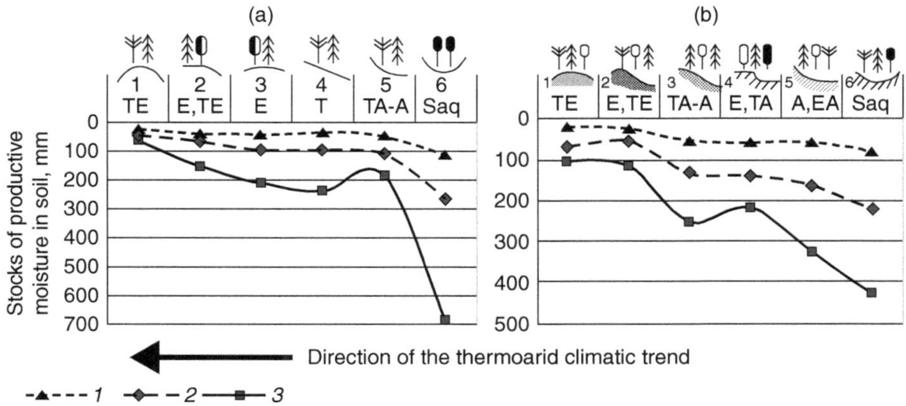


**FIGURE 9.2** Distribution of taxonomic rates of July mean stocks of productive moisture in soil layers (1) 0–20 cm, (2) 0–50 cm, and (3) 0–100 cm of different groups of forest biogeocoenoses in the system of their catenary organization consistent with the thermo-arid climatic trend. The Middle Volga Eco-regions (see Table 5.3): (a) the northern edge of the broadleaf sub-zone of the Volga Upland (Urochishche "Green Town"); (b) the southern band of the sub-taiga zone in the Lowland trans-Volga basin (Kerzhenets reserve). Types of locations: E and TE, eluvial and trans-eluvial; T and TA, transit and trans-accumulative; A and AE, accumulative and eluvial–accumulative.

the real spectrum of possible directions of functional–structural transitions between ecosystems.

In line with the pattern of projected climatic changes, *vector series* were constructed by the *degree of aridization of topo-ecosystems* and their individual components for each regional ecosystem (Kolomyts 2022). This concerned not only the quantitative parameters of their state but also the qualitative attributes themselves. In the binary ordination plots, all states of the objects under consideration were arranged from left to right along the *X* axis (for phenomenon *A*) and from top to bottom along the *Y* axis (for factor *B*) in such a way that the change of the states themselves had a vector character and would be adequate for the general trend of aridization of local landscape–environmental conditions (Figure 9.2). The degree of hydro-edaphic aridization of biogeocoenoses was estimated by Equation (9.1).

## 9.5 THE PHENOMENON OF BIOCLIMATIC POLYZONALITY AT THE LOCAL LEVEL

It is a question of revealing the regularities whereby local geomorphological and edaphic factors refract the background bioclimatic conditions and the formation of certain categories of topo-ecosystems as some kind of representatives of different types of the zonal geographical environment, often quite well spaced. Local landscape–environmental structures emerge, forming *regional systems of localized natural zonality*. For their part, these regional–topological systems, being objects of the effect of background climatic signals, may simulate the main directions and scales of local ecosystem reconstructions, thus making an empirical basis for prognostic constructions. Moreover, it seems practicable to assess the extent to which the structure–function transformations of local ecosystems and the corresponding shifts of natural boundaries are able to reorganize the whole zonal–regional background of a given territory under the effect of this or that climatic trend. This aspect of landscape–ecological prognosis is particularly significant because geographical zonality, according to V. V. Dokuchaev (1899, 1949), is the highest form of interaction of natural components and its features are directly reflected in the structure and functioning of geo(eco)systems.

The landscape mosaic of any territory is made up of areas of both local upland geosystems, reflecting in the most "pure form" the zonal–regional background of a given territory and being its representatives, and extra-zonal geo-topes, which arose as a result of various types of refraction of this background under the influence of local factors (Vysotskiy 1960). In upland areas, *indigenous landscape facies (biogeocenoses)* are formed, which represent a certain equifinal dynamic state of elementary geosystems, corresponding to the "harmonious (optimal) combination of their components" (Sochava 1978, p. 294). In the 1930s, I.P. Gerasimov expressed this pattern in a capacious aphorism: "zonality is facies" [cited from: Sochava (1978, p. 183)].

Within the given ecoregion, a hydrothermal series of zonal and extra-zonal biogeocenoses is constructed in the direction of their humidity or, conversely, aridity, representing this series (based on the self-similarity property of the operating system) as a certain analogue of the corresponding background climatic changes (trend). Thus, according to the vector spectrum of *topological polyzonality*, it is possible to outline real scenarios of the reaction of local geosystems to certain shifts in the regional climatic system and to outline the corresponding chains of local landscape–ecological transitions.

Thus, they provide an empirical basis for *predictive constructs based on the principle of actualism*. Moreover, it seems realistic to assess the extent to which the functional–structural transformations of local ecosystems and the corresponding shifts in natural boundaries can transform the entire zonal and regional background of a given territory under the influence of a particular climatic trend.

Let us consider this situation using the Oka-Terrace Biosphere Nature Reserve as an example. We know that the zonal geographical shape of forest phytocoenosis is reflected in the vegetation cover as an indicator of *biologically equivalent habitats*, according to Sukachev (1972), which are formed by a certain combination of abiotic factors. The present-day forest communities of the reserve are secondary and are in various stages of recovery, but all its biogeocoenotic groups have already acquired quite distinct zonal features (Figure 9.3a). The latter, in turn, correlate quite closely with three abiotic edificator factors: with the type of location (Figure 9.3b), with the depth of the groundwater table, and with the associated surface moistening of the edaphotope (Figures 9.3c and d).

The regional system of local zonality in the reserve is characterized by *a high degree of contrast even on the flat-interfluve types of locations* (see Table 5.3). The boreal taiga

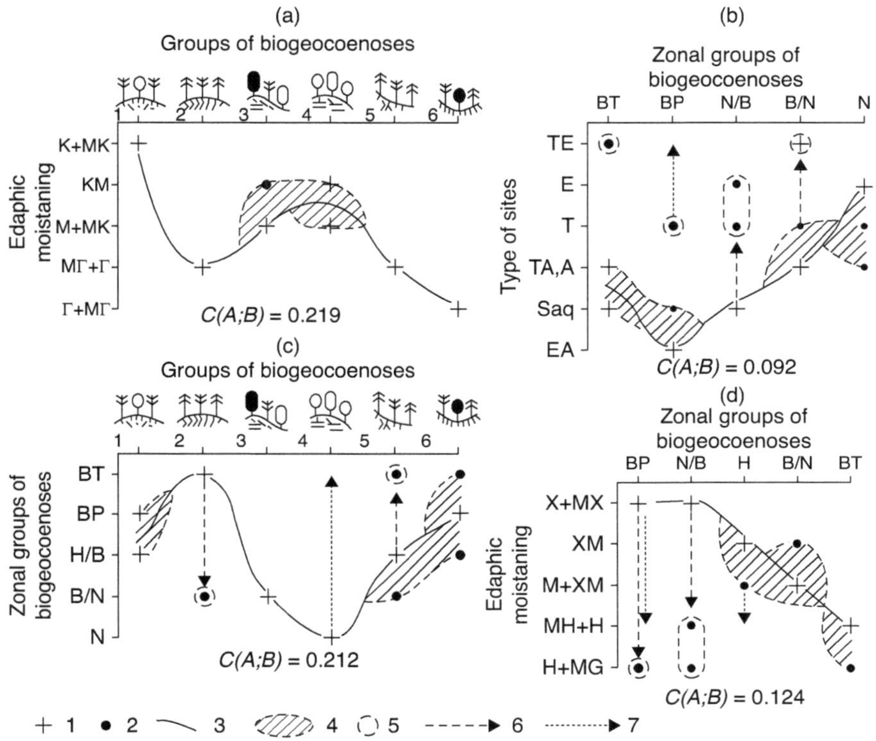


**FIGURE 9.3** Binary ordination of groups of biogeocoenoses of the Oka-Terrace Biosphere Reserve and their zonal groups by types of locations and hydro-edaphic conditions. 1. Environmental dominant; 2. "fuzzy" part of the environmental niche; 3. trajectory connecting environmental dominants; 4. space of the environmental niche; 5. enclave; 6. direction of possible transgression of this phenomenon away from its environmental dominant; 7. region of sporadic occurrence of the phenomenon within the given gradations of the factor. Zonal groups of biocoenoses: BT, boreal taiga; BF, boreal forest; N/B, nemoral–boreal; B/N, boreal–nemoral; N, nemoral. Environmental groups of biocoenoses: X, xerophytic; MK, meso-xerophytic; M, mesophytic; MH, mesohydrophytic; H, hydrophytic. For symbols of groups of biogeocoenoses, see Table 5.3.

group of biogeocoenoses (E and TE spruce and spruce–pine forests, group 2) neighbors the zonal representatives of the nemoral forest sub-zone (TE and TE oak–lime–birch forests, group 4), as well as with sub-taiga-type ecosystems (pine–lime–oak forests, group 3). Meanwhile, all three groups of biogeocoenoses are found in the same ranges of groundwater table depths and gradations of edaphic moistening. There is then a significant overlap of their ecological niches, which should provoke fierce competition and, as a consequence, a high potential capacity for mutual functional and structural transitions under one climatic trend or another. Such transitions are quite clearly revealed under the projected thermo-arid environmental scenario even in the moderate climate model E GISS. The affinity of hydro-edaphic conditions of eluvial–accumulative coniferous–leaved waterlogged forests (biogeocoenosis group 6) to meso–hydro-morphic flat-interfluve spruce forests (group 2) and to some mesomorphic lime–birch forests (group 4) led to their assignment to the same boreal–taiga zonal group. Accordingly, here we also predict the most significant functional transitions of group 6 into groups 2 and 4.

Thus, the vector spectrum of *topological polyzonality* makes it possible to reproduce some real scenarios of the response of local natural complexes to certain shifts in the climatic system and to outline the corresponding chains of local landscape–environmental transitions. Our proposed method of *empirical simulation of the bioclimatic trend* as an initial procedure for predictive landscape–environmental constructions is a realization of one of the declared methods of geographical forecasting, the method of space–time analogies (Puzachenko and Skulkin 1981).

## 9.6 MODELS OF CRITICAL STATES OF FOREST COMMUNITIES AT THE SOUTHERN BOUNDARY OF THE FOREST ZONE

To forecast the ecological effects of climate-genic changes in the hydrological regime of soils, it was necessary to determine the critical values in summer stored soil moisture ($W_{crit}$) leading to transitions of zonal types or sub-zonal subtypes of vegetation mantle from one to another. For this purpose, we used the previously obtained correlation relationships of the zonal and sub-zonal borders in different sectors of the Volga River basin with the atmospheric humidity factor (Kolomyts 2005). It was calculated based on the atmospheric moisture coefficients $F_{hum}(1)$. According to the data from agrometeorological and water balance stations of the Volga basin (for forest areas), fairly close relationships were established between the $F_{hum}(1)$ parameter and the average July reserves of productive moisture in the soil layers of 0–20 cm ($W$–20), 0–50 cm ($W$–50), and 0–100 cm ($W$–), with $R^2$=0.79–0.81; see Equations (7.7)–(7.9). The transition from the boundary (critical) values of $F_{hum}(1)$ to the respective values of $F_{hum}(2)$ is performed by the empirical dependence ($R$=0.90; $R^2$=0.81):

$$F_{hum}(2) = 0.409 - 0.5697\, F_{hum}(1) + 1.399\, F_{hum}(1)^2. \tag{9.3}$$

The values of $F_{hum}(2)$ generally proved to be 1.5–2 times higher than $F_{hum}(1)$. That explains why forest communities can show sustainable existence under the evident deficiency of atmospheric humidity.

Using the correlation relationships between the July stored soil productive moisture in upland forest ecosystems and $F_{hum}(2)$, we obtained two series of critical stored moisture (taxonomic norms and minimums) for the three main zonal boundaries in the territory of the Middle Volga Region: (1) the transition from the boreal to the sub-boreal belt; (2) the shift from broadleaf forests to typical and southern forest–steppe; and (3) the transition from southern to northern forest–steppe. According to Armand (1989), the critical state of the geo(eco)system results in its qualitative reorganization, including structural changes. The July critical stored soil moisture given in Table 9.1 was taken from the plots of statistical connections used to obtain formulas Equations (7.7)–(7.9) and (9.1). The $W_{crit}$ values characterize particular states of hydro-edaphotopes favoring the above-mentioned most abrupt (stepwise) functional and structural changes in the geo(eco)systems of the upland group, representing, as has been mentioned earlier, the zonal–regional background of the given territory.

On the basis of these correlation links, two series of critical moisture reserves (taxonomic norms and minimum values) were obtained for the three main zonal boundaries in the territory of the Middle Volga region: (1) the transition from the boreal to the sub-boreal belt; (2) the replacement of broadleaf forests by typical and southern forest–steppe; and (3) the transition from the southern forest–steppe to the northern steppe (Table 9.1). The $W_{crit}$ values characterize specific states of hydro-edaphotopes at which the above-mentioned most abrupt functional–structural changes in the geosystems of the upland series, reflecting the zonal–regional background of the given territory, should occur.

It should be noted that the range of $W_{crit}$ values is comparatively narrow between the first and second levels, i.e., between the northern and southern boundaries of the nemoral–forest sub-zone. This means that with the intensive development of the thermo-arid trend, the process of steppe formation and subsequent disintegration of the forest cover can equally affect both broadleaf and mixed formations, creating not only small-leaf oak but also coniferous typical forest–steppe. Judging by the critical values of summer moisture reserves, the hydro-edaphotopes of upland maple–elm–linden–oak

**TABLE 9.1** The average and minimum (in brackets) critical values of the July reserves of productive moisture in the soil, at which zonal changes of placor plant communities occur in the Middle Volga Region

| *LANDSCAPE–ZONAL CONDITIONS (ECOREGION)* | *SOIL LAYER, CM* | *LEVELS OF CRITICAL RESERVES OF PRODUCTIVE MOISTURE IN THE SOIL, MM* | | |
|---|---|---|---|---|
| | | I – for mixed forests (zonal transition to broadleaf forests) | II – for a continuous nemoral forest cover (zonal transition to a typical and southern forest–steppe) | III – for island park forests and open woodlands (zonal transition to the northern steppe) |
| Sub-zone of broadleaf forests (high plain by-Volga region near city Nizhny Novgorod, Kud'ma) | 0 – 20<br>0 – 50<br>0 – 100 | 46 (37)<br>108 (87)<br>215 (180) | 40 (32)<br>92 (74)<br>185 (147) | 30 (24)<br>62 (48)<br>122 (102) |
| Sub-zone of typical forest–steppe (By-Sura) | 0 – 20<br>0 – 50<br>0 – 100 | 42 (34)<br>96 (78)<br>190 (150) | 38 (29)<br>85 (70)<br>170 (132) | 29 (23)<br>57 (43)<br>110 (87) |
| Sub-zone of southern forest–steppe, (Zhiguli) | 0 – 20<br>0 – 50<br>0 – 100 | 38 (32)<br>84 (70)<br>170 (130) | 36 (30)<br>79 (65)<br>155 (120) | 28 (22)<br>52 (38)<br>100 (75) |

forests will, by the middle of the century (according to the scenarios of both HadCM3 and GISS models), reach the second critical line and will find themselves in conditions that almost exclude the existence of continuous broadleaf forests. In their place, island forests and sparse forests with steppe meadows should form.

It is noteworthy that there is a relatively short range of $W_{crit}$ values between the first and second levels, i.e., between the northern and southern boundaries of the nemoral forest sub-zone. This indicates that steppification and further breakup of forest cover, under the conditions of intensive development of the thermal-arid trend, can equally involve both broadleaf and mixed formations, creating not only small-leaf oak groves but also typical coniferous forest–steppe.

It is also notable that continuous nemoral forest cover and insular forests with open tree formations at their southern outposts (in this case, on the Zhiguli plateau) reach a critical state at a higher summer stored soil moisture compared to the "core of typicality" of the broad-leaved forest sub-zone (on the farewell rocks of ancient peneplanation planes in the By-Sura). The forest and forest–steppe ecosystems of the uplands in the sub-zone "core" have higher reserves of homeostaticity compared to its southern periphery. Therefore, with the same hydrothermal signal, forest communities should reach the critical state first at the zonal ecotones of forest and steppe and only then in the more hinterland regions of forest zones. This conclusion is confirmed by the results of the analysis of prediction models for landscape–ecological transitions.

## 9.7 THE PRINCIPLE OF OPTIMALITY IN THE FUNCTIONING OF ECOSYSTEMS

High emphasis was placed on the integral estimation of autotrophic biogenesis using dimensionless parameters CE, CA, and CR (see Section 5.4), which serve as characteristics of internal phytomass turnover in a forest community (Odum 1983). The parameter CE = PC/BC is the most generalizing as a measure of ecological efficiency of the phytoproduction process. The value of this parameter was assessed in terms of the known principle of optimality, which is expressed by *the allometric low* (Rosen 1969). According to the latter, there is a certain constant value of parameter CA = log(PC)/log(BC) for each of the zonal types of landscapes, in spite of the significant fluctuations of productivity and biomass within this type (Perel'man 1975). On the zonal–regional scale (for flat interfluve types of location), this coefficient monotonously increases from 0.53–0.54 in forest–tundra and northern taiga to 0.60–0.62 in the broadleaf forests of plains and mountains and then more abruptly increases to 0.84–0.88 at the transition to meadow and then dry steppe of the sub-boreal belt. Along with the above, each zonal sub-division of the geographical environment has CA values that are higher than in flat interfluves and correspond to plant communities of more humid habitats.

The latter regularity, relatively weak at the level of zonal types (or subtypes) of landscapes, develops and becomes a phenomenal event at the local level. The deviations from flat interfluves towards lower CA values in the upper, xeromorphic link of landscape couplings are becoming increasingly distinct. The CA values for flat interfluve forest phytocoenoses of the Middle Volga Region, on the whole, meet the above zonal norms. However, xeromorphic links of the local hydro-edaphic series of biogeocoenoses, standing as representatives of more southern bioclimatic conditions, in every ecoregion are also characterized by lower CA values (over the Middle Volga Region as a whole $\leq$0.40–0.35). At the same time, the groups of facies simulating more northern zonal conditions have coefficient values of up to 0.75–0.80. The above indicates that the ecological efficiency of production as a whole increases from dry to humid and then wet habitats, reaching its maximum in paludal forests. Consequently, the thermo-arid trend must induce a certain decline in the degree of optimality of the biological cycle, which will imply a general decrease in the level of functioning of forest topo-ecosystems. Thus, the edaphic humidification factor creates a fundamentally different pattern of spatial distribution of parameters CE and CA at the local level compared to the atmospheric humidification factor at the zonal–regional level.

Aridization of soils in the middle and southern forest–steppe of the High-plain by-Volga Region under global warming must cause a general decrease in the CA parameter from the base taxonomic level of 0.54–0.56 (maximal for the given zonal–regional conditions) to 0.43–0.45. At the same time, the maximum decline in production efficiency will take place in the forest communities of eluvial (flat interfluve), TA, and A locations, with the initial values of *(W-50)*$\geq$100 mm. Here, with a decrease in *(W-50)* to 60–80 mm, the *CA* parameter will drop from 0.56–0.60 to 0.48–0.50. This may happen in the middle of the 21st century according to GISS or may have already occurred by 2030 according to HadCM3. By the above dates, the total productive potential of broadleaf forests of the Middle Volga Region, which are the most valuable in ecological and economic respects, may have decreased by 2–2.5 times.

## 9.8 TIME LIMITATIONS OF THE EMPIRICAL SIMULATION METHOD OF LANDSCAPE–ECOLOGICAL FORECASTING

Let us pay attention to a certain limitation of the empirical simulation method of landscape–ecological forecasting, based, as already mentioned, on the fundamental property of ergodicity of the natural environment. The essence of these limitations is that the forecast can be realistic only for the period of maintaining equilibrium or, at least, stationary processes (Puzachenko 1998). The adequacy of the models will be significantly reduced if the role of environmental factors begins to change dramatically during the forecast period. In this case, the forecasted geo(eco)system, being initially linear, will move into the category of non-linear systems, with a disproportionality of their response to a particular value of the disturbing signal (Zeidis et al. 2001). For the forest zone of the Volga basin, a similar situation will develop by 2050, when the state of the atmosphere goes beyond the current interannual variability of its state parameters (Paris Agreement 2015).

The above limitations are largely "removed" if, at each discrete step of forecasting, we accept, according to Puzachenko (1998), the trajectory of the expected functional changes of the geo(eco)system as a stationary nonequilibrium process that has the properties of reversibility. Under external disturbance, the system moves from the state of equilibrium to a nonequilibrium stationary regime. In this case, the merger of the previous stable equilibrium with the unstable stationary regime causes a certain jump-like transition to a new regime (Svirezhev and Logofet 1978). Thus, each new step of the forecast trajectory represents a jump in the functional changes of the geo(eco)system, expressed to a greater or lesser degree.

## 9.9 MAIN INFERENCES

The development of scientific and methodological bases of spatial functional monitoring of forests is presented, based on empirically established local and regional landscape–ecological connections, which are considered as the mechanisms of metabolic responses of forest ecosystems to particular climatic trends. Empirical simulation predictive models describe the future scenarios of biogeocoenological systems according to the laws of their basic spatial organization, in accordance with the fundamental properties of ergodicity of the natural environment.

The boreal bioclimatic ecotone of the Volga River basin is used as an example to carry out the experience of analytical and cartographic prognostic scenario modeling of the near and distant future for the given ecoregion. The mechanisms of local and regional response of forest ecosystems to the anticipated global warming in the 21st and 22nd centuries have been assessed.

Certain time limitations of the empirical simulation method of landscape–ecological forecasting are noted, with a recommendation to use a more meaningful interpretation of the results of forecast calculations.

In light of the forthcoming and already beginning global warming, along with the progressing development of soil–plant cover aridization in the temperate latitudes of continents, the described empirical –simulation problems under consideration are particularly urgent. It is necessary to assess the state of forest and forest–steppe ecosystems and predict their impending changes in order to develop an ecological basis for the preservation and reproduction of forest resources and to arrange a network of conservation and recreation territories (nature reserves, national parks), which is especially important for regions with critical conditions for the existence of forest plants, including the zone of contact between forest and steppe.

# Regional Landscape–Ecological Prognoses

# 10

## 10.1 ANALYTIC METHODS OF LANDSCAPE–ECOLOGICAL PROGNOSES

### 10.1.1 Evaluation of Probabilities of Changes in the Functional States of Ecosystems

The mechanism of estimating the probability of ecosystem transitions (for the predictions) is illustrated by Venn diagram (Figure 10.1). Here, $A_0$ and $A_1$ are respectively the initial (contemporary) and final (future) values of the row vector of the ecological niche of object *A* (either landscape kind or facies group), which must be absorbed by another object *B* if their niches intersect in the final state more than in the initial state. The landscape–ecological transition proceeds in the direction $A \rightarrow B$, and the absorption proceeds in the opposite direction $B \rightarrow A$. The shaded area $a+b+c$ on the diagrams reflects the total value of the transformation of object *A*, which can be expressed in terms of the following measure of inclusion by Andreev (1980):

$$a + ß = A_0 \cap A_1; \;\; ß = A_0 \cap B_0; \;\; ß + c = A_1 \cap B_0, \qquad (10.1)$$

where "$\cap$" is the sign of intersection (logical product) of two sets. The elements of the intersection "zone" belong to two sets simultaneously.

It is supposed that any object *A* (reducent) must be absorbed by another object *B* (adsorbent) if their niches intersect in the final state more than in the initial state. Initial (contemporary) $A_0$ and final (future) values $A_1$ of the row vector of the climatic niche of object *A* are considered. In each researched pair of objects *A* and *B*, two parameters of

DOI: 10.1201/9781003733119-10

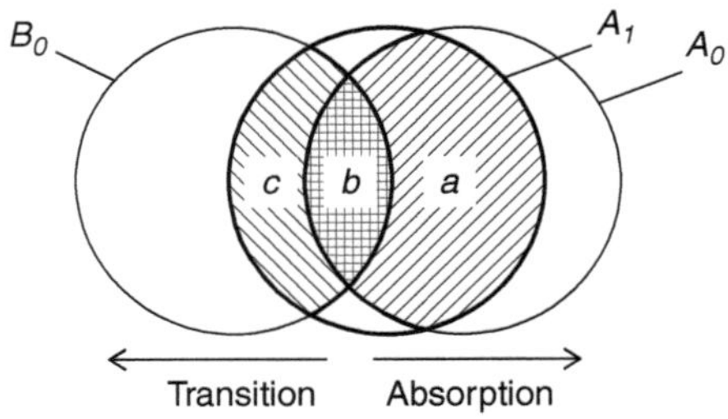


**FIGURE 10.1** Venn diagram illustrating the mechanism of probability estimation of landscape–ecological transitions (explanation in the text).

transformation are calculated: the probability measure $P_{ii}$ of object $A$ stabilization: $K(A_0 \rightarrow A_1)$; and the measure $P_{ij}$ of its absorption by object $B$: $K(A_0 \rightarrow B_0)$. These parameters are expressed in the following measures of inclusion:

$$K(A_0 \rightarrow A_1) = \frac{A_0 \cap A_1}{A_0}; \tag{10.2}$$

$$K(A_0 \rightarrow B_0) = \frac{A_1 \cap B_0 - A_0 \cap B_0}{A_0}. \tag{10.3}$$

Since the operations are performed with descriptive fuzzy sets, the operating formulas have the following shape:

$$K(A_0 \rightarrow A_1) = \frac{\sum_1^{NP} \min\left[L_i(A_0), L_j(A_1)\right]}{\sum_1^{N} L_i(A_0)},$$

$$K(A_0 \rightarrow B_0) = \frac{\sum_1^{PQ} \min\left[L_j(A_1), L_k(B_0)\right] - \sum_1^{NQ} \mathrm{Min}\left[L_i(A_0), L_k(B_0)\right]}{\sum_1^{N} L_i(A_0)}. \tag{10.4}$$

Here, $i$, $j$, and $k$ are the order numbers of partial association coefficients in row vectors describing the ecological niches respectively $A_0$, $A_1$, and $B_0$.

By analogy, we calculate the values of transitions $K(A_0 \rightarrow C_0)$, $K(A_0 \rightarrow D_0)$, etc. For the following state, we have $K(B_0 \rightarrow B_1)$, $K(B_0 \rightarrow C_0)$, $K(B_0 \rightarrow D_0)$, etc. For the next stage of prediction, its own measures of transition are calculated: $K(A_0 \rightarrow A_2)$, $K(A_0 \rightarrow B_0)$, $K(A_0 \rightarrow D_0)$, …; $K(B_0 \rightarrow B_2)$, $K(B_0 \rightarrow C_0)$, $K(B_0 \rightarrow D_0)$,

Thus, a specific quadratic matrix of probabilities of stabilization of each object (diagonal elements of the matrix, $P_{ii}$) and its transitions into other objects ($P_{ij}$) is obtained for each prediction period. Here, zero and negative values of probabilities are possible. The former indicate the absence of transitions, and the latter show the intensification of the contrast between the functional states of objects (at $A_1 \cap B_0 < A_0 \cap B_0$). Further operations are executed with the transition probability matrices. All negative transitions are provisionally replaced by zeros.

If objects *A* and *B* are first-order neighbors, then the value of displacement of the boundary between them in the $A \rightarrow B$ direction is taken as proportional to the calculated measure of transformation of object *A*. In turn, object *B* can be transformed into the following objects (*D*,*C*, etc.). The evolution of ecosystems is usually multivariate; hence, all possible transitions have to be considered for each of them, revealing the maximum measures of transformation, which will indicate the most probable directions of transformation of phytocoenological structures as a whole. For first-order landscape neighborhoods, this process will be rather distinct if $P_{cn}(B) > P_{cn}(A)$, i.e., given that the absorbing object *B* has a higher power of climatic niche $P_{cn}$ compared to the absorbed object *A*.

## 10.1.2 Algorithm of Predictive Calculations

The procedure of the stage of own prognosis analysis consisted of the following operations (Kolomyts 2022).

1. Using the information analysis of connections on each object (plant formation) with various climatic characteristics, two matrices of partial association coefficients ($C(a_i/b_j)$) were obtained: matrix $T_0$ for the base period and matrix $T_1$ for the given prognostic dates. On the columns of each matrix, the graduations of the given factor are placed, and on the rows, the gradations of the given object are. Because the regional climatic prognosis is differentiated with respect to temperature and precipitation, we have formed four pairs of matrices: on average January and July temperature ($T_0(1)$ and $T_1(1)$; $T_0(2)$ and $T_1(2)$) and on the quantity of atmospheric precipitation of cold and warm periods ($T_0(3)$ and _$T_1(3)$; $T_0(4)$ and $T_1(4)$).
2. The formulas above serve to calculate four square matrices *T*(1), *T*(2), *T*(3), and *T*(4) of measures of stabilization (resistant stability) $P_{ii}$ of each *i*-th object (diagonal matrix elements) and measures of its transformation $P_{ij}$ into other *j*-th objects. Zero values of matrix elements show the absence of a transition in the given pair of objects, and negative values point to the intensification of contrast between them ($A_1 \cap B_0 < A_0 \cap B_0$).

   For simplification of predictive calculations, the above condition $P_{cn}(B) > P_{cn}(A)$ is not taken into consideration at the given stage of described algorithm development, which certainly reduces the reliability of results.
3. Account was taken of simultaneous changes in all four (or three) cited climatic factors. Then, on the basis of matrices *T*(1)–*T*(4), the first factor-weighted average matrix $T_{wa}$ of object transformation was calculated. The "weighting" coefficients for its calculation were values of parameter *C*(*X*;*Y*), which were normalized once again to obtain, as per Puzachenko and Skulkin (1981), a linear polynomial of dependencies between the distributions of objects and

the geophysical factors under consideration. For the groups of plant formations, prognostic calculations for matrix $T_{wa}(1)$ were based on the linear polynomial with the following coefficients:

$$T_{wa}(1) = 0.131\,T(1) + 0.388T(2) + 0.147T(3) + 0.334T(4). \tag{10.5}$$

For paleo-reconstructions, matrix $T_{wa}(1)$ was created using the following linear polynomial:

$$T_{wa}(1) = 0.161T(1) + 0.477T(2) + 0.362T(3). \tag{10.6}$$

4. At this stage of operations with the matrices, it is necessary to take into account the probable presence of "residual" transitions of some objects into other prototype objects absent in our operational system (sampling), i.e., to extra-sampling objects. For some facial groups, these transitions seem to be dominant and even unique. Such transitional "residue" $P_{ij}(x)$ for each object (i.e., for each line of the weight-average transition matrix) can be easily found from the ratio:

$$P_{ij}(x) = 1 - P_{ii} - \sum P_{ij}. \tag{10.7}$$

The $P_{ij}(x)$ values are entered as an additional column into matrix $T_{wa}$. If the "residue" is negative, it means that all predicted transitions of the given object are within the limits of the considered set of prototype objects and there are no other transitions. Negative "residues" of transitions are changed to zeros.
5. Comparable values of the probabilities of transition of some objects into different prototype objects can be obtained if the sum of these probabilities in each line of the matrix is equal to 1. After normalizing each of the five $T_{wa}$ matrices (with "residual" columns) by lines, we will obtain the second weight-average matrices $T_{wa}(2)$. The lines of the matrix with positive transition "residues" are normalized already at the preceding (4th) stage of calculations.
6. In conclusion, the correcting procedure is performed. Matrix $T_{wa}(2)$ contains quite a number of elements with very low (below 1%–2%) probabilities going beyond the limits of measurement and calculation accuracy. It is necessary to calculate the guaranteed minimum of probabilities of landscape–ecological transitions: $M_{min}$. The simplest method of finding the guaranteed minimum of elements is based on the application of a one-sided criterion of their significance, $t=\sigma$, which provides the 5% level of significance (Mitropol'sky 1971). All values of $M$ transitions or deviations ($P_{ii}$ or $P_{ij}$) below the difference of $M - \sigma$ must be eliminated. For this purpose, we calculate $M_{min}=M - \sigma$ and drop all vector elements below this difference.

Thus, each object acquires a much shorter transition vector. Residual vector elements are normalized again and, thus, we obtain the third matrix of transition probabilities weighted by the given geophysical factor: the $\overline{T}_{wa}(3)$

matrix. It provides a clear notion of the potentials for the transformation of various objects or, conversely, their ability to resist external influence. The oriented graphs of probabilities of functional transitions between the objects are plotted on the basis of $T_{wa}(3)$ matrices (see Figures 10.4). These graph analytical models are the main tool for landscape–ecological prediction. They provide the most general notion of the exogenous succession dynamics of ecosystems as whole natural–territorial complexes.

### 10.1.3 Calculations of the Rates of Ecosystem Transformations

The second stage of the analysis included the following procedure: calculation of the potential rates of transitions and the average times of full or partial absorption of the hydrothermal niche of one examined object (either landscape or facie groups) by the niches of other objects. For this purpose, we used the mathematical apparatus of Markov chain theory. Markov models are used in geology (Harbaugh and Bonham-Carter 1974), ecology (Jeffers 1981), physical geography (Viktorov 1986), etc. They describe natural processes as random events, determining the influence of each preceding state on each following state, which aligns with the peculiarities of the dynamics of geo(eco)systems. Using known methods, it has been previously established that the chains of geo(eco) system transitions we have studied may possess Markov properties. By applying the methods of the theory of finite Markov chains, the potential rates of transformations of climatic niches of geosystems were determined, enabling predictive estimates with a given lead time. For the simulation, discrete first-order Markov chains with a short "memory" of one prediction step, during which the process is considered stationary, are used.

The rates of landscape transitions were calculated using the model of W. Krumbein (Harbaugh and Bonham-Carter 1974). The rates $h_{ij}$ (and consequently, times) of transitions are defined through a non-complicated algebraic transformation of the initial normalized matrix of transition probabilities, where both the relative transition probabilities ($P_{ij}$) and the a priori probabilities ($P_i$) are considered. The total rates of transition $M_i$ from the given state in one step, that is, in the time interval $d=1$, are calculated first. In general,

$$M_i = -\text{In}P_{\text{ii}}/d \text{ and by } d = 1.\ \ M_i = -\text{In}P_{\text{ii}}. \tag{10.8}$$

The values of $M_i$ are placed in the next additional column of the matrix. Here, $d$ is the number of steps. In this case, we have three steps: 1985–2050, 2050–2100, and 2100–2200. The initial year of the forecast, taking into account modern trends in global warming (Houghton et al. 1996), was 1985.

The expected total time of residence of the system in state $P_{ii}$, that is, up to the moment when it will give way to all other states, is given by

$$\tau(P_{\text{ii}}) = 1/M_i. \tag{10.9}$$

Here, $\tau(P_{ii})$ is expressed in conventional units: the number of steps. For example, it may be expressed as the degree of plant transformation or in the parts of the area of a given object falling at one step of reduction of this area. By $\tau(P_{ii}) \leq 1$, it is evident that the given area would be absorbed entirely just in the course of the first step. By multiplying the $\tau(P_{ii})$ value by the time equivalent of the first step (20–100 years in our case), we obtain the expected absorption time $T(P_{ii})$ (in years), which conceptually represents the final time of full transformation of the given object.

Further, two square matrices are created for the relative and absolute rates of transitions of objects into each other. In case $M_i=1$, the relative partial rates of transitions $m_{ij}$ from $i$-th state to $j$-th state are found by the correlation:

$$m_{\mathrm{ij}} = P_{\mathrm{ij}}/P_{i}. \tag{10.10}$$

Next, it is essential to calculate $h_{ij}$ – the absolute partial rate of transition of the $i$-th object to the $j$-th object (as a conditional probability). This rate is equal to

$$h_{\mathrm{ij}} = m_{\mathrm{ij}} \cdot M_{i}. \tag{10.11}$$

Values $h_{ij}$ are calculated on all lines of matrix $P_{ij}$ (except for its diagonal elements). The $h_{ij}/M_i$ ratio will show us that part of the total one-step reduction of the area of the $i$-th object (or the part of any other attribute of its transformation) that accounts for the transgression of the $j$-th object. Since Markov's process is stationary, this part remains constant across all transition steps. It is thereafter an easy matter to calculate what part of the area of the given object will be absorbed by areas of other objects at time intervals (steps) of 20, 40, 60 years, etc.

The matrices of probabilities of mutual transitions $P_{ij}$ of the objects have been constructed for all prognosis intervals. However, in further estimations of transition rates on Markov's chains, the duration of the intervals themselves, that is, the number of steps ($d$), was taken into consideration. In particular, when calculating $h_{\mathrm{ij}}$, it is conventional that $d=1$ for the first prognosis interval, $d=2$ for the second prognosis interval, $d=3$ for the third prognosis interval, etc.

# 10.2 PROGNOSES OF FUNCTIONAL CHANGES OF PLANT FORMATIONS

## 10.2.1 Main Features of Climatic Prognosis in the Region

According to the GISS model, the temperature rise expected in the next few decades in the temperate belt of the Russian Plain will be more prominent in the warm rather than cold season (Table 10.1). By the year 2050, the average July temperature will

**TABLE 10.1** Taxonomic norms of average temperatures in January and July for zonal types and groups of plant formations of the Volga River basin and its surroundings for the base period (1881–1980), as well as deviations from these norms for the projected periods of 2050, 2075 and 2100, according to the GISS-1993 model

| *ZONAL AND SUB-ZONAL TYPES OF FORMATIONS* | *GROUPS OF PLANT ASSOCIATIONS*[a] | *AVERAGE* $T_{JAN}$, °C | | | | *AVERAGE* $T_{JULY}$, °C | | | |
|---|---|---|---|---|---|---|---|---|---|
| | | *BASIC PERIOD* | *2050* | *2075* | *2100* | *BASIC PERIOD* | *2050* | *2075* | *2100* |
| Middle and south taiga | 1 | −12.2 | 0.6 | 1.3 | 3.7 | 16 | 1.3 | 2.3 | 3.2 |
| | 2 | −15.6 | 1.4 | 3.6 | 5 | 16.7 | 1.1 | 1.9 | 3 |
| | 3 | −11.6 | 0.5 | 2.1 | 4 | 17.2 | 0.8 | 2 | 2 7 |
| | 4 | −11.9 | 0.6 | 2.1 | 3.9 | 17.4 | 0.7 | 2 | 2.7 |
| Sub-taiga | 5 | −10.2 | 0.2 | 1.4 | 3.3 | 17.3 | 0.7 | 2.3 | 3 |
| | 6 | −14.7 | 0.9 | 2.9 | 4.8 | 18 | 0.8 | 2 | 2.6 |
| | 7 | −13.9 | 0.5 | 2.7 | 4.5 | 18.6 | 0.8 | 2.1 | 2.6 |
| | 8 | −12.2 | 0.5 | 2.3 | 4.1 | 18.8 | 1.1 | 2 | 2.8 |
| | 9 | −12 | 0.3 | 1.7 | 3.8 | 19.2 | 1.1 | 2.2 | 2.8 |
| Broadleaf forest | 10 | −11 | 0.1 | 1.5 | 3.7 | 18.7 | 1 | 2.1 | 3 |
| | 11 | −11.5 | −0.2 | 1.9 | 3.3 | 19.6 | 1.1 | 2.6 | 3 |
| Typical forest–steppe | 12 | −12.5 | 0.3 | 2.4 | 4.3 | 20.2 | 1 | 2.4 | 3 |
| | 13 | −11.1 | 0.1 | 1.6 | 3.5 | 20.5 | 1.3 | 3.4 | 3.3 |
| South forest–steppe | 14 | −12.3 | 0.4 | 2 | 3.9 | 21.6 | 0.9 | 2.2 | 3 |
| | 15 | −14.9 | 0.9 | 3.2 | 4.6 | 20.9 | 0.9 | 2.4 | 2.7 |
| North steppe | 16 | −14.3 | 0.4 | 2.4 | 24.4 | 22.2 | 0.8 | 2 | 2.9 |

[a] For conventional meanings of groups of plant associations, see Table 5.2.

generally increase by 1.1°–1.3° C, and in some areas of the sub-taiga and broadleaf forest sub-zones, even by 1.5°–1.7°.

The duration of the vegetation period will be increased everywhere as well. The average January temperature, by contrast, will become only 0.2°–0.5° higher. This is of great ecological significance because, as we have already shown before (Kolomyts 2003), the structural changes in vegetation and soils are determined mostly by their functional shifts during the vegetation period. The initial prognosis has already been partially justified: in 1995–1999, which turned out to be the warmest of the instrumentally monitored periods, the deviation of mean annual temperature from the 20-year norm of the "pre-industrial period" (1886–1905) on the European territory of Russia was 0.7°–0.8°to 1.9°(Hansen et al. 1988).

The rate of warming after 2050 will remain generally the same, with the monthly average temperatures exceeding the corresponding baseline values by 2.2°–2.3°. However, by the year 2050, the rise of the average January temperature will finally become prevalent, reaching 4.5°–5.0°. Thus, considerable shifts in the thermal energy supply to ecosystems will take place in all natural zones of the Volga River basin, with the ecological effect of these shifts increasing in a north–south direction. Taiga and nemoral forests will remain under climatic conditions of the temperate belt. Formations of the southern forest steppe and, especially, the northern steppe will transition from the sub-boreal to the subtropical thermal–radiation category according to classification (Volobuev 1974); i.e., they will be in a different climatic belt. The anthropogenic warming will be accompanied by a general increase in precipitation. By 2050 and 2100, the annual amounts of precipitation in the middle and southern taiga sub-zones will increase by 100–120 and 200–270 mm, and those in the forest–steppe with broadleaf forests and in the northern steppe will increase by 60–80 and 100–150 mm, respectively (Table 10.2). More abundant precipitation in the warm season will account for 70%–80% of this increase. Natural climate fluctuations may periodically intensify or weaken the predicted process of total aridization of natural ecosystems in the Volga River basin caused by anthropogenic enhancement of the atmospheric greenhouse effect. However, it seems impossible to quantitatively assess this correction at each stage of prediction. One can only provide the following estimates and considerations. The amplitude of super-century and century fluctuations of the global and regional mean annual air temperature reached 1°–2° in the past. Consequently, in the future, it may be considered commensurable to anthropogenic warming only until the beginning of the 30s of the 21st century.

We have analyzed the materials on climatic changes that occurred during the recent 30-year period (1975–2004) in the Middle Belt of European Russia (Sirotenko and Abashina 1998). The analysis showed a generally satisfactory adequacy of the climate prediction of the GISS model being used at the initial stage of global warming. During that period, the mean July temperature rose by 2.7° on average in the Upper Volga Region and by 2.2°–2.5° in the Middle Povolzhye (Middle Volga Region). As we can see, the actual thermal shift even advances the predicted one: it would have set in only in 2030–2040 according to the GISS model (see Table 10.1).

As regards the conditions of atmospheric humidification, the recent climatic trend also fully corresponded to the GISS prediction. By 2004, the sum of biologically active

**TABLE 10.2** Basic taxonomic norms of annual precipitation amount and $F_{hum}(1)$, as well as their predicted deviations in the Volga River basin, according to the GISS-1993 model

| *ZONAL AND SUB-ZONAL TYPES OF FORMATIONS* | *GROUPS OF PLANT ASSOCIATIONS[a])* | *ANNUAL PRECIPITATION AMOUNT, MM* | | | | *$F_{HUM}(1)$* | | | |
|---|---|---|---|---|---|---|---|---|---|
| | | *BASIC PERIOD* | *2050* | *2075* | *2100* | *BASIC PERIOD* | *2050* | *2075* | *2100* |
| Middle and south taiga | 1 | 768 | 14 | 142 | 172 | 1.80 | –0.14 | –0.13 | -0.24 |
| | 2 | 697 | 122 | 123 | 250 | 1.42 | –0.16 | 0.08 | 0.07 |
| | 3 | 759 | 94 | 166 | 229 | 1.73 | –0.13 | –0.34 | -0.21 |
| | 4 | 737 | 59 | 106 | 193 | 1.47 | –0.02 | –0.07 | -0.08 |
| Sub-taiga | 5 | 763 | 62 | 134 | 196 | 1.54 | –0.01 | –0.19 | -0.18 |
| | 6 | 679 | 54 | 93 | 170 | 1.23 | 0.06 | –0.07 | -0.03 |
| | 7 | 630 | 44 | 92 | 158 | 1.07 | 0.04 | –0.07 | -0.04 |
| | 8 | 650 | 56 | 93 | 148 | 1.21 | –0.09 | –0.18 | -0.13 |
| | 9 | 652 | 50 | 110 | 178 | 1.12 | –0.02 | –0.09 | -0.15 |
| Broadleaf forest | 10 | 655 | 64 | 102 | 156 | 1.07 | –0.10 | 0.01 | -0.19 |
| | 11 | 602 | 56 | 65 | 134 | 0.94 | –0.02 | –0.09 | -0.10 |
| Typical forest–steppe | 12 | 556 | 81 | 76 | 121 | 0.80 | 0.05 | –0.10 | -0.08 |
| | 13 | 536 | 109 | 96 | 158 | 0.90 | –0.06 | –0.38 | -0.27 |
| South forest-steppe | 14 | 507 | 77 | 25 | 126 | 0.72 | –0.02 | –0.20 | -0.26 |
| | 15 | 509 | 48 | 38 | 81 | 0.62 | 0.06 | –0.01 | -0.03 |
| North steppe | 16 | 438 | 66 | 61 | 81 | 0.57 | –0.11 | –0.23 | -0.26 |

[a] For conventional meanings of groups of plant associations, see Table 5.2.

temperatures in the Povolzhye region had increased by 30°–100°, with almost unvaried annual precipitation; therefore, the Vysotzky–Ivanov atmospheric humidification factor decreased by 0.07–0.10 (see Table 10.2). This means that the hydrothermal regime of broadleaf forests shifted to the southern boundary of their area, approaching the conditions of typical forest steppe (trend $F_{hum}(1)=1.37 \rightarrow 1.30$), while mixed forests of the southern belt of the sub-taiga zone turned out to be in the state of neighboring nemoral–forest communities (trend $F_{hum}(1)=1.29 \rightarrow 1.19$).

Comparison of both GISS scenarios (1988 and 1993) by a number of landscape–geophysical parameters with the scenarios of other models of the AOGCMs family (Houghton et al. 1996): GFDL and UKMO models, shows that they generally predict an unambiguous, though expressed in varying degrees, lowering of the atmospheric humidification factor by 2050 in the wide range of natural zones of the Russian Plain: from south taiga to middle steppe. This undoubtedly indicates the dominance of a thermo-arid climatic trend. At the same time, as compared to GFDL and particularly UKMO models, GISS-93 gives quite moderate climate aridization, especially in the forest and forest–steppe zones of the Russian Plain.

It is also necessary to take into account significant seasonal differences in the natural climate fluctuations. In the period of 1965–1990, an increasing positive temperature trend was noted for the latitudes of the middle band of the Russian Plain for January (reaching 1.0°–1.2°), while the smoothed curve of the mean July temperature had zero deviations. Generally, the change of epochs of natural cooling and warming relative to the summer season is rather weakly manifested in temperate latitudes; deviations from the norm are no more than ±0.2°–0.6°. For this reason, one can hardly consider the correction introduced by the natural climate fluctuations into the ecological effect determined by the increasing atmospheric greenhouse effect as essential. Considering the changes in the mean January temperature and winter precipitation, the ecological role of which is comparatively low (Volobuyev 1974), as the most significant changes, one may suppose that the reduction of this effect will be no more than 25%–30% of the predicted value.

### 10.2.2 Forecasting Changes in the Functional States of Plant Formations

Although the rise of summer and winter temperatures will be accompanied by a certain increase in precipitation, particularly during the warm season, the increment of precipitation will be insignificant and will not be able to neutralize the temperature increase. Therefore, the annual and summer humidity factors will start to decrease almost everywhere and quite significantly (see Table 10.2). Besides, the *progressive summer aridization of soil* will begin: a common event for the moderate belt of continents (Wetherald and Manabe 1995).

According to HadCM3, the mean July temperatures in the Middle Povolzhye will have increased by 2100 by more than 7°–8°, and the neutralization of such an increase will require a minimal increment of annual precipitation of about 1,000 mm. Such a situation is not predicted for this region by any of the AOGCM models (see Budyko et al. 1991; Albritton et al. 2003). It is not observed in paleoclimatic analogs either. For example, the annual precipitation in the Mikulino interglacial period exceeded the modern norm by no more than 100 mm (Velichko et al. 1983). Even for the Pliocene, as an analog of climatic conditions of the second half of the 21st century, the annual precipitation increment in the Middle belt of the Russian Plain is below 200 mm (Borzenkova 1992).

Thus, up to the end of the 21st century, *the thermo-arid climatic trend* will develop in the territory of the Volga River basin and its surroundings (Figures 10.2 and 10.3). This is evidenced by the negative values of changes in the atmospheric humidity factor dominating at all stages of the forecast period, as well as the July reserves of productive moisture in soil (see Table 10.2). The anticipated global anthropogenic warming will cause *general aridization of the regional bioclimatic system*, and this aridization will become even deeper and more total over time. The most significant aridization is expected in the first decades of the 21st century, when the increment of atmospheric precipitation will be especially low compared to the temperature increase. HadCM3 predicts two "leaps" of aridization: (1) at the first stage (1985–2050) and (2) at the beginning of the fifth stage (2075–2100). The first leap can already be clearly observed.

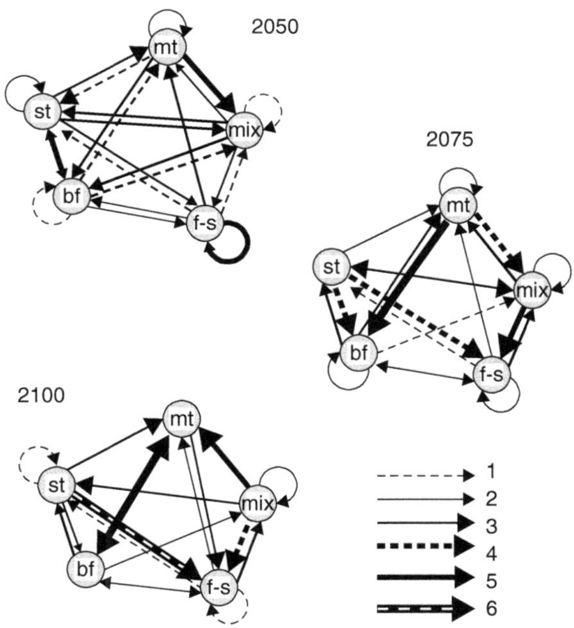


**FIGURE 10.2** Directions and probabilities of functional transformation of the landscape–zonal systems in the territory of the Volga River basin and its surroundings for different prognostic dates, according to the GISS model. Landscape–zonal systems: Mid T – middle taiga; Sou T – southern taiga; Mix F – mixed forest; BL F – broadleaf forest; F/S – typical and southern forest–steppe. Probabilities of landscape–zonal transitions: 1 – 0.10–0.20; 2 – 0.21–0.30; 3 – 0.31–0.40; 4 – 0.41–0.50; 5 – 0.51–0.60; 6 – 0.81–0.90.

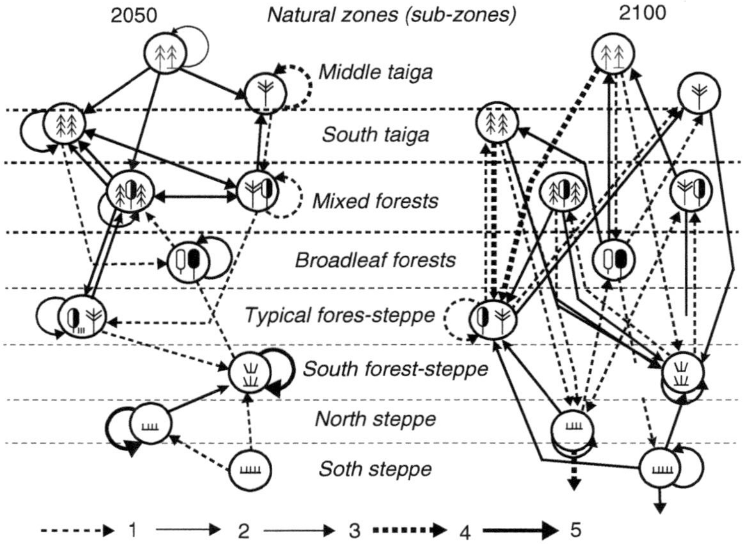


**FIGURE 10.3** Vector graphs of the probabilities of functional transitions between the plant formation groups of the headwater of the Volga River basin (see in Supplement II) for prognostic periods up to the years 2050 (a) and 2100 (b), according to the GISS model. Probabilities of inter-formation transitions: 1 – 0.20 and less; 2 – 0.21–0.30; 3 – 0.31–0.40; 4 – 0.41–0.50; 5 – 0.51–0.60.

In the European territory of Russia, during the period of 1995–1999, which proved to be the warmest over the entire time of meteorological observations, the mean annual temperature deviated from the 20-year norm of the "pre-industrial period" (1886–1905) by 0.7°–0.8° to 1.9° (Izrael 2001). During the next rather long interval (2050–2075), one can expect a certain weakening of the tempo of summer aridization against the background of the lasting tendency for enhanced winter mildness.

In accordance with the described climate scenario for all three forecast periods, the constructed maps and or-graphs (see Figure 10.3) indicate a consistent northward shift of the zonal and sub-zonal boundaries of the functioning of geo(eco)systems and the corresponding transformation of plant formations and soil cover under the influence of the thermo-arid bioclimatic trend.

The dominant and sub-dominant objects, according to the functional and structural appearance of which the majority of landscape–zonal systems will be transformed, will be exclusively sub-boreal natural complexes – typical and southern forest–steppe and northern steppe. At the same time, the middle and southern taiga pine forests will find themselves in the position of reduced ecosystems. They will progressively lose their features, transforming first into forest–steppe, and then even into dry steppe formations. Steppe formation of light coniferous taiga will occur mainly in the easternmost continental sector of the basin. At the third forecast stage (2075–2100), southern taiga fir–spruce forests will join the pine forests, which will also be subject to aridization.

**TABLE 10.3** Generalized characteristics of landscape transitions in the Volga River basin

| *PARAMETERS (SEE SECTION 10.1.3)* | *CLIMATE FORECAST VARIANT* | | |
|---|---|---|---|
| | *2050* | *2075* | *2100* |
| $P_{ij}$ | 0.505 | 0.462 | 0.378 |
| $M_{i1}$ | 0.684 | 0.386 | 9.325 |
| $M_{i2}$ | 1.368 | 0.772 | 0.649 |
| $M_{i3}$ | 2.052 | 1.158 | 0.974 |
| $t_{(steps)}$ | 1,462 | 2,590 | 3,081 |
| $t_{(years)}$ | 29,234 | 51,806 | 61,627 |

*Note:* $P_{ij}$ is the measure of stabilization of a given landscape type for the forecasted timeframe; $M_{i1}$, $M_{i2}$, and $M_{i3}$ are the total rates of transition of a given landscape type to other types by 2050, 2075, and 2100, respectively; $t_{(steps)}$ and $t_{(years)}$ are the total time required for the complete transformation of landscapes of a given type expressed in steps and years, respectively.

Along with this, another group of reducers will appear, consisting of mixed and broadleaf forests, which will be in the sphere of action of the opposite bioclimatic process – humidification, with a certain tendency of transformations towards dark coniferous–taiga ecosystems. The humid trend will also manifest itself in the southern dry steppe, which will begin to transform into nemoral, and then into mixed forests.

The most significant shift of landscape–zonal conditions towards aridization is expected in the first (to 2050) stage of the forecast period (Table 10.3), when the increase in atmospheric precipitation will still be small compared to the increase in temperature, which will cause a fairly sharp decrease in the moisture coefficient – first in the middle and southern taiga, and then in the typical and southern forest–steppe. The initial signal of anthropogenic warming will thus be ecologically the strongest and territorially the most effective, which we have already noted earlier (Kolomyts et al. 2012). Further shifts in the boundaries of zonal types of functioning of geo(eco)systems will be much smaller (see Table 10.1), due to the increasing, especially by 2050, increase in the amount of precipitation, which will stop the decrease in the $F_{hum}(1)$ parameter. This will contribute to the relaxation and functional stabilization of natural ecosystems subjected to transformations, their gradual transition to a new stable state.

By 2050, the heat and moisture ratios in the region will develop in such a way that the most noticeable functional shifts will occur in the zonal subdivisions of the boreal belt, while the northern strip of the sub-boreal belt will still remain relatively stable. This is indicated by both the expected sizes of the shift in zonal boundaries (see Figure 10.2) and the probabilities of phytocoenological transitions. However, the typical and southern forest–steppe will already be in the role of weakly expressed subdominants – prototypes of states for the entire galaxy of forest natural complexes.

During this period, the zonal transitions or-graph for 2050 (see Figure 9.1a) indicates a very intensive (up to 60%) potential transformation of middle taiga plant communities into sub-taiga ones. Middle and southern taiga pine forests, as reduced ecosystems, will have a tendency to transform into typical forest–steppe landscapes by almost a third of their appearance.

For its part, the southern taiga functional space will also be significantly crowded out by mixed forest (transition probability up to 40%), which will lead in the future to a noticeable narrowing of the entire southern taiga sub-zone. The mixed forest zone, in turn, will begin to transform near its southern boundaries, acquiring the functioning regime of the neighboring nemoral forest sub-zone, although not in such a significant latitudinal range and with much less intensity (up to 30%) than the middle or southern taiga. Broadleaf forests will overcome the well-known oro-climatic barrier: the latitudinal segment of the Oka, Volga, and Kama valley system, and will be able to transgress to the north. Warming will sharply increase the competitiveness of oak and other broadleaf species (Izrael 1992), which will begin to actively penetrate into coniferous forest phytocoenoses.

By 2100, the forest–steppe appearance will increasingly be acquired by the soil and vegetation cover not only in the sub-taiga but even in the southern and middle taiga sub-zones. The probability of steppe formation in the southern taiga is especially high (up to 80%–90%), since both typical and southern forest–steppe will develop simultaneously on its territory.

In the western sector of the Volga River basin, the forest–steppe type of functioning will significantly approach the current position of the Main Landscape Boundary of the Russian Plain, according to Milkov (1981), and in the lower reaches of the Vyatka and Kama rivers, it will "step over" it.' (In the middle of the 21st century, due to a significant increase in annual precipitation and relative stabilization of the humidity coefficient, a counter transgression of the sub-zones of the southern and even middle taiga into the territory of broadleaf forests is possible. As a result, the mutual transformation of the middle taiga and nemoral forest formations will be 50%–60%.

Thus, the thermo-arid bioclimatic trend should disrupt the stability of forest biogeosystems in the entire transition zone from forest to steppe on the Russian Plain, already from the first half of the 21st century of anthropogenic warming. In general, a progressive encroachment of the steppe on the forest is expected here over the entire 100–200 km space of the south of the forest zone. Such a landscape–ecological forecast is very unfavorable for forestry and will inevitably exacerbate the problem of preserving forest, and with it water resources in the vast transition zone from forest to steppe.

## 10.3 PROGNOSTIC SCENARIOS OF ZONAL STRUCTURE OF THE BASIN TERRITORY

As a whole, the orienting graphs (see Figures 10.2 and 10.3) and maps (Figure 10.4) plotted for both prognostic dates (2050 and 2100) demonstrate a gradual northward shift of zonal and sub-zonal boundaries and the corresponding changes in the types of ecosystem functioning under the action of the thermo-arid climatic trend. Objects-dominants and sub-dominants, whose appearance will be a prototype for transformations of most landscape–zonal systems, will be exclusively sub-boreal natural complexes: typical and southern forest–steppe and northern steppe (see Figure 10.4). By 2100, the features of

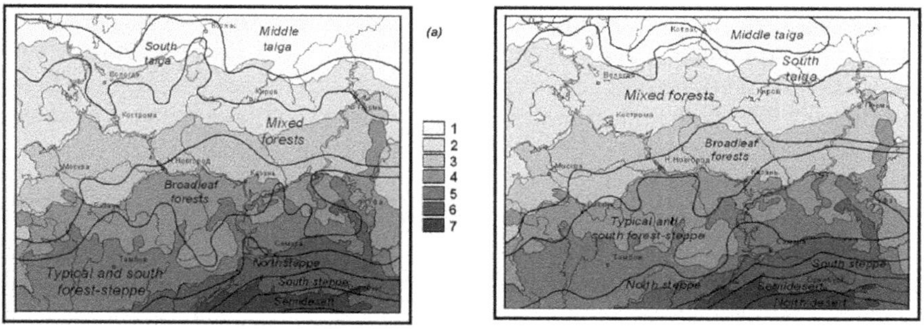


**FIGURE 10.4** The zonal landscape and environmental conditions of the Volga River basin and its environment, predicted for the years 2050 (a) and 2100 (b) according to the climate model GISS -1993. Areas of modern natural zones (sub-zones): 1 and 2 – middle and southern taiga; 2 – sub-taiga; 4 – broadleaf forests; 5 – typical and southern forest–steppe; 6 – northern steppe; 7 – southern (dry) steppe.

forest–steppe will have become increasingly characteristic of not only the sub-taiga but also the south and middle taiga sub-zones of the Volga River basin.

The most significant shift of landscape–zonal conditions towards aridization is outlined at the first and second stages of the forecast period, when the increase in atmospheric precipitation will still be low compared with the temperature increase, resulting in a rather abrupt fall of the coefficient of humidification – first in the middle and south taiga and then in the typical and southern forest–steppe. Thus, the starting signal of the warming will be the most effective ecologically and territorially. Already in 2050, the average taiga regime of ecosystem functioning will have almost completely moved to the north, beyond the boundaries of the Volga River basin. Further shifts of landscape–zonal boundaries will be much less pronounced due to increasing precipitation, particularly in 2100, slowing down the decrease of $F_{hum}(1)$. For this very reason, a counter-transgression of the sub-zones of south and even middle taiga into the territory of broadleaf forests is possible.

Thus, the following equifinal transformations of the zonal structure of the Volga River basin may be anticipated. First, two zonal types of the environment will be most developed territorially: mixed forests, on the one hand, and typical and southern forest–steppe, on the other hand. This forest–steppe belt will become broader overall. As we have shown previously (Kolomyts 2005), the southern boundary of the south taiga sub-zone is the main climatic border between boreal and sub-boreal belts, because the most abrupt

increase in the substance–energy basis of natural complexes is observed here. Consequently, by the middle of the 21st century, almost the entire territory of the main catchment area of the Volga River basin will have been under sub-boreal climatic conditions. At the same time, if in the first 30 years the transgression will manifest mostly in the mixed and, to a lesser extent, broadleaf forests, by the year 2075 the whole forest–steppe zone will be of no less priority and will become the predominant absorbing state by the end of the forecast period.

Second, the nemoral sub-zone will abruptly become narrower and change its location, having moved beyond the known oro-climatic barrier – a latitudinal segment of the systems of valleys of the Oka, Volga, and Kama rivers. The territory of modern broadleaf forests will be completely absorbed by typical forest–steppe formations (meadow steppe), which will partially expand even over the pre-Ural taiga. The oak–lime communities, meanwhile, will show transgression first to the south and then to the middle taiga, creating new mixed phytocoenotic structures and thus extending the total area of the sub-taiga zone.

Finally, the bioclimatic contrasts in the southeast of the region (in its most continental part) will sharply intensify, resulting in a considerable closing in of zonal and sub-zonal boundaries in the south of By-Ural, up to the appearance of the zone of semi-deserts at the latitude of Orenburg. The forest–steppe zone will gradually lose its stability: the degree of its stabilization from the first to the third forecast stage would decrease from 50%–60% to 10%–20%. Accordingly, it will start to transform into steppe in its southern and particularly southeastern regions.

The predicted thermo-arid bioclimatic trend is in harmony with the global scenarios of natural zonality obtained from various models, including the GISS model, for double $CO_2$ concentrations in the atmosphere, i.e., for the year 2100. The maps and schemes plotted by the authors show that the forest–steppe and steppe formations should intrude deeply into the forest zone of Eurasia, reaching the latitude of the city of Kirov in the east of the Russian Plain and the mouths of the Ob and Yenisei rivers in Western and Central Siberia. At the same time, the farthest expansion of "steppe areas" on the European territory of Russia is anticipated along the Central Russian and Privolzhskaya (By-Volga) Uplands.

The extreme climatic signal completely disrupts the functional stability of all biogeocoenotic groups (see further Table 12.2 and Figure 12.4b). In the sub-taiga Kerzhenskoye Trans-Volga region, the process of thermo-aridization of forest communities will proceed at such a significant rate that after 2050 they will pass through the nemoral forest stage of transformation, and by the middle of the 22nd century, almost 60%–70% will transition into the state of northern meadow steppes. Ultimately, in place of the sub-zone of broadleaf forests and the southern strip of the sub-taiga zone of the Middle Volga region, coniferous (pine) forest–steppe will develop as a new (for the Russian Plain) zonal ecotone of direct contact between boreal and steppe plant formations. "Fragmentation of the vegetation cover is a fundamentally new factor"; it "entails the development of ecotoneization of the boundaries of natural ecosystems, the formation of wide transitional strips" (Tishkov 2012, pp. 39, 40). This will probably be the eastern version of the general process of "savannization" of broadleaf and mixed forests on the Russian Plain under current global warming.

Prognosis results, quite similar to our model, have also been obtained for the territory of the former USSR (Kobak et al. 2002). This concerns, in particular, significant (2.5–3.4-fold) cutting of the taiga zone anticipated by the 10–30s of the 21st century. Moreover, a 10-fold extension of the area of broadleaf forests seems unlikely. We could also agree that by 2050, when the value of global warming will have reached 1.4°, the landscape–ecological conditions of mixed forests will prevail in the territory of the Volga River basin. According to the calculations (Velichko 1991; Velichko 2010), the zonal conditions of coniferous–broadleaf forests are expected to extend to 62°–63° N in the east of the Russian Plain already in 2010, which goes far beyond the northern boundaries of the Volga River basin. Our scenario, as a matter of fact, assumes such a possibility, but only after the middle of the 21st century.

## 10.4 DYNAMICS OF BIOLOGICAL CYCLE PARAMETERS IN ZONAL–REGIONAL ECOSYSTEMS

It is significant to evaluate the possible changes in primary productivity and the velocity of forest litter mass decomposition in terrestrial ecosystems as global warming continues. The prevailing transitions of northern landscape–zonal systems into southern systems under the effect of global warming will be accounted for by the acceleration of the biological cycle throughout the boreal ecotone, from the middle taiga to the southern forest–steppe. Both the rate of decomposition of dead organic matter, including forest litter, and the level of autotrophic biogenesis will become higher. Therefore, the net primary production of phytocoenoses will increase, entailing additional consumption of soil moisture and mineral nutrients.

Now the south-taiga forests have maximum annual $P_{prim}$ (10.5–11.5 t/ha per year), and north-steppe formations have a minimum one (6–7 t/ha).

According to prognosis calculations, primary bio-productivity may rise everywhere by 1.5–2 t/ha as early as in the 2010s due to the increase in both temperature and precipitation, and by the year 2050, even in dry steppe, it will reach 9.5–10.0 t/ha.

The effect of these functional changes will expectedly reach a peak in the eastern sector of the taiga zone (Tables 10.4 and 10.5). By the year 2050, the annual rate of increase in the zonal norms of *Pp* in this sector will be 1.8–2.5 t/ha per degree of

**TABLE 10.4** Specific increase rates of taxonomical standards of plant formation productivity for the Volga River basin and its encirclement at various predictive intervals, according to GISS model

| *NATURE ZONES AND SUB-ZONES* | *GROUPS OF PLANT FORMATIONS* | *PREDICTIVE INTERVALS* | | |
|---|---|---|---|---|
| | | *1985–2050* | *2050–2075* | *2075–2100* |
| Middle and south taiga |  1 | 2.2/2.3 | 2.0/3.6 | 1.8/3.3 |
| |  2 | 1.8/1.6 | 1.5/2.4 | 1.6/2.0 |
| |  3 | 1.6/1.5 | 1.8/2.2 | 1.9/2.0 |
| |  4 | 2.3/2.7 | 1.7/2.9 | 1.7/2.4 |

*(Continued)*

**TABLE 10.4 (*Continued*)** Specific increase rates of taxonomical standards of plant formation productivity for the Volga River basin and its encirclement at various predictive intervals, according to GISS model

| *NATURE ZONES AND SUB-ZONES* | *GROUPS OF PLANT FORMATIONS* | *PREDICTIVE INTERVALS* | | |
|---|---|---|---|---|
| | | *1985–2050* | *2050–2075* | *2075–2100* |
| Mixed forests | 5 | 1.8/2.1 | 1.4/2.5 | 1.3/2.0 |
| | 6 | 2.5/3.7 | 1.5/1.7 | 1.7/2.6 |
| | 7 | 1.8/3.2 | 1.0/2.4 | 1.4/2.3 |
| | 8 | 0.9/1.8 | 1.0/2.0 | 1.1/2.0 |
| | 9 | 1.5/3.2 | 1.0/1.9 | 1.1/1.7 |
| Broadleaf forests | 10 | 0.8/0.8 | 0.9/1.9 | 1.1/2.1 |
| | 11 | 1.8/3.6 | 1.0 / 3.8 | 1.2/2.7 |
| Typical forest–steppe | 12 | 1.3/1.6 | 0.3/0.9 | 1.0/2.5 |
| | 13 | 1.2/1.7 | 0.7/2.5 | 0.9/1.9 |
| South forest–steppe | 14 | 1.9/2.2 | 1.1/10.0 | 1.1/2.5 |
| | 15 | 0.8/1.4 | 0.2/1.3 | 0.7/2.2 |
| North steppe | 16 | –1.6/– | –0.6/– | 0.4/1.4 |

*Note:* Specific increase of productivity: in numerator – at the expense of average July temperature (t/h·year for 1°C); in denominator – at the expense of increase of total precipitation per year (t/h·year for 100 mm of precipitation). Conventional meanings of plant formation groups see in Table 5.2.

temperature rise. The increase in precipitation will have an effect mainly in the sub-taiga zone of this sector, where the pluvial gradient of annual *Pp* growth will reach 3.2–3.7 t/ha per 100-mL increment in the annual amount of precipitation. The initial acceleration of metabolism in the western sector will be lower, but both meridional and latitudinal (zonal) differences in the rates of functional activation of boreal forests will gradually level off. Under the effect of increasing temperature in the warm season, parameter CY in boreal forests will sharply drop already at the first stage (before 2010). This will

**TABLE 10.5** Forecasted changes of the litter-fall index (CY) in groups of plant formations for the dates 2050 (+), 2075 (×), and 2100 (*) as compared with base period (•)

| Nature zones and subzones | Groups of plant formations | Gradations of *CY* parameter | | | | | |
|---|---|---|---|---|---|---|---|
| | | 0.8–1.5 | 1.6–5 | 6–10 | 11–15 | 16–20 | 21–50 |
| Middle and south taiga | 1 | | * ← | × ← | —— | +• | |
| | 2 | | *× ← | + ← | —— | —— | • |
| | 3 | | *× ← | + ← | —— | • | |
| | 4 | | *× ← | + ← | —— | • | |
| Mixed forests | 5 | * ← | × ← | +• | | | |
| | 6 | * ← | × ← | + ← | • | | |
| | 7 | *× ← | + | ← —— | • | | |
| | 8 | *× ← | + ← | • | | | |
| | 9 | | • → | + → | × → | * | |
| Broadleaf forests | 10 | | | •+× | → | * | |
| | 11 | | •+× | → | * | | |
| Typical forest-steppe | 12 | | •+× → | * | | | |
| | 13 | •+ → | × → | * | | | |
| South forest-steppe | 14 | | *× +• | | | | |
| | 15 | ×+• → | * | | | | |
| North steppe | 16 | *×+• | | | | | |

*Note:* Conventional meanings of plant formation groups see in Table 4.2. Arrows show the transitional directions.

indicate a significant increase in the efficiency of the detrital segment of the biological cycle, which is its main limiting component.

Boreal forests of the Volga basin will generally become more productive (in particular, due to the increasing rate of forest litter decomposition). Their annual net production will exceed 13–14 t/ha by the year 2075 and reach 15.5–16.5 t/ha by the year 2100. Following Budyko (2002), it can be figuratively stated that, by the mid-21st century, these forest formations will exemplify the return of the boreal biosphere belt to a more fertile geological epoch in which the concentration of greenhouse gases in the atmosphere was much higher than today.

The calculations showed that already for the period of 2050–2075, the predicted fields of $R_{ann}$ and $I_{arid}$ will go significantly beyond the limits of their contemporary combinations on the territory of the Russian Plain. The same could be referred to the optimum of the Mikulino interglacial period. Within the Volga River basin and adjoining regions, there are at present no combinations of high values of $R_{ann}$ (more than 1,900–2,000 MJ/m$^2$) and very low $I_{arid}$ (below 0.7–0.8), when the productivity of vegetation must exceed the value of 14 t/ha per year. Such hydrothermal conditions are typical of contemporary humid forest formations of the extreme south of the sub-boreal belt and even the subtropical belt, e.g., Colchis and Lenkoran lowlands, where the annual productivity of forests exceeds 20 t/ha.

The productivity of boreal forests in the Volga River basin will reach a level that, according to published data (Dumitrashko 1966; Bazilevich and Rodin 1971), is equivalent to the lower limit of the present-day productivity of mesophilic meadow steppes and forest–steppes in some well-moistened regions in the extreme south of the Russian Plain (the Azov–Kuban Lowland and the Terek delta) and humid–subtropical liana forests of Transcaucasia. The productivity of broadleaf forests in the middle Volga region, especially that of more xerophilic oak forests, will also approach this level.

These data were used for extrapolating the prognosis of primary productivity to the years 2050 and 2100 for the natural ecosystems of the west and northwest regions of the Volga River Basin. The combinations of predicted $R_{ann}$ and $I_{arid}$ values here exceed the limits of the nomogram in Figure 6.5.

A certain growth of net production by the year 2050 (from 8–9 to 11–12 t/ha per year), resulting mainly from the increase in annual precipitation, is also expected in phytocoenoses of typical and southern forest–steppes. It will be especially evident at the next stage (between 2050 and 2075) in the western, more humid sector. In the extreme southeast of the region, in northern steppes, tendencies in biological production will be ambiguous. In the steppes of the Transvolga and Orenburg regions, a drop in *Pp* is expected due to the unfavorable combination of high values of the annual radiation balance (more than 2,050 MJ/m$^2$) and the index of aridity (1.7–1.8).

The predicted thermo-arid trend will definitely entail a significant decrease in soil moisture content during the growing period throughout the Volga River basin (see Table 10.4). By the year 2030, the average soil moisture content in July will drop below the lower limit of its current interannual variation. By the mid-21st century, this parameter in all middle and southern taiga forests will be five to seven times lower than the minimum water capacity of soils (today, the difference is by a factor of no more than 1.5–1.7). The extent of soil aridization in the mixed forest zone will be virtually the same. With respect to moisture deficiency in summer, soddy podzolic soils will approach the current state of not only gray forest soils of broadleaf forests but even leached and podzolized chernozems of the typical forest–steppe. Aridization of gray forest soils will bring them to the state of recent meadow chernozem soils of the southern forest–steppe, as well as typical and ordinary chernozems of the northern forest–steppe, which, in turn, will be gradually converted into semidesert chestnut soils.

# 10.5 MODERN REGIONAL ANALOGUES OF PREDICTED LANDSCAPE–ECOLOGICAL SITUATIONS UNDER GLOBAL WARMING

Analogues of equifinal results of the process of aridization of boreal and nemoral forests are easy to find in the extra-continental plain and mountain regions of Siberia and Central Asia. Thus, in the south of Central Siberia, forest–steppe landscapes that arose as a result of the direct interaction of southern taiga and steppe formations are widespread in very large areas (Krauklis 1991). At the same time, in contrast to the European forest–steppe, taiga and steppe communities here are more clearly differentiated by types of locations. For example, the forest–steppe landscapes of the Kansk, Nazarovo, and Achinsk–Bogotol basins, located at the northern foothills of the Eastern Sayan and Kuznetsk Alatau, are a combination of boreal (birch–larch and pine) park-type forests on flat watersheds and valley bottoms, with meadow–steppe and steppe associations on steep, predominantly sunny slopes and hilltops. Small-bunch grass steppe grass stands also penetrate into the ground cover of park light coniferous forests, an example of which is the Yenisei forest–steppe pine forests. Even in the spruce forests near streams of the Kansk Depression, shrubs are quite abundant, and the grass stand is represented not so much by boreal as by meadow–steppe species (reed grass, aconite, delphinium, kakalia, smilacina, etc.).

The main reason for the territorial differentiation of the vegetation cover in the south of Central Siberia is considered to be the uneven distribution of winter precipitation and, accordingly, the depth of the snow cover (Krauklis 1991). Forest–steppe landscapes arise, as a rule, in conditions of low-snow winters, despite the fact that, according to the annual moisture coefficient, they should belong to the southern taiga. Low-snow and cold winters contribute to the steppe formation of the territory. This fact, which is currently characteristic of the entire chain of forest–steppes in the south of Siberia and the Far East, is directly related to the above-described forecast climate scenario for the Volga River basin. The transition of vast territories of the southern taiga and sub-taiga to the category of medium- and low-snow regions should significantly accelerate the process of steppe formation of boreal forests. According to Alekhine's "landscape anticipation rule" (cited in Milkov 1981), this process should manifest itself first on well-drained, sunny slopes and watersheds.

Landscape–ecological analogues from the plains of Transcaucasia and highland regions of Central Asia, which are located in the most pronounced thermo-arid climatic conditions, may turn out to be quite representative. Such are, in particular, the subtropical hemic–serophytic sparse forests in the Kura River basin in eastern Transcaucasia (Dumitrashko 1966; Gulisashvili 1980). Here, trees and shrubs are scattered in disarray and are located at a considerable distance from each other against the background of herbaceous vegetation. The climatic conditions of this territory ($t_{\text{July}}$=23°–24°, $r_{\text{year}}$=500–600 mm, mild winters with little snow) are very close to those predicted for the eastern sector of the broadleaf forest sub-zones and typical forest–steppe in the Volga River basin by 2100.

The mid-mountain forest–meadow–steppe belt of the Western and Central Tien Shan is also of interest, occupying absolute altitudes from 1,200–1,500 m to 2,600–3,000 m. Against the background of the general high aridity of this mountainous country, these regions are distinguished by increased atmospheric humidity. Nevertheless, intra-belt bioclimatic contrasts are expressed quite sharply here. In conditions of deep erosion–tectonic dissection of the ridges, forest communities (deciduous in the lower part of the altitudinal belt and coniferous from the Tien Shan Shrenk spruce in the northern part) do not form a continuous cover but are scattered in separate park-type massifs, mainly along the northern slopes of mountain gorges, descending to the bottoms of the valleys. The slopes of the southern exposure are occupied mainly by meadow steppes and shrub thickets. Such a clear phytocoenological differentiation is closely related to the distribution and regime of snow cover in the region. According to snowfall characteristics, the northern and nearby slopes of the forest–meadow–forest belt approach the overlying alpine belt, and the southern ones approach the underlying desert steppe belt. As we can see, this example also demonstrates the decisive importance of reducing winter snowfall in the steppization of the territory.

In general, as shown by floro-genetic and geobotanical studies (Morozov 1949; Agakhanyants 1981; Gulisashvili 1980; Korotkov 1978; etc.), arid phytocoenogenesis has the following features:

- A decrease in site quality and the total amount of phytomass, with the movement of most of it into the underground sphere and a decrease in the productivity of communities.
- An increase in antagonistic relationships between forest and herbaceous vegetation, leading to an increase in the contrast of the species that make up the community and to an exceptionally high ability of trees and shrubs to regenerate by means of coppice; the latter significantly helps tree and shrub species to compete with herbaceous species.
- Increased xeromorphism of the green parts of plants and the development of a powerful closed root system that provides water and mineral nutrition to plants in conditions of a dry growing season.
- Replacement of forests by sparse savannah-type forests (as a way for the root system of tree stands to combat the lack of soil moisture) and sparseness of permanent layers of vegetation cover, indicating a decrease in the biological activity of soils.
- Typological vagueness of communities, expressed by their deep interpenetration into each other; hence, the formation of mixed types of vegetation and increased instability of aridized communities.
- A decrease in the vertical capacity of ecosystems (narrowing of the life layer) and an increase in their spatial mosaicism and micro-zonality; these processes are territorially differentiated by geomorphological and soil–edaphic factors.

The listed homeostatic mechanisms of adaptation of forest communities to conditions of intensive aridization of the environment serve as a clear illustration of the equifinal state to which the above-described functional and structural changes in the natural ecosystems of the Volga River basin can lead under the influence of the predicted thermo-arid climatic trend. They confirm, in particular, the expected phenomenon of transgressive development of mixed-forest phytocoenoses not only in the south of the boreal belt but also in its "core of typicality" – in the middle taiga. Already at the first stages of the forecast period, one can also assume a certain "savannization" of shady broadleaf forests and their merging with the general mosaic complex of sparse forests, meadows, and steppes of the typical forest–steppe.

A further transformation of the closed cover of mixed and even coniferous forests into sparse forest massifs of the park type, separated by areas of meadow–steppe communities – analogues of modern elani – is also more clearly imagined. The latter are formed, as is known, by long-term human impact on coniferous forests (Murzaev 1984). In this case, a certain differentiation of the formations indicated by types of locations in the meso-relief should be expected. Finally, we can speak with complete confidence about the increasing mobility of geobotanical areas and the instability of zonal boundaries.

## 10.6 MAIN INFERENCE

Aridization of bioclimatic conditions will inevitably aggravate the situation regarding the conservation and reproduction of forest resources and crop cultivation in the area of the present-day forest–steppe zone and broadleaf forest sub-zone. Based on available data on phytoecological soil hydrology (Verigo and Razumova 1973; Volobuev 1963), it may be expected that, as early as 2010, natural reproduction of linden–oak phytocoenoses will already be impaired not only in the northern steppe but also in the typical and southern forest–steppes. By 2030, the same situation will develop in broadleaf forests themselves, facilitating their steppification. The July soil moisture content in these forests will drop to the level at which the viability of mature trees sharply decreases. Within the next two decades, the conditions for forest reproduction will become critical even in the mixed forest zone.

Agroclimatic conditions will also become unfavorable. By 2010, the soil hydrologic regime in July, at the phase of the most active vegetative development of grain crops (leaf tube formation and flowering), will deteriorate throughout the entire forest–steppe zone so that crop productivity will inevitably become 20%–30% lower. In 2050, the mixed forest zone will be in the same situation, with crop productivity in the forest steppe decreasing by 40%–65%. It may be expected that, due to global warming, significant additional energy expenditures for irrigation projects will be necessary in the next few decades to maintain crop productivity in the middle part of the Russian Plain.

# 11 Cartographic Models of Regional Landscape–Ecological Forecasting

## 11.1 IDEOLOGICAL FOUNDATIONS OF PREDICTIVE RESEARCH

The given section presents one of the results of the author's scientific research on cartographic modeling of regional landscape–ecological forecasts – the creation of forecast maps displaying the governance and rates of transformation of the landscape–zonal structure of the region based on the expected (according to this geographical forecast) global anthropogenic climate changes. The regional landscape–ecological forecast, carried out on the example of the boreal ecotone of the Volga River basin according to the scenario of one of the global climate models, includes establishing the mechanisms of climat-genic transformations of zonal types of regional geosystems, indicated, according to V.N. Sukachev (1975), by indigenous plant formations; calculating and mapping interzonal transitions of geosystems; and obtaining a general picture of the transformation of the landscape texture of the region; assessing the limiting role of morphogenetic barriers caused by the lithogenic basis of landscapes. In the processes of landscape–zonal restructuring, changes in the structure of the phytobiotic block of landscapes are considered, mainly, as the most sensitive to external influences and the most dynamic. Such an approach in geographical forecasting is quite justified (Simonov


DOI: 10.1201/9781003733119-11

1982;) however, it is limited due to the complication of the exogenous dynamics of phytocoenoses by their internal interrelations.

The initial cartographic material for the regional forecast analysis was the landscape map of the Volga basin compiled by V.P. Yunina at a scale of 1:2,500,000 (Kolomyts 2005). Chapter 5 (Section 5.3.1; Figure 5.1) describes the procedure for "greening" the landscape map so that the zonal–regional features of the background bioclimatic system would be more clearly reflected in the appearance of the main operational units (landscape types).All types of landscapes of the Volga River basin (61 types, see Figure 4.1) were combined into 20 typological groups of landscapes (see Figure 5.2) as the main natural–territorial units of landscape–ecological forecasting.

Two features were used to identify them – zonal affiliation and regional type of location: eluvial (E), trans-eluvial (TE), transit (T), trans-accumulative (TA), and accumulative (A).According to taxonomic rank, landscape groups correspond to regional units of vegetation – classes and groups of formations (Sochava 1979). Landscape species and groups were also classified according to the degree of lithomorphism–hydromorphism as a "directly acting" ecological factor that links together the landscape-forming effect of zonal and azonal factors. This feature fits quite clearly into the coordinate system of genetic genera and landscape groups (Table 11.1).

The general process of phytocoenological transformations in natural complexes is conventionally divided into two groups. The first of these includes trans-boundary interactions of phytocoenoses belonging to different landscape areas in conditions of first, second, etc. order neighborhoods. This is, in essence, a phenomenon of phytocoenotic induction, or guidance, caused primarily by the dispersal of seeds and their survival (Udra 1988) at a given stage of exogenetic successions (Sukachev 1975). Such a process of frontal phytocoenotic transformations can be called inductive.

The second group is formed by the processes of focal or area transformation of phytocenoses, when the decisive role in the change of dominants is played by the presence of a ready seed bank in the soil or the existence (in other life forms or at least in a depressed state) of certain bio-groups, capable of occupying a dominant position in the phytocoenosis under favorable conditions (Sukachev 1975; Rabotnov 1978). This process of phytocoenological transformation of the landscape occurs as if from within. It is least of all associated with the overcoming of certain natural boundaries by biogenic flows. The speed of its manifestation depends mainly on the severity of the competitive relations of old and new coenotic groups, their individual elements, and the corresponding rates of change of dominants.

The process of focal transformation of phytocoenoses is, apparently, the leading one. It is more effective than frontal interaction and is capable of causing a much faster transformation of the landscape–zonal structure of the territory, often with abrupt shifts of zonal boundaries. In this case, the effect of frontal interactions of phytocoenoses should be considered as part of the general, predominantly focal, process of landscape restructuring.

For all forecast periods for each $i$-th group of landscapes of the Volga River basin, the following were calculated (see Sections 10.1.1 and 10.1.3): the measure of its stabilization $P_{ii}$ and the total a priori probabilities $P_i$ of its transition to all other groups; conditional probabilities $P_{ij}$ and the total rates of transitions $M_i$ from its niche $i$ to the

**TABLE 11.1** Ecological classification of landscapes of the main catchment area of the Volga River basin and its surroundings

| *LANDSCAPE GENERA (SUBSTRATE)* | *LANDSCAPE GROUPS (LOCATION TYPES)* | | | | |
|---|---|---|---|---|---|
| | *TE* | *E* | *T* | *TA* | *A* |
| Erosion-denudation, E-D (loamy) | 2/2 (S/L)<br>3/2 (L)<br>4/4 (L)<br>5/2 (L)<br>6/1 (L) | 2/1 (S/L)<br>3/1 (S/L)<br>4/1 – 4/3 (L) | | | |
| Erosion, E (loamy) | 3/3 (L) | 4/5, 4/6 (L) | 5/4 (L) | | |
| Morainic, M (loamy, loamy sand) | | 2/5 (G-L)<br>3/7 (G-L) | | | |
| Denudation-outwash, D-Out (loamy–sand) | | | 2/3, 2/4 (G-L)<br>3/4 – 3/6(G-L)<br>4/7, 4/8 (S/L) | | |
| Morainic-Outwash, M-Out (sand–loamy) | | 2/6 (S/G) | 3/8 (S/G)<br>4/9 (G-L) | | |
| Lake-Alljuvial, L-Al (loamy) | | | 3/11 (G)<br>4/12 (S/L)<br>5/3 (S/L) | 6/4 (G-L) | 2/9 (G) |
| Outwash, Out (sand, loamy sand) | | | | 2/7 (S/G)<br>3/9, 3/10 (S/G)<br>4/10 (G-L) | |
| Valley-Outwash, V/Out (sand, loamy sand) | | | | | 2/10 (G)<br>3/12 (G)<br>4/14 (S/G) |

*Note:* 2/1, 2/2, 3/1, etc. – designations of landscape types (see the legend to the landscape map, Figure 4.1); In brackets – designations of soil properties: G, hydromorphic; S/G, sub-hydromorphic; G-L, hydro-lithomorphic; S/L, sub-lithomorphic; L, lithomorphic.

niches of other groups *j*; absolute $h_{ij}$ and relative $m_{ij}$ partial speeds of its transition to the niche of the *j*-th group (see Table 10.1);time $T(P_{ii})$ of the object's stay in state $P_{ii}$, expressed in the number of steps, i.e., until the moment when it is replaced by all other states. Multiplying the value $T(P_{ii})$ by the duration of the step, we express the time of the object's stay in the number of years.

Climate scenarios for 2050, 2075, and 2100 are specific forecast variants, and each of them assesses changes in landscape–ecological conditions for the same stages – the forecast periods themselves. Climate forecast options differ significantly in their landscape-dynamic characteristics.

## 11.2 PREDICTIVE MAPPING OF FOCAL LANDSCAPE TRANSFORMATIONS

The processes of multiple transformations of natural complexes consist of a whole series of mutual transitions of some landscape (more precisely, phytocoenological) niches to others, not only at the level of zonal types and subtypes of landscapes according to the climatic trend but also between different genera, i.e., morphogenetic units associated with the features of the lithogenic basis. Thus, the background climatic signal causes a certain "landscape-generic drift" of the phytocoenological core of geosystems within a given natural zone (sub-zone) and is potentially no less significant than the zonal shift of the boundaries of the range of a particular type of landscape. In the given example (Table 11.2), according to the GISS and EGISS models, with a thermo-humid trend by 2050 and 2075, the climatic niches of sub-taiga spruce forests of hilly–moraine plains (landscape type 3/7) will be replaced by niches of, on the one hand, coniferous, mixed, and even broadleaf forests currently growing on erosion–denudation and erosion interfluves (types 2/1, 3/1–3/4, 4/5, 4/6), and on the other hand, spruce–pine and coniferous–broadleaf complexes of valley sanders and lake–alluvial plains (types 2/10, 3/12, 4/12).

To map the landscape–ecological conditions of the future, it is necessary to have an integral picture of the transformation of climatic niches of geosystems. For this purpose, a generalization, with a cumulative total (Kolomyts 2022) of the $h_{ij}$ values for each landscape type in a two-dimensional coordinate system, was carried out by the bioclimatic zonality feature and by the generic feature. Such a generalization for landscape type 3/7 is given in Table 11.2. The sum of the transition rates by rows characterizes the rate of change of climatic niches, which determines the interzonal focal phytocoenological transformation of landscapes of a given genus and species, and the sum by columns characterizes the rate of "absorption" of a given niche by niches of other landscape genera, regardless of the zonal and species affiliation of the latter. The final sum $\Sigma\Sigma$ $h_{ij}(2)$ (in the lower right corner of the matrix) is an analogue of the $M_i$ value, while taking into account the restrictions imposed on the speed of landscape transitions by the factor of the generic affiliation of landscape types. The assessment of such limitations was as follows.

Interzonal transformation of a given landscape type into another type will be most probable and rapid within the same genus of geosystems or close genera. The "generic barrier" should negatively affect the speed of landscape transitions. A sharp change in the physical and mechanical properties of soil-forming rocks, the form of flat relief, and the type of location serve as significant obstacles to the mutual transformation of neighboring landscapes of different types, located on both sides of the lithogenic boundary. Such natural boundaries will shift last and at the lowest speed; therefore, the height of the "generic barrier" must be taken into account in the procedure of further adjustment of the obtained calculated speeds $h_{ij}$ of landscape transitions.

The height of the "generic barrier" was taken into account using the correction coefficient $C_{gb}$, which generally characterizes the relative generic contrasts between different types of landscapes. The approach proposed in Ivashutina and Nikolaev (1972)

**TABLE 11.2** Matrices of absolute partial measures $h_{ij}(2)$ of the rates of transitions of the niche of landscape type 3/7 (see Figure 5.1) into niches of other types belonging to various natural zones (sub-zones) and landscape genera

| *NATURE ZONE (SUB-ZONE)* | *LANDSCAPE GENERA (SEE LEGEND TO FIGURE 5.1)* | | | | | | | | *SUM* |
|---|---|---|---|---|---|---|---|---|---|
| | *E-D* | *E* | *M* | *D-O* | *M-O* | З | *L-AL* | Д-О | |
| *a*) **Forecast for 2050** | | | | | | | | | |
| Middle taiga | 0.350 | | | | 0.093 | | | | 0.443 |
| Soth taiga | 0.138 | | 0.039 | 0.088 | | 0.011 | | | 0.276 |
| Sub-taiga | 0.022 | | | 0.030 | | 0.050 | | | 0.107 |
| Sum | 0.510 | | 0.039 | 0.118 | 0.093 | 0.066 | | | 0.727 |
| *b*) **Forecast for 2075** | | | | | | | | | |
| Middle taiga | 0.035 | | 0.026 | 0.093 | | | 0.028 | | 0.182 |
| South taiga | 0.138 | | 0.039 | 0.088 | | | 0.011 | | 0.276 |
| Sub-taiga | 0.053 | 0.025 | | 0.080 | | | 0.055 | 0.004 | 0.217 |
| Broadleaf for. | | 0.045 | | | 0.015 | 0.012 | | | 0.072 |
| Forest-steppe | 0.017 | 0.022 | | | | | | | 0.039 |
| Sum | 0.243 | 0.092 | 0.065 | 0.261 | 0.015 | 0.012 | 0.094 | 0.004 | 0.786 |
| c) **Forecast for 2100** | | | | | | | | | |
| Middle taiga | 0.035 | | 0.057 | 0.110 | | | 0.047 | | 0.249 |
| South taiga | 0.165 | | 0.095 | 0.112 | 0.016 | 0.010 | 0.021 | 0.006 | 0.425 |
| Sub-taiga | 0.053 | 0.025 | | 0.080 | | | 0.055 | 0.004 | 0.217 |
| Broadleaf for. | | 0.075 | | | 0.025 | 0.012 | | | 0.112 |
| Forest-steppe | 0.017 | 0.035 | | | | | | | 0.052 |
| Sum | 0.270 | 0.135 | 0.152 | 0.302 | 0.041 | 0.022 | 0.123 | 0.010 | 1.055 |

was used to assess the contrast of the landscape structure of the territory. To calculate $C_{gb}$, all eight landscape genera were located, according to Glazovskaya (1964) and Mil'kov (1981), in one morphogenetic (paragenetic) series in the general direction from watersheds to valley bottoms:

$$E-D \rightarrow E \rightarrow M \rightarrow D-O \rightarrow M- \rightarrow L/Al \rightarrow \text{Outwash} \rightarrow V-O.$$

The further two types of landscapes are separated from each other in this series, the higher the lithogenic contrast between them is, and, accordingly, the slower the process of inter-generic transitions of landscape types will be in the event of the intersection of their climatic niches.

There are eight intervals (steps) in the morphogenetic series, which are proposed to be called the paragenetic distance, or the index $C(j,i)$ (Kolomyts 2005; Kolomyts 2018). For example, in the trans-boundary bundle of landscape types 2/1(E)–1/3(T), this distance is equal to one step; in the pair 2/5(E)–1/9(TA) – two steps; and for the most distant types, 4/5(TE) and 3/12(A), there are four steps. We conditionally accept that each $i$-th interval (step) has a "weight" $\Delta Q_i=0.1$ (for lake–alluvial landscapes, $\Delta Q_i=0.2$).For the maximum interval, the total sum of "weights" $\Delta Q_i=0.9$.This applies to E-D and valley–outwash landscapes, which occupy extreme positions in the morphogenetic series and therefore have the minimum possible mutual transformation. The correction "generic" coefficient, as a value inverse to the paragenetic distance, is equal to

$$C_{gb} = 1 - n \cdot (\Delta Q_i), \tag{11.1}$$

where $n$ is the number of intervals separating the landscape type under consideration in the morphogenetic series from the type into which it should be transformed. Thus, the absolute speed of landscape transitions $h_{ij}$ (2), corrected for the value of the "generic barrier", is equal to

$$h_{ij}(2) \ = C_{gb} \cdot h_{ij}. \tag{11.2}$$

The matrix of measures $h_{ij}$ (2) describes quite fully the zonal–regional diversity of the rates of landscape transitions, allowing us to estimate at each step both the total number and prevailing directions of transformations of niches of a given landscape type, as well as the partial and total rates of these processes. For example, for mixed-forest moraine spruce forests (landscape type 3/7, see Table 11.2) during the forecast period, the following will be characteristic: a tendency for the diversity of their transitions to other types to increase; preservation of the prevailing transformation into neighboring E-D and lower down the slopes valley–outwash geosystems; dominance of changes in sub-taiga phytocoenoses by southern and middle taiga ones; increased transformation rates at the beginning and end of the forecast period and their obvious decrease at the second step; achievement of a complete change of climatic niches by 2100.

The $h_{ij}$ (2) values summed up by the rows of the matrix, excluding transitions within "their" zone, allow us to estimate the rates of interzonal transformation of ecological niches of the considered type of landscapes. Thus, for sub-taiga geosystems of type 3/7, such total rates of transition to niches of other zones are, for three forecast periods, 0.404, 0.358, and 0.828, respectively.

According to the matrices of rates of landscape–ecological transitions $h_{ij}$ (2), calculated for all types of landscapes, the transformation of modern landscape areas into areas of future equilibrium states of geosystems was carried out (Figure 11.1). Let us emphasize that the forecast map characterizes not the distribution of landscape areas in a given forecast period but a mosaic of those landscape–ecological conditions that will develop by this date in a given territory and in accordance with which, first of all, shifts in the functioning of landscapes will occur (Kolomyts 2008). That is why the forecast map, being a static forecast model, is called landscape–ecological and not landscape. The primary subject of forecast mapping, shown in Figures 11.1a and b, is the zonal transformation of geosystems, expressed by the sum $h_{ij}$ (2) along the rows of the matrix of climatic niche transition rates.

The forecast maps (see Figure 11.1) show two dynamic characteristics: (1) the prevailing trend of transitions of a niche of a given landscape type from "its" natural zone

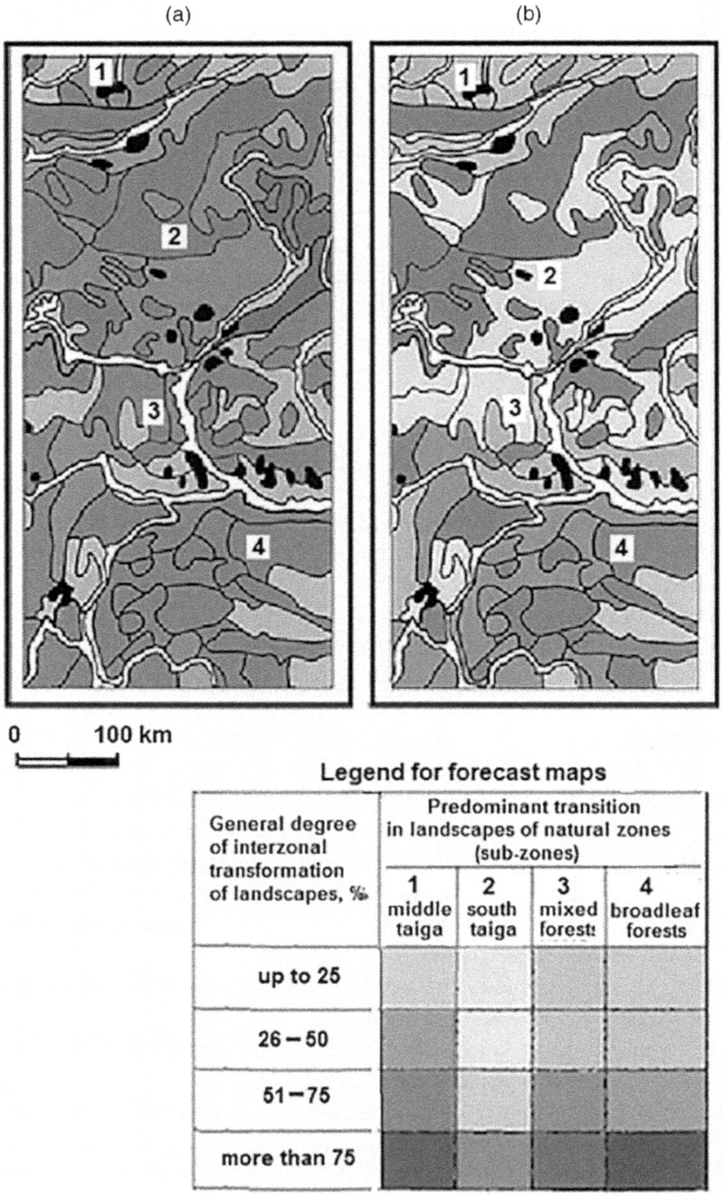


**FIGURE 11.1** Fragments of maps of directions and measures of focal transformation of landscape types in the Volga River basin for the forecast periods of 2050 (*A*) and 2100 (*B*) according to climate scenarios, based on the GISS and E GISS models. (Conventional meanings of landscape groups are shown in Figure 5.2.)

(sub-zone) to a niche of another zone – by the largest sum $h_{ij}$ (2), obtained by the rows of the matrix (Table 11.3); (2) the general degree of interzonal transformation of landscapes of a given type – by the total values $h_{ij}$ (2), taken by the rows and columns of the matrix, minus the sum $h_{ij}$ (2) by the row corresponding to "its" natural zone. Both of these characteristics are reflected in the legend of the forecast maps of landscape–ecological conditions (see Figure 11.1). The following gradations of the degree of total transformation of landscapes are distinguished: weak (less than 25%), average (26%–50%), increased (51%–75%), and high (more than 75%).

Along with this, it was important to take into account the entire spectrum of directions of zonal differentiation of each type of landscape, which is reflected in the legend to the forecast maps. For example, mixed-forest landscapes of type 3/7 during the first step (up to 2050) will acquire a predominantly southern taiga appearance, with a total value of interzonal transformations of their climatic niches within 25%–50% (see Figure 10.1).This trend will continue in subsequent steps with a further growth of the overall measure of transformation.

The legend to the forecast landscape–ecological maps reflects a rather sharp and multidirectional interspecific divergence of the climatic areas of the geosystems. Thus, the southern taiga E-D landscapes of type 2/1 will initially tend to transition to sub-taiga (Table 11.3), but a significant proportion of broadleaf elements is also expected in them. In the depressions and on the northern slopes, conditions will be created for the introduction of boreal middle taiga phytocoenotic groups. Finally, in the phytocoenoses of the tracts of the southern slopes, the appearance of forest–steppe representatives is expected.

In many cases, it is not so much the zonal bioclimatic type (subtype) that will change in landscapes, as the edaphic moisture. An example is the sub-taiga landscapes of type 3/2 in the lower reaches of the Kama and Vyatka (Table 11.4). At all stages of their transformations, these landscapes will remain sub-taiga with interzonal transformation up to 30%–50%.The main changes will consist of increasing edaphic moisture. This trend is favored by the presence of hydro-lithomorphic and sub-hydromorphic tracts in modern landscapes of type 3/2 (they occupy about 10% of the area).It can be expected that in these geosystems, at all predicted stages of the thermo-arid trend, swamping will increase, and the role of boreal vegetation elements will increase.

**TABLE 11.3** Integral measures (in %) of predicted inter-zonal transformation of southern taiga landscapes of type 2/1 (erosion-denudation, eluvial, hydro-lithomorphic, with fir-spruce forests)

| *ABSORBING ZONAL (SUB-ZONAL) TYPES (SUB-TYPES) OF LANDSCAPES* | *PREDICTED TERMS* | | |
|---|---|---|---|
| | *2050* | *2075* | *2100* |
| Middle taiga | 15 | 15 | 18 |
| Sub-taiga | 28 | 39 | 49 |
| Broadleaf forests | 23 | 45 | 53 |
| Middle and south forest-steppe | 12 | 16 | 21 |

**TABLE 11.4** Measure $h_{ij}(2)$ of intergeneric transformation of sub-taiga landscapes of type 3/2 (erosion-denudation, transeluvial, lithomorphic, with broad-leaved-fir-spruce forests), in %

| *LANDSCAPE GENERA SIMULATING ABSORBING STATES* | *PREDICTIVE TERMS* | | |
|---|---|---|---|
| | *2050* | *2075* | *2100* |
| Moranic | 5 | 9 | 11 |
| Denudation-outwash | 18 | 18 | 22 |
| Moranic-outwach | 11 | 12 | 14 |
| Lake-alluvial | 2 | 5 | 6 |
| Outwash | 4 | 4 | 5 |
| Valley-outwash | 1 | 1 | 1 |

## 11.3 ANALYTICAL AND CARTOGRAPHIC ASSESSMENT OF INDUCTIVE LANDSCAPE TRANSFORMATION

Of no less interest is the direct trans-boundary interaction of neighboring geosystems and the resulting possible shifts in natural boundaries as a result of the inductive transformation of phytocoenosis niches. The methodology for these assessments is as follows. The probability matrices of landscape–ecological transitions (see Table 11.1) represent only the first approximation to the real picture, since they are built on the basis of only the spatio-temporal diversity of climatic niches of landscape types and do not take into account the influence of enhancing or, conversely, limiting factors associated with the landscape texture of the region itself. We are talking, first of all, about the types of landscape boundaries and the order of landscape neighborhood (Simonov 1982; Preobrazhensky 1986; Mamay 1992; Rodoman 1999; Viktorov 2006); therefore, the applied methods for calculating the rates of niche change, using finite Markov chains, make it possible to estimate the potentially possible, rather than close to reality, rates of phytocoenotic transformation of natural complexes.

It can be assumed that the elements of the transition probability matrix and all subsequent calculated transition parameters ($M_i$, $P_{ij}$, $m_{ij}$, $h_{ij}$) characterize a certain hypothetical set of averaged rates of transformation of landscape species niches under the following conditions: ubiquity of membrane (semi-permeable) boundaries (by definition (Preobrazhensky 1986)) and the presence of only first-order neighbors for specific binary relations of landscape areas. The last condition is fulfilled in the case when the area of each landscape type is surrounded by only two neighbors, which ensures unambiguous directionality and uniformity of the chain of landscape transitions.

Meanwhile, the landscape texture should significantly refract such a hypothetical system of geosystem transformations. The very type of natural boundary will make adjustments to the process of landscape transformation, making it multifaceted and ambiguous. Connection (functional-contact) boundaries are formed under conditions of the migration type of landscape junctions (Viktorov 2006) and, in general, should facilitate the transformation of contacting geosystems. Barrier boundaries (divergent ridge and convergent trough) will, on the contrary, hinder this (Rodoman 1999). In addition, in the presence of barrier boundaries, the probability of transforming one type of landscape into another should be, apparently, the lower the order of their neighborhood, i.e., the greater the number of intermediate landscape types separating them.

Based on these assumptions, to calculate the inductive component of the potential phytocoenological transformation of landscapes, the author proposed to use the correction "textural coefficient" $C_m(j,i)$, which is equal to

$$C_m(j,i) = \prod_1^n C(j,i) + 1, \tag{11.3}$$

$$C_m(j,i) = \prod_1^n 2P(i,j) / P(i). \tag{11.4}$$

There $C_m(j,i)$ is the paragenetic distance (index) between the $i$-th ($A$) and $j$-th ($B$) landscape types (see above); $P(i,j)$ is the probability of their areas occurring together; $P(i)$ is the a priori probability of the $i$-th landscape type's neighborhood, $\prod$ is the product sign, and $n$ is the order of landscape neighborhood (in this case, the number of boundaries separating the areas of these types). Obviously, for $n=1$, i.e., for a first-order neighborhood, the $i$-th landscape type is $A$, and the $j$-th landscape type is $B$. If $n>1$, then types $A$ and $B$ participate only as factors in the n-th order neighborhood. The product of paragenetic indices (for $n>1$) yields a value similar to the conditional probability, which reflects a sharp drop in the probability of mutual transformation of landscape types separated by several barrier-type boundaries. Adding 1 to $C_m(j,i)$ in the calculation formula allows us to avoid negative values of the texture coefficient. A deviation of the texture coefficient from 1 in either direction means a change from the membrane (intermediate) type of landscape boundary to a connection type or, conversely, to a barrier type. Connection boundaries will correspond to the coefficient $C_m(B,A)>1$, thus increasing the probability of landscape transitions, and barrier boundaries will correspond to the coefficient $C_m(B,A)<1$, reducing this probability.

Calculations have shown that it makes sense to assess the induction component of the transformation of geosystem areas only for first-order landscape neighborhoods, since at the second-order neighborhood the texture coefficient becomes, as a rule, negligible. Thus, for landscapes of type 3/7, there is the following chain of induction transformation into other types, with the corresponding texture coefficients:

$$3/7 \xrightarrow[0.16]{} 2/6 \xrightarrow[0.33]{} 1/9.$$

The final coefficient of trans-boundary landscape transitions of sub-taiga moraine complex spruce forests (3/7) into mid-taiga outwash pine forests (1/9) is 0.16 · 0.33 = 0.05. Such an insignificant value of the coefficient can already be neglected.

Figure 11.2 shows the forecast estimates for 2100 of the induction transformation of landscape areas – first-order neighbors. The measure of the shift of the landscape boundary $h_{ij}$ (3) was calculated (with a cumulative total of three steps) as

$$h_{ij}(3) = C_m(j,i) \cdot h_{ij}. \tag{11.5}$$

The vector of natural boundary shift is indicated by arrows, and the relative potential rates of absorption of one landscape area by another (in %) are indicated by a system of lines of varying thickness passing along this natural boundary. For the first two forecast periods, the pattern of boundary shift is almost identical. For many landscape neighborhoods, changes will be noticeable only by 2100.

## 11.4 BRIEF ANALYSIS OF FORECAST SCENARIOS

The vector of shifting natural boundaries is indicated by arrows, and the relative potential rates of absorption of one landscape area by another (in %) are shown by a system of lines of different thicknesses passing along this natural boundary. For the first two forecast periods, the picture of the shifting boundaries is almost identical. For many landscape neighborhoods, changes will become noticeable only by 2100. According to the GISS model, for the central zone of the Russian Plain in the first decades, a higher temperature increase is expected not in the cold, but in the warm period of the year. By 2050, in most of the Volga basin, the average July temperature will increase by 1.1°C–1.3°C, and in the sub-taiga zone and the sub-zone of broadleaf forests, in places by 1.5°C–1.7°C. The average January temperature will increase much less – generally no more than 0.2°C–0.5°C (in the Upper Volga region, the increase is up to 1.4°C). In 2075, the warming rate will generally remain the same, but the temperature increase will level off and exceed 2.2°C–2.3°C compared to the base period. In the taiga of the Upper Volga region, winter warming will already be predominant (by 3.6°C), but in most areas, summer priority will remain. Finally, by 2100, the increase in the average January temperature will become dominant (up to 4.5°C–5°C in the southern and middle taiga zones and in broadleaf forests).

Anthropogenic warming will be accompanied by an increase in precipitation, but this growth will slow down significantly from north to south. By 2050 and 2100, in the middle and southern taiga, the annual precipitation will increase by 100–120 mm and 200–270 mm, respectively, and in the forest–steppe and northern steppe – by 60–80 mm and 100–150 mm, respectively. As a result, the annual Vysotsky–Ivanov moisture coefficient $F_{\text{hum}}(1)$ for the forest zone of the Russian Plain as a whole will decrease very slightly by 0.05, 0.10, and 0.12 by 2050, 2075, and 2100, respectively, which allows us

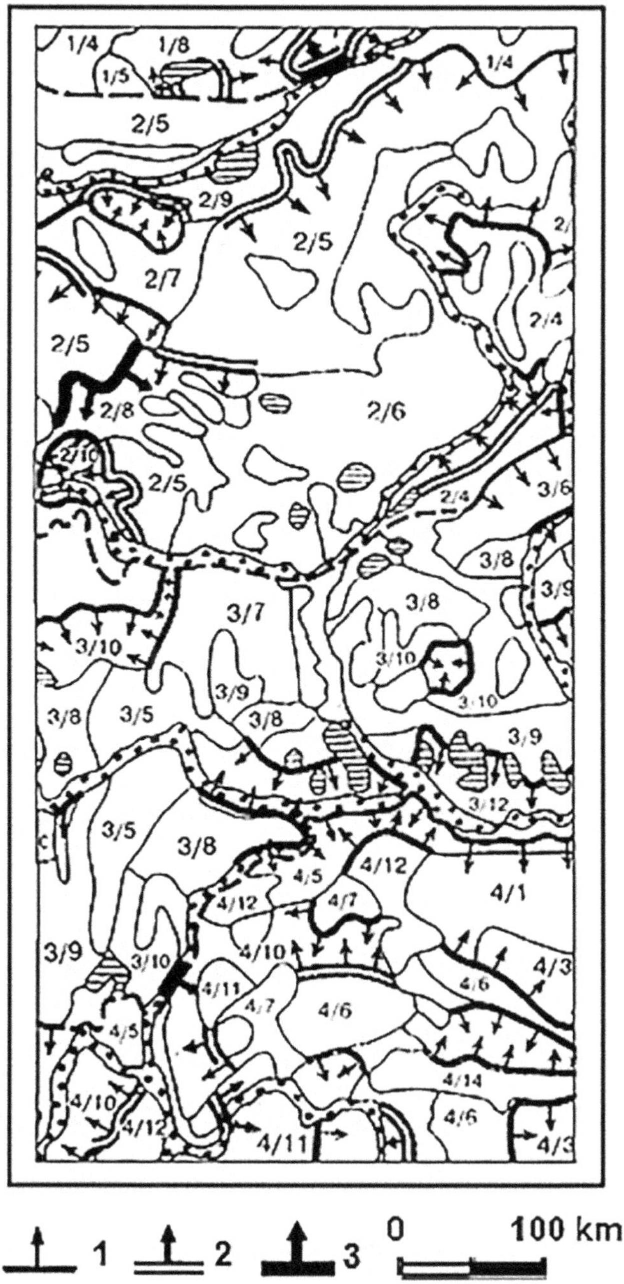


**FIGURE 11.2** Fragment of the map of induction transformation of landscape types in the Volga River basin for the forecast period of 2100, according to the climate scenario of the E GISS model. Measure of shift of landscape boundaries hij(3): 1 – 0.1–0.05; 2 – 0.06–0.104; 3 – 0.11–0.15. Conventional meanings of landscape groups and other objects are shown in Figure 5.2.

to classify this climate trend as thermo-arid. The later EGISS model generally gives a cold–humid trend up to 2050 with a decrease in summer and winter temperatures by 0.5°C–1.9°C and an increase in the humidity coefficient by 0.06–0.25. Only by 2100 will a slight increase in average July temperatures (up to 0.3°C–0.4°C) cause a decrease in the coefficient by 0.02–0.13.

The expected increase in temperature and growth of annual atmospheric humidity will lead to an inevitable expansion of the ecological space in each natural zone, which will cause a corresponding complication of the zonal structures of the region and a general increase in the degree of its ecotonization. Conditions will be created for the shift of many landscape boundaries under the influence of focal and frontal transformations of hydrothermal niches of phytocoenoses (see Figures 11.1 and 11.2). The dominant exogenous successions will be, firstly, the replacement of grassy (steppe and meadow–steppe) associations with woody (forest) ones, and secondly, the replacement of broadleaf forest phytocoenoses with mixed and coniferous forest ones. This is explained by the higher sensitivity of zonal vegetation types of the Russian Plain to an increase in the amount of atmospheric precipitation. Mass transgression of boreal elements into the broadleaf forest sub-zone and even into the forest–steppe zone is one of the phenomena predicted by these models in the boreal ecotone of the Volga River basin, especially in its western part.

Phytocoenoses of northern steppe landscapes will experience the least influence from forest natural zones. Zonal transitions were not found here at any of the forecast stages. Nevertheless, from 2075 coniferous species and taiga elements of the ground cover will begin to penetrate here. Pine forests of sandy landscapes will largely acquire a middle taiga appearance. In the sub-zone of the northern forest–steppe, absorption of broadleaf forests by southern and even middle taiga "aliens" will occur in all landscapes, with the exception of lacustrine–glacial and erosion–denudation. In the latter, on the contrary, steppe and forest–steppe associations will appear.

Mixed forests will be actively absorbed by the southern taiga. In the denudation–outwash, outwash, and lake–alluvial landscapes, with their low substrate trophicity, even the penetration of middle taiga vegetation can be expected. In trans-eluvial and eluvial landscapes, the introduction of forest–steppe phytocoenoses and the strengthening of the positions of broadleaf forests are most likely. In erosion–denudation and lake–alluvial loamy landscapes, southern taiga vegetation will be replaced by broadleaf forest, while middle taiga elements will penetrate landscapes with a sandy substrate. In the middle taiga, the transformation of almost all types of landscapes (except for moraine complexes) into mixed and broadleaf forest, and partly into southern taiga, is expected. Finally, the introduction of broadleaf forest and forest–steppe elements into the boreal zone is likely. By 2050, this will occur in the middle taiga zone, as well as in the southern taiga of the Vyatskiye Uvaly and the Verkhnekamskaya Upland. At the second step, the introduction of nemoral and forest–steppe species and entire associations into the southern taiga and sub-taiga is expected. At the third step, this process will weaken.

The most pronounced transgression of zonal geo-spaces from north to south will manifest itself in the western semi-humid sector of the Volga Basin, and their transgression from south to north will occur in the eastern semi-arid sector. Therefore, the landscape–zonal structure of the boreal ecotone, formed by the end of the forecast period, will have a more pronounced meridional asymmetry than at present.

## 11.5 MAIN INFERENCES

The presented material outlines one of the working versions of the concept of regional landscape–ecological forecasting, developing the known provisions of the theory of biogeographical and landscape ecotones and leading to the construction of analytical and cartographic forecasting models for a large region. Forecast mapping is differentiated by territory, taking into account the properties of the landscape structure and texture – morphogenetic complexes, types and kinds of boundaries, and the direction and strength of paragenetic connections. This made it possible to adjust the rate of shift of the landscape boundaries themselves and to identify the relative time step of transformation of specific landscape areas.

The conducted scientific research aimed to enrich one of the fundamental provisions of complex physical geography – geographic zonality – with regional forecast aspects. The study was conducted using the example of the boreal ecotone of the Volga River basin as a macrosystem with the most pronounced ecological effects of natural interactions.

It is necessary to emphasize the obvious limitations of the mapped and described forecast scenarios. They mainly reflect changes in the structure of the vegetation cover as the most sensitive and dynamic natural component. Meanwhile, the very diverse dynamics of vegetation is composed not only of its exogenous, but also endoeco-genetic successions (Sukachev 1975), which often complicate the identification of environmental consequences of external influences. In this regard, it is advisable to use another, no less important, forecast indicator – soil, as a more inert, but also more stable component of the landscape. As is known (Mamay 1992), the main nodal point in the irreversible transformation of the landscape is the change in the type of soil formation.

The presented scientific and methodological developments can be included in the methodological arsenal of geosystem monitoring by creating regional models of global changes in the natural environment. The results obtained show that regional predictive landscape–ecological analysis opens up one of the promising ways to solve global environmental problems of our time.

# Local Landscape–Ecological Prognoses

# 12

## 12.1 THE IDEOLOGY OF LOCAL ENVIRONMENTAL FORECASTING

Predictions of the ecological consequences of the expected anthropogenic global warming (Budyko 1980; Emanuel et al. 1985; Budyko et al. 1992; Izrael 1992; Izrael 2001; Dahl 1993; Houghton et al. 1996; Aber et al. 2001) pay the least attention to local (topological) responses to these climatic changes. However, the mechanisms of the responses of natural ecosystems to external perturbations are known to be triggered at the level of biogeocoenoses as elementary chorological units of the biosphere (Timofeev-Resovsky and Tyuryukanov 1966). The *Global Changes at the Local Level* concept should be created as a scientific and methodological basis for local geo-ecological monitoring of global changes.

The productive significance of this concept in solving global environmental problems lies in the fact that biogeocoenosis, as an elementary cell of the biogeochemical work of the biosphere (Timofeev-Resovsky 1970), is "the primary apparatus of energy and material exchange" (Sochava 1974, p. 5). Consequently, the initial mechanisms of manifestation of the systemic commonality of various natural components are concentrated at the topological level. Due to their wide distribution, local processes acquire a global character (Kotlyakov 2001).

The following predictive modeling of the main directions and rates of functional and structural changes in local ecosystems is based on the representation of the latter as integral natural–territorial units, following the organismic concept of plant communities of Kostychev–Clements–Tansley (Mirkin et al. 1989; Rozenberg et al. 1999; Rozenberg 2004). This concept allows, firstly, consideration of biogeocoenoses of a particular group as integrated systems with discrete spatial boundaries and their equally


DOI: 10.1201/9781003733119-12

clear changes over time, and secondly, to more definitively judge the stability of plant communities to external disturbances and their sensitivity to climate signals. With a discrete approach to the analysis of the behavior of topo-ecosystems in the space of changes in local hydrothermal fields, a certain framework of the predicted pattern of changes in the functional states of ecosystems is established, which makes it possible to immediately embrace various local variants of such changes within each ecoregion and then proceed to broader territorial generalizations.

The trajectories of climate-genic transformations of local ecosystems themselves can be considered, according to Sukachev (1972), as a biogeocoenotic process. This thesis expounds on the main statements of the original topo-ecological concept of prediction: "Global Changes on the Local Level". Predictions based on one of the latest extreme prognostic models, the Hadley Centre Model HadCM3 (version A2), have been performed using an original method of discrete empirical statistical modeling of ecosystems.

For plain territories of the temperate zone, it has been empirically established that the main channel of connections of regional and local ecosystems with the background climatic system passes through summer soil moisture content, which is a rather reliable geophysical indicator of the state of ecosystems on the one hand, and the most powerful local ecological factor on the other hand (see Chapter 8). Probabilistic scenarios of the expected changes in summer soil moisture content until the mid-22nd century and the respective structural and functional changes in forest ecosystems for different zonal–regional conditions of the Volga River basin are described.

As known, "all the time there is a process of destruction of some relationships and the creation of other relationships. This ultimately results in a continuous restructuring of biogeocoenoses, in the destruction of some … and the creation of new biogeocoenoses" (Sukachev and Dylis 1964, p. 2). At the same time, new types of connections appear, characteristic only of these newly formed integral formations. The local landscape–ecological forecast is inherently experimental in nature, since it is based on a deliberate search for mechanisms for the local display of global changes, with the identification of transfer and transforming functions in landscape connections of various hierarchical levels. At the same time, the forecast scenarios themselves are described by empirical–statistical models, for the construction of which the materials of targeted large-scale landscape–ecological surveys are used (see Section 5.3.4).

When searching for ways to develop the aforementioned concept, we found that, to date, (1) the responses of ecological (hydrothermal) niches of topological–ecological systems (biogeocoenoses and their groups) in a given ecoregion to the behavior of the background climatic system remain unclear, and (2) the transmitting mechanisms of this response have not been studied. None of the available Russian or foreign sources discusses the transformation of global–regional climatic signals to the local level in sufficient detail. The predicted hydrothermal trend is usually described by three parameters: the mean January and July temperatures and the annual precipitation, whose relationships with biogeocoenoses are weak and often statistically insignificant. We attempted to determine the causative mechanisms and consistent patterns of the local response to global–regional climatic signals through finding transmitting functions in the landscape relationships that transform these signals.

The predicted dynamics of forest ecosystems under changing climatic conditions remain a poorly developed field. Existing simulation models of the response of forests to climatic factors (Bonan et al. 1992; Aber et al. 2001; etc.) are only true for narrow ranges of habitat conditions; therefore, the results of this modeling are insufficient for predicting the state of the forest community as a whole and do not encompass the spatial diversity of succession changes under exposure to the same background factor. Our local landscape–ecological prediction is based on discrete empirical statistical models of natural ecosystems, which make it possible to deal with relatively few but most informative characteristics and obtain results that, though not so accurate, offer a higher spatial resolution than simulation modeling.

## 12.2 PREREQUISITES OF RESEARCH

We addressed this section by studying forest and forest–steppe ecosystems of the main catchment area of the Volga River basin (Kolomyts 2009). The conservation and reproduction of forest resources at the southern limit of the boreal zone, where forest communities are approaching a critical state (Kolomyts 2003), is a fundamental and urgent ecological problem.

As already stated (see Section 8.5), for lowland territories, the main channel of relationships between regional and local ecosystems and the background climatic system is the summer soil moisture content, which is a reliable geophysical indicator of the state of an ecosystem, on the one hand, and the strongest ecological factor determining their spatial organization, on the other hand. We have found that the July productive moisture reserves in the soil levels of 0–20 cm (*W*–20), 0–50 cm (*W*–50), and 0–100 cm (*W*–100) are closely related to Vysotsky–Ivanov's atmospheric humidity factor $\_F_{\_hum}(1)$ in the Volga River basin (see Equation 6.1).

We also postulate that the *directional change of ecosystem functioning*, i.e., shifts in the minor ecological cycle (production and decomposition of phytomass), rather than the structural evolution of ecosystems, should be the priority object of landscape–ecological prediction within a century. These shifts occur within the first several years in the taiga zone and within 1 year in the sub-zone of broadleaf forests (Kolomyts 2008). These characteristic times of functional relaxation are comparable with those of exogenous changes in the carbon contents of forest phytomass (live and dead) and soil mobile humus (Osipov 2004).

Predictive interpretation of hydrothermal ordination of different parameters of the minor biological turnover is uncertain in its extreme indices. In the polygon, the ecological gradients of thermo- and hydro-edaphotopes are not high enough to describe all possible future states of edaphotopes by the example of initial states. The xeromorphic and arid branches of the thermo- and hydro-edaphic series are particularly insufficiently represented. Hence, extrapolation is inevitable: a hypothetical stretch of the available sampling towards more arid zonal ecosystems, which makes this part of the prediction rather "fuzzy".

The range of the greatest scattering of the mean summer soil (and air) temperatures between the extreme facies groups in their topological order is about 40. Such an interval almost completely corresponds to the maximal deviations of July temperature according to GISS for the end of the 21st century, but for HadCM3 it is adequate to deviations only for the period until 2050. Environmental prediction by this model for the second half of the century is entirely of extrapolation character.

## 12.3 EMPIRICAL MATERIALS

For prognostic analysis, we used the results of large-scale landscape–ecological surveys that we performed according to a specially developed technique from 1987 to 1998 in eight test sites in the following areas of the middle Volga River basin (see Section 5.3 and Figures 5.1 and 5.4). The test sites covered a wide spectrum of transitional zones from mixed forests to northern steppes in the Russian Plain. As already mentioned (see Section 2.6), we termed this spectrum the *boreal ecotone*, a planetary-scale zonal boundary. Each site characterized a specific *regional ecosystem*. Hereinafter, we will refer to them as Kerzhenets, Vyksa, Shchelokovsky Farmstead, Green Town, Kud'ma, By-Oka-Terrace Reserve, By-Sura, and Zhiguli.

Prognostic estimation was performed in the system of two coordinates: the levels of basal functioning (or the location and habitat types) and zonal–geographical groups of biogeocoenoses (see Figures 12.1–12.3). The measures of the inclusion relationships (see Supplement) of 23 parameters of the functional phytocoenotic unit of topological ecological systems were used to determine their *ecological groups* (*ecotypes*) corresponding to different functional levels, including both phytomass production and phytomass decomposition (see Section 5.4). The functional states of forest phytocoenoses are rather contrasting in each ecoregion, which should result in diversity of their responses to the same background hydrothermal signals.

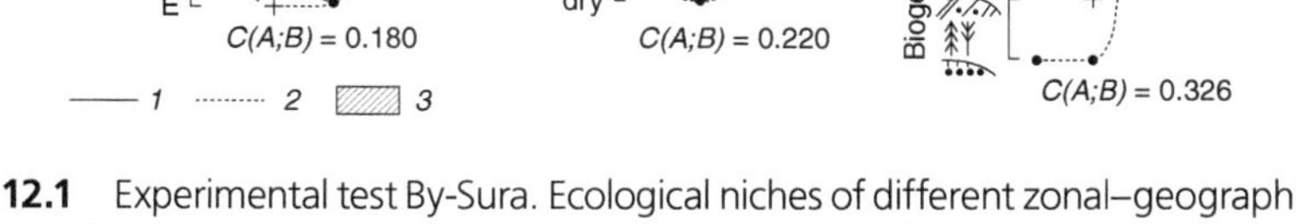


**FIGURE 12.1** Experimental test By-Sura. Ecological niches of different zonal–geographic groups of coniferous forest ecosystems in relation to abiotic factors and the groups of facies themselves. Zonal–geographical groups: 1 – boreal taiga; 2 – boreal pine forest; 3 – mixed boreal–nemoral; 4 – nemoral Eurasian; 5 – nemoral European. Trajectories of ***1*** – ecological dominants; ***2*** – ecological niche boundaries. *C*(*A;B*) – values of the normalized coefficient of interrelation (see Equation 4.7 in Chapter 4). For other conventional meanings, see the text (Chapter 4).

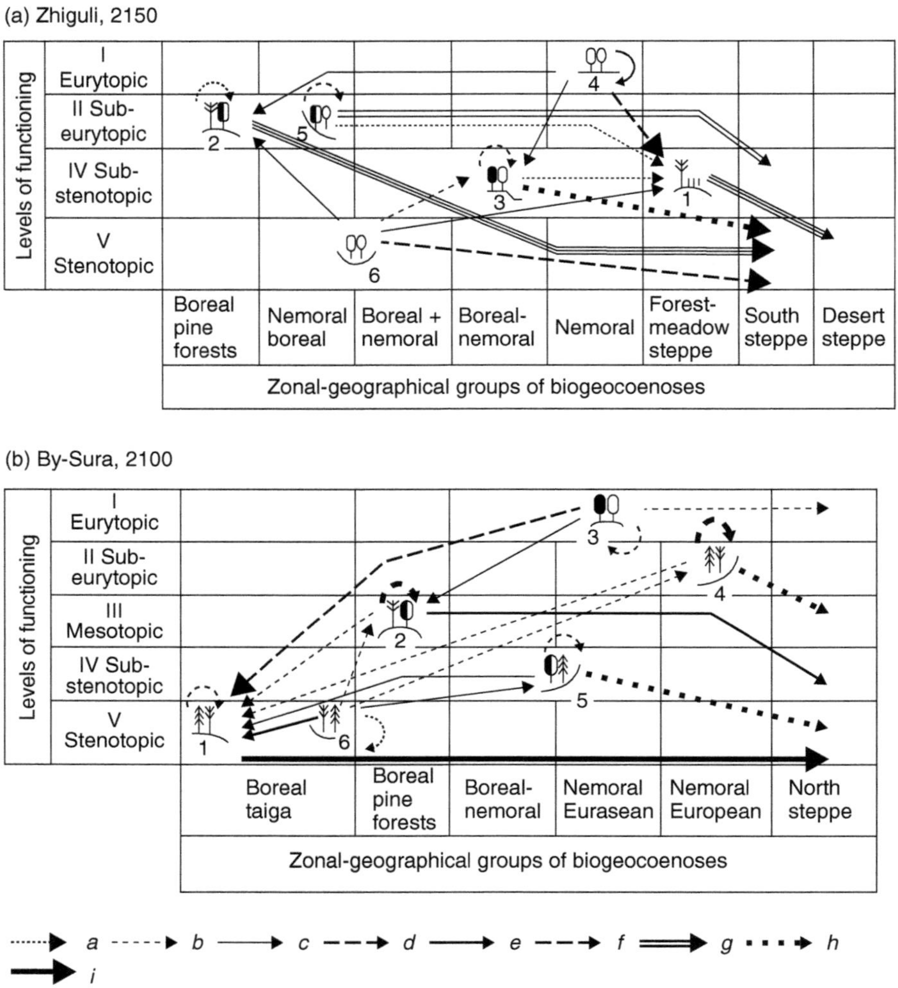


**FIGURE 12.2** The vector graphs of functional transitions between facies groups: *a* – on the zonal forest and steppe ecotone (the Zhiguli low-mountain range); *b* – on the sub-taiga–forest–steppe ecotone (high right-bank By-Sura), according to the HadCM3 model. Probabilities of interface transitions: *a* – 0.10 and less; *b* – 0.11–0.20; *c* – 0.21–0.30; *d* – 0.31–0.40; *e* – 0.41–0.50; *f* – 0.51–0.60; *g* – 0.61–0.70; *h* – 0.71–0.80; *i* – 0.81–0.90. 1–6 – groups of landscape faces (see Table 5.2). Levels of biogeocoenosis functioning according to the classification of Odum (1971).

The zonal–geographical differentiation of topological–ecological systems is also distinct. Local geomorphological and edaphic factors modify the background biological and climatic conditions and form, in addition to terrace ecosystems (Vysotskiy 1960) reflecting the zonal–regional background of different parts of the boreal ecotone, many extra-zonal categories (Sprygin 1986) as representatives of other zonal types of the geographical environment, both neighboring and remote. This is the essence of the

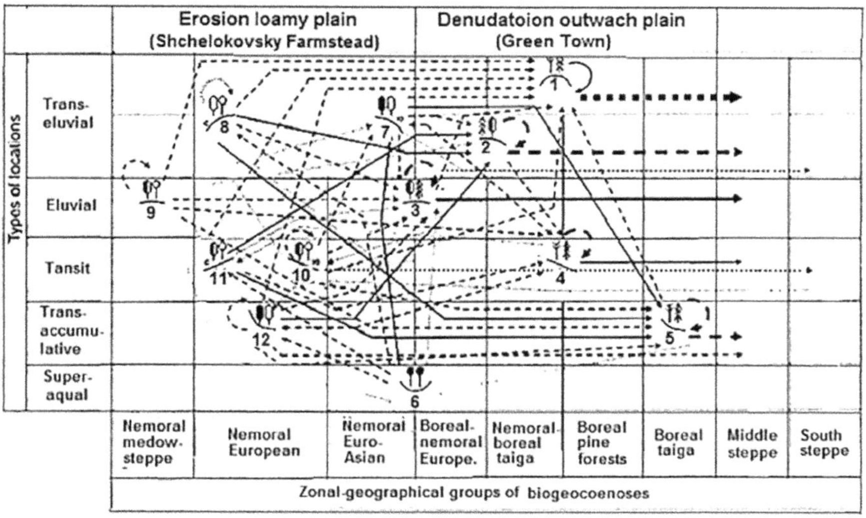


**FIGURE 12.3** The vector graphs of functional transitions between facies groups on the boreal–nemoral forest ecotone – coniferous, mixed, and broadleaf forests of the high-plain By-Volga Region for the prediction period up to 2100, according to the HadCM3 model. Conventional meanings are as in Figure 12.2.

multi-zonal biological and climatic character of topological–ecological systems as a form of their response to global climatic change (see Chapter 6). *Regional systems of local zonality* are formed; they imitate the main directions and scales of local ecosystem rearrangements, thereby providing the empirical basis for quantitative predictions and correction of their results.

## 12.4 POLYZONALITY OF LOCAL ECOSYSTEMS AS A WAY OF THEIR RESPONSE TO CLIMATE SIGNALS

It is of great significance for the knowledge of local mechanisms of global–regional changes in the environment to disclose the patterns of refraction of background bioclimatic conditions by local geomorphological and soil–edaphic factors and the formation of certain categories of topo-geo(eco)systems as representatives, in a sense, of different types of zonal geographical environments. This is particularly important for understanding the role of local landscape–ecological structures in the creation of *regional systems of localized natural zonality* (see above Section 9.5). From their part, these regional–topological systems, being objects of the effect of background climatic signals, may simulate the main directions and scales of local ecosystem reconstructions, thus making an empirical basis for prognostic constructions. Moreover, it seems practicable to assess the extent to which the structure–function transformations of local ecosystems and the corresponding shifts of natural boundaries are able to reorganize the whole zonal–regional background of a given territory under the effect of this or that climatic trend.

In light of the forthcoming and already beginning global warming, along with the progressing development of soil–plant cover aridization in the temperate latitudes of continents (Albritton et al. 2003; Kolomyts 2005), the problem under consideration is particularly urgent. It is necessary to assess the state of forest and forest–steppe ecosystems and predict their approaching changes in order to develop an ecological basis for the preservation and reproduction of forest resources and to arrange a network of conservation and recreation territories (nature reserves, national parks), which is especially important for regions with critical conditions for the existence of forest plants, including the zone of contact between forest and steppe.

Let us note that the landscape mosaic of any territory is known to comprise the areas of flat-interfluve local ecosystems, which reflect the zonal–regional background of the territory in the "purest" form and are its representatives, as well as so-called extra-zonal natural complexes formed as a result of various refractions of this background under the influence of local geomorphological and hydro-edaphic factors. Within any ecoregion, we can construct a hydrothermal series of extra-zonal topo-ecosystems (along with zonal representatives) leading to their humidity or, conversely, aridity. This series may be presented (on the basis of the self-similarity of operational systems) as a certain analog of the corresponding background climatic changes (trend). Thus, using the vector spectrum of *topological polyzonality*, it is possible to present realistic scenarios of local ecosystems' responses to various shifts in the regional climatic system and to outline the corresponding chains of local landscape–ecological transitions.

It would be advisable to examine the local systems of natural zonality using the example of zonal ecotones in the south of the boreal belt, particularly the forest–steppe ecotone as a band of interaction and interpenetration of the most contrasting bioclimatic media. The experimental test site, By-Sura (see Section 5.3.5), has been chosen to address the given problem. The ordination of facial groups was conducted using two categories: biogeocoenotic groups and their zonal–geographical groups.

The phytocoenotic and, accordingly, landscape diversity of Prisurye is determined by the above peculiarities of its ecotone geographical position: (1) in the transition band between the sub-zones of broadleaf forests and typical forest–steppe; (2) in immediate proximity to the meridian border of continentality, which separates the more humid Privolzhskaya (By-Volga) Upland from the less humid Lowland Zavozhye (Trans-Volga region), forming two contrasting geobotanic provinces (Lavrenko and Isachenko 1976); and (3) at the turn of a vast lowland Prisurye outwash plain – the conductor of taiga plants shifting southwards – and loamy upland watersheds of the Volga and Sura Rivers of nemoral–forest appearance.

The confinement of taiga landscape representatives (pine–spruce sphagnum–long-moss forests) almost solely to accumulative, eluvial–accumulative, and super-aquatic sites, with sod-podzolic, deeply podzolized loamy soils, points to a clearly extra-zonal character of their presence in this region of the Privozhskaya Upland (Figure 12.1). This is also indicated by the hygrophyte nature of their edaphotops. The drier spruce–pine blackberry–green moss forests, with elements of Eurasian herbage, ascend the surrounding slopes from valley bottoms. At the same time, pine–forest phytocenotic complexes occurring merely in a lowland–outwash region relate to xerophytes and meso-xerophytes and gravitate towards the opposite "pole" of landscape connections: trans-eluvial and eluvial types of location, with sod-shallow-podzol sabulous and sandy soils.

Gently sloping interfluves are occupied by (1) meso-hygrophyte communities of the sub-taiga zonal group (ecological optimum) in the lowland–outwash region and (2) mesophyte oak–lime forests with European-type herbage in the upland–watershed region. Finally, stably mesophyte nemoral–grass spruce–lime oak forests and their derivatives, pine–small-leaf associations, are widespread on the steeper loamy transit slopes facing river valleys. All of them have the appearance of nemoral Eurasian forests (Table 12.1).

**TABLE 12.1** High right-bank By-Sura region. Standardized partial association coefficients $C(x_i/y_j)$ and corresponding ecological optimums (•) and secondary ("diffused") parts of ecologic niches (×) for groups of biogeocoenosis in relation to their zonal–geographical groups

| | | | | | *C(A;B)=0.326* |
|---|---|---|---|---|---|
| *GROUPS OF BIOGEOCOENOSES* | *ZONAL–GEOGRAPHICAL GROUPS OF BIOGEOCOENOSIS* | | | | |
| | *BOREAL TAIGA* | *BOREAL PINE FOREST* | *BOREAL–NEMORAL* | *NEMORAL EURO-ASIAN* | *NEMORAL EUROPEAN* |
| 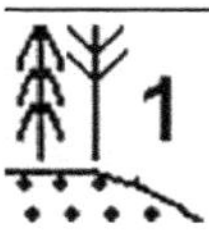 1 | 0.36 × | 0.18 × | 0.12 | | 0.09 |
|  2 | | **0.82** | | | |
| 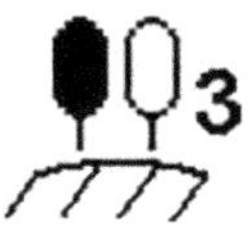 3 | | | 0.33 • | 0.26 × | 0.34 × |
|  4 | | | 0.18 | 0.59 • | 0.14 |
|  5 | | | 0.37 • | 0.15 | 0.42 • |
|  6 | 0.64 • | | | | |

⟶ *Direction of the thermo-arid climatic trend*

*Note:* For conventional meanings of biogeocoenosis groups, see Table 5.3 in Chapter 5. The optimum values of parameter $C(x_i/y_j)$ are showed by half-rich print. Bold dots indicate ecological optima.

Thus, there is no single trend of zonality in the system of local landscape connections, which results in an explicit "overlapping" in the distribution of forest communities of different zonal appearances by the forms of microrelief. This in itself must predetermine a rather diverse picture of phytocoenotic transformations of topo-ecosystems and the shifting of their boundaries at one and the same background hydrothermal signal. Climatogenic transformation of ecosystems will be of multi-vector character; therefore, there may be concurrent shifts of interfacial boundaries both up and down the chain of topological landscape interfaces.

## 12.5 THE MAIN CHARACTERISTICS OF CLIMATE PREDICTION

Hydrothermal trends for the period until the year 2150 were determined using the HadCM3 model (version A2). This is one of the most extreme models predicting the hydrothermal conditions that may go beyond the limits of tolerance of the main forest-forming species in the boreal ecotone of the Russian Plain. We repeat that we divided the prediction period into five stages (steps): 1985–2025–2050–2075–2100–2150. The period of instrumental meteorological surveys from 1881 to 1985, at the end of which the current global climate warming began (Houghton et al. 1996), served as the basic period.

The HadCM3 model predicts a 5.5°C–7.0°C rise in the mean July temperature in the Middle Volga region by the late 21st century (Table 12.2). The annual precipitation will change insignificantly; hence, the atmospheric moisture coefficients will decrease throughout the region. The $F_{hum}(1)$ is expected to decrease significantly during the

**TABLE 12.2** Prediction of values of main climatic parameters and corresponding landscape–zonal conditions of Middle and Upper Volga regions, according to prognosis climate model HadCM3

| *CLIMATIC PARAMETERS* | *BASIC PERIOD* | *PROGNOSTIC DATES* | | | | |
|---|---|---|---|---|---|---|
| | | *2025* | *2050* | *2075* | *2100* | *2150* |
| a. Zonal ecotone of forest and steppe (Zhiguli low-mountain range) | | | | | | |
| $t_{Jan}$ | –4.0 | –11.2 | –8,1 | –6.5 | –8.0 | –7.0 |
| $t_{July}$ | 20.1 | 23.3 | 24,3 | 25.6 | 27.2 | 29.0 |
| $r_{year}$ | 629 | 591 | 622 | 682 | 579 | 650 |
| $E_0$ | 659 | 1009 | 1115 | 1340 | 1609 | 1950 |
| $F_{hum}(1)$ | 0.95 | 0.59 | 0,56 | 0.51 | 0.36 | 0.33 |
| Zonal belonging | Typical forest–steppe, at its north boundary | Middle steppe, at its northern boundary | Middle steppe | Boundary between middle and south steppe | South steppe | At southern boundary of south steppe |

*(Continued)*

**TABLE 12.2 (*Continued*)** Prediction of values of main climatic parameters and corresponding landscape–zonal conditions of Middle and Upper Volga regions, according to prognosis climate model HadCM3

| *CLIMATIC PARAMETERS* | *BASIC PERIOD* | *PROGNOSTIC DATES* | | | | |
|---|---|---|---|---|---|---|
| | | *2025* | *2050* | *2075* | *2100* | *2150* |
| *B*. Boreal–nemoral forest ecotone, high plain of the By-Volga region (Green Town) | | | | | | |
| $t_{Jan}$ | –11.8 | –8.6 | –6.7 | –4.4 | –4.8 | –4.0 |
| $t_{July}$ | 18.2 | 21.5 | 22.5 | 23.5 | 24.4 | 26.0 |
| $r_{year}$ | 660 | 594 | 667 | 705 | 658 | 650 |
| $E_0$ | 511 | 796 | 910 | 1035 | 1159 | 1400 |
| $F_{hum}(1)$ | 1.29 | 0.75 | 0.73 | 0.68 | 0.56 | 0.46 |
| Zonal belonging | Mixed forests, at its south boundary | North steppe, at its northern boundary | | North steppe | Middle steppe | South steppe, at its northern boundary |
| c. South strip of sub-taiga zone, the moraine–outwash forested lowland of Trans-Volga region (Kerzhensky biosphere reserve) | | | | | | |
| $t_{Jan}$ | –2.4 | –9.3 | –7.8 | –5.2 | –5.3 | –5.0 |
| $t_{July}$ | 18.1 | 20.3 | 22.3 | 22.2 | 23.7 | 25.0 |
| $r_{year}$ | 685 | 609 | 687 | 725 | 694 | 700 |
| $E_0$ | 500 | 921 | 983 | 1,131 | 1,160 | 1,250 |
| $F_{hum}(1)$ | 1.37 | 0.66 | 0.70 | 0.64 | 0.60 | 0.56 |
| Zonal belonging | South region of mixed forests | | North steppe | | Middle steppe | Middle steppe, at south boundary |
| *d*) Middle left-bank lowland region of By-Oka River, south boundary of sub-taiga zone (By-Oka-Terrace biosphere reserve) | | | | | | |
| $t_{jan}$ | –10.6 | –7.9 | –7.4 | –5.6 | –2.7 | –4.1 |
| $t_{july}$ | 17.5 | 19.7 | 21.5 | 21.6 | 23.6 | 22.6 |
| $r_{year}$ | 712 | 700 | 616 | 651 | 730 | 681 |
| $E_0$ | 468 | 624 | 796 | 807 | 1,048 | 921 |
| $F_{hum}(1)$ | 1.52 | 1.12 | 0.77 | 0.81 | 0.70 | 0.74 |
| Zonal belonging | Mixed forests (in its middle strip) | Broadleaf forests | South forest–steppe | | North steppe | |

*Note:* Average temperature: $t_{Jan}$ – January, $t_{July}$ – July; $r_{year}$ – total of precipitation per year, mm; $E_0$ – annual evaporativity; $F_{hum}(1)$ – Vysotsky–Ivanov's annual atmospheric humidify factor.

century from 0.95–1.07 to 0.36–0.45 in the forest–steppes of the Zhiguli and Sura river basins and from 1.37 to 0.65 in the left-bank Volga–Kerzhenets sub-taiga region (see Table 12.2). The climatic conditions of the entire southern boreal zone of the Russian

Plain will become sub-boreal. Accordingly, progressive drying of the soil in summer will begin. According to our estimation, an increase in annual precipitation of at least 1,000 mm will be necessary to compensate for a 7°C rise in the mean July temperature in the Samara part of the Volga River basin; none of the AOGCMs predicts such an increase in precipitation in this region (Houghton et al. 1996). Thus, there will be a *thermal arid climatic trend* throughout the boreal ecotone of the Volga River basin until the mid-22nd century. The expected climate warming will cause the general *aridization of regional biological climatic systems*, which will gradually enhance over time. The general picture of the predicted changes in summer productive moisture reserves in the soil in the Middle Volga Region is described in Section 6.5.

Meteorological data were used to estimate the representativeness of soil hydrothermal parameters obtained utilizing a single landscape–ecological survey of the experimental sites. The soil moisture reserves measured in the test sites were normalized to the mean July moisture reserve averaged over many years, which served as a basic reference value.

## 12.6 PREDICTION PROCEDURE

The algorithm of ecological prognostic calculations had four stages.

I. *Preliminary estimation of zonal–regional relationships of the July productive soil moisture reserves with the atmospheric annual humidity factor.* For this purpose, we used data from 120 agro-meteorological stations of the Volga River basin and the neighboring regions, as well as five water balance stations (for the transition from agricultural fields to forested areas). For example, the relationships of the July moisture reserves with $F_{\text{hum}}(1)$ in a 1-m layer of field soil in the Zhiguli and By-Sura regional ecosystems were as follows:
   a. on sands and sandy loams,

$$(W-100)=\exp\left[3.451+0.907\ F_{hum}(1)\right],\ R=0.83;\ R^2=0.69; \tag{12.1}$$

   b. on medium loams,

$$(W-100)=\exp\left[2.821+0.401\ F_{hum}(1)\right],\ R=0.93;\ R^2=0.86; \tag{12.2}$$

   c. on clayey loam,

$$(W-100)=\exp\left[2.67+1.59\ F_{hum}(1)\right],\ R=0.88;\ R^2=0.78. \tag{12.3}$$

The following equations were used for the transition from field ($W_{\text{field}}$) to forest ($W_{\text{forest}}$) plots:

1. for Zhiguli,

$$W_{\text{forest}}=0.991W_{\text{field}}+7.2,\ R=0.98,\ \ R^2=0.97; \tag{12.4}$$

2. for the Sura River basin,

$$W_{forest} = 1.01W_{field} + 7.7, \quad R = 0.92, \quad R^2 = 0.85. \tag{12.5}$$

II. *Prognostic calculations of the changes in summer soil moisture reserves in different topological–ecological systems at the end of each of the five stages.* The calculations were based on the fact that the summer moisture content of a 1 m soil layer in forest and forest–steppe ecosystems of the regions studied was no higher than 150 and 100 mm, respectively; i.e., it is always lower than the critical moisture content according to Budyko (1971). Under these conditions, evaporation of soil moisture is determined not only by the evaporating capacity but also by the amount of soil moisture itself. Therefore, with the same decrease in $F_{hum}(1)$, the moisture reserves of different biogeocoenoses should decrease in proportion to their initial moisture content at the given stage of prediction: the higher this initial level, the greater the absolute loss of soil moisture by the end of the stage. Only the relative decrease in moisture reserves is the same for all facies. The validity of this approach is confirmed by the results of long-term biogeocoenological studies in pine forests of the southern forest–steppe zone (Osipov 2004).

III. *Estimation of the critical levels of summer soil moisture reserves* for transitions between zonal types of vegetation: from mixed forest to broadleaf ones and from these two types of forests to forest–steppe formations, i.e., to a system of island forests and meadow steppes. For this purpose, we used the _*local edaphic humidity factors* $F_{hum}(2)$ whose critical limits are shown in Table 7.5.

IV. *Local landscape–ecological prediction*: estimation of the probabilities and rates of mutual transitions in a system of biogeocoenoses according to the previously determined trends of climate-induced changes in the July soil moisture reserves. Graphical–analytical models (Figures 12.2–12.4) show the probabilities of landscape–ecological transitions. By transition, we mean a change in the functional and, then, structural state of an object such that the object assumes (at a certain probability) characteristics of other prototype objects because, as its ecological (in the given case, hydro-edaphic) niche changes, it increasingly overlaps the niches of these objects.

The forecasting procedure itself is detailed in Chapter 9. Only two main characteristics of local prediction should be noted here. First, the moisture reserves of the 0–20, 0–50, and 0–100 cm soil levels, with the corresponding weight coefficients expected for each period (see above), serve as predictors of prognostic calculations. Second, the probability of "residual" transitions of an object into prototype objects that are absent in the given operational system (sample) is taken into account. For marginal forest ecosystems of the Zhiguli ecosystem, these transitions are dominant or even the only possible ones (Figure 12.2). Square matrix probabilities of transitions of biogeocoenosis groups into one another and the probabilities of stabilization of the groups (the diagonal elements of the matrices) were the main results of prognostic calculations at each step. The latter parameter is regarded as a measure of the resistance of biogeocoenoses to a given hydrothermal signal. These matrices were used to construct the or-graphs of transition probabilities (see Figures 12.2–12.4), describing the overall pattern of the response of forest topological–ecological systems of a given ecoregion to global climatic changes.

# 12.7 PROGNOSTIC SCENARIOS OF TRANSFORMATION OF FOREST ECOSYSTEMS

## 12.7.1 Changes in Soil Moisture Content

Prediction of changes in the state of hydro-edaphotopes and their ecological consequences has been described in Section 7.5. First of all, note that the range of summer soil moisture content between the first and second critical values, i.e., between the northern and southern limits of the nemoral sub-zone, is relatively narrow (see Table 7.5). This means that, in the case of a strong thermal arid trend, steppification and subsequent breakup of the forest cover may spread over broadleaf and mixed forests to equal degrees, leaving in their place not only typical small-leaf oak forest steppes, but also typical conifer forest steppes. In general, considerable aridization of soil is expected in forest ecosystems of the Middle Volga Region. The predicted changes in summer water content suggest inevitable deterioration of site conditions throughout the transitional zone between forests and steppes in the Russian Plain in the course of global climate warming.

The HadCM3 model is used to predict very high tempos of aridization of soils and forest ecosystems (see Section 9.6 and Figure 9.1). The predicted thermo-arid trend will initiate an unequivocal and very significant reduction of productive soil moisture resources in the middle of the vegetation period across the whole territory of the Volga River basin. Soil drying during global warming will cause the transitions of hydro-edaphotopes through second and third critical levels (see Table 7.5) for the majority of forest communities across the whole spectrum of natural zones of the boreal ecotone – from mixed forests to southern forest–steppe.

It will be caused by quite a number of reasons: reduction of fore-spring snow resources, the earlier disappearance of snow cover, and finally, which is most important, increased evapotranspiration at a decreasing $F_{\text{hum}}(1)$ (see Table 12.2). At the third prognostic stage (2050–2075), the average multi-year resource of productive moisture in soil will go beyond the lower level of its modern year-to-year variation. By the end of the 21st century, the July moisture resource in all middle- and south-taiga forests will be 5–7 times less than the normal (field) moisture capacity (currently, this ratio does not exceed 1.5–1.7). In the zone of mixed forests, this resource will fall to the same level. Due to the deficiency of summer moisture content, sod-podzol soils will considerably approach the modern condition of not only gray forest soils of shadowy broadleaf forests but also leached and podzolized chernozems of typical forest–steppe. In broadleaf formations, the drying out of gray forest soils will bring them to a condition similar to modern meadow–chernozem soils of the southern forest–steppe, as well as typical and common chernozems of the northern steppe.

### 12.7.2 Climate-Genic Dynamic of Forest Biogeocoenoses

A considerable initial "saltation" of changes in the state of forest ecosystems due to a drastic decrease in summer soil moisture content is expected across the entire spectrum of natural zones and sub-zones of the boreal ecotone as early as the first prediction step. The functional stability of boreal and nemoral forest communities may be substantially upset. The increase in the intensity of landscape–ecological transitions will be accompanied by an increase in selectivity: the functional shift of an ecosystem will be not only increasingly great (compared to the basic state) but also increasingly definite.

In the southern forest–steppe, specifically in marginal forests of the zonal ecotone of forests and steppes (the Zhiguli ecosystem), biogeocoenoses of two relatively moist "poles" of landscape systems should undergo the most diverse changes: first, terrace mesomorphic oak–linden forests (facies group 4); second, mesomorphic–hydromorphic deciduous forests of the bottoms of small valleys, as well as the adjacent oak–linden and aspen forests of the lower and middle parts of non-insolated slopes (groups 3 and 5; Figures 12.2a). In the first half of the 22nd century, 40% of terrace oak forests of the Zhiguli region will already have turned into xerophytic forest–meadow systems, which are currently confined to steep insolated slopes.

The thermo-arid signal will mainly affect the state of Zhiguli forest–meadow–steppe systems (facies group 1), 90% of which may be transformed into desertified steppes as early as the mid-21st century. By that time, watershed pine–broadleaf forests (group 2) will have reached the state of the southern (dry) steppe.

In the Sura River basin moraine–outwash plains, where the pleiad of boreal formations penetrates deep into the nemoral sub-zone and reaches the boundary with the typical forest–steppe, pine forests with a spruce admixture and pine–broadleaf forests (facies groups 1 and 2) will begin a rapid transformation, first into southern forest–steppe and then into the northern steppe.

In turn, forest communities of this pleiad will serve as the prototypes of the future states of almost all other facies groups, including terrace oak–linden forests of the neighboring elevated loamy plains (Figure 12.2b). In general, the widest spectrum of functional transformation may be expected for shaded broadleaf forests of elevated loamy plains in the northern half of the By-Volga Upland. There are several possible paths for their transition, with a relatively low probability (no higher than 30%–40%), to xerophytic–mesophytic forest communities, which are currently common in neighboring sandy watersheds.

Coupled prognostic analysis of a two-level (regional–local) system of landscape systems in the right-bank part of the Nizhny Novgorod region (Figure 12.3) has demonstrated a consistent trend of functional transitions of forest ecosystems from the upper geomorphological level (erosional loamy uplands) to the lower one (accumulative moraine–washout lowlands). Conversely, landscape–ecological transitions within each of these levels are directed upwards in the system of local catenae, from trans-accumulative locations to eluvial and, especially, trans-eluvial ones, i.e., in the order of increasing aridity of biogeocoenoses.

In general, the aridization of broadleaf forests in the course of global warming should progressively accelerate in the following exedynamic succession series: purely oak forests and spruce–oak forests, pine forests and spruce–pine mixed forests, typical coniferous and small-leaf forest–steppes, northern and middle steppes, southern (arid) steppes.

Note that the steppification of oak forests always includes an intermediate stage of boreal, mainly pine, forest and open forest associations. The subsequent decomposition of forest communities and their replacement by meadow–steppe and, later, steppe associations will be more rapid. In the sub-taiga depressed moraine–outwash plains on the left bank of the Volga River (the Kerzhenets ecosystem), coniferous forests of the first links of catenae, namely, xerophytic and mesophytic–xerophytic pine forests and spruce–pine forests (facies groups 1 and 2) will be the first to turn into steppes (Figure 12.4). Under these local conditions, thermo-aridization of forest communities

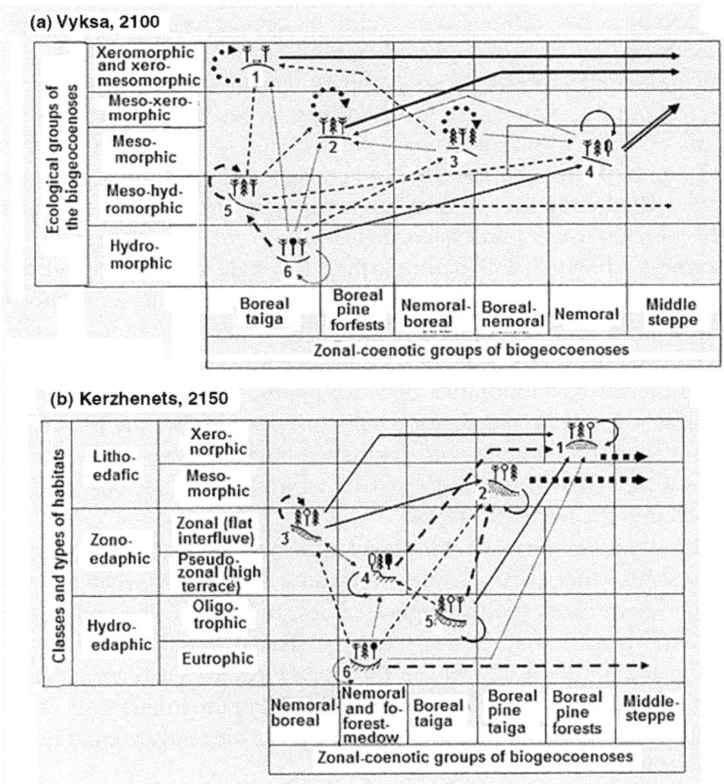


**FIGURE 12.4** The vector graphs of functional transitions between facies groups of the outwash forested lowlands in the southern regions of the mixed forest sub-zone, according to the HadCM3 model. Conventional meanings are as in Figure 12.2.

will be so rapid that they will pass the nemoral stage of transformation soon after the year 2050, and almost 60%–70% of them will become northern meadow steppes by the mid-22nd century. In turn, terrace coniferous and mixed forests (groups 3 and 4) will begin to assume the basic xerophytic form of these taiga pine forest communities, with a transition probability as high as 50% by the end of the second step of the prediction.

The mixed forest zone differs from the forest–steppe zone in the higher resistance of forest biogeocoenoses to global warming. However, according to the extreme forecasting and climatic model HadCM3, near the southern borders of the sub-taiga zone (the By-Oka-Terrace biosphere reserve), one can expect that a thermo-arid climate trend will begin to develop in this ecoregion ($F_{hum}(1) = 1.52 \rightarrow 0.70$, see Table 12.2d). In the future, by the end of the 21st century, pine forests with spruce and pine–broadleaf forests will begin to transform, first into the typical, and then into the southern forest–steppe (Figure 12.5). At the same time, the swampy *A* and CA coniferous small-leaf forests will be replaced by meso-hydromorphic *E* and TE spruce and pine forests. These, in turn, will begin to transform at the same rate into xeromorphic birch–pine forests in similar locations. In the areas of the latter, there will be a significant introduction of the already nemoral northern forest–steppe communities.

Coniferous (pine) forest–steppes, as a new (for the Russian Plain) zonal ecotone of direct contact between boreal and steppe plant formations, with a mosaic pattern of light coniferous and oak–small-leaf park forests, meadows, and steppes, will eventually

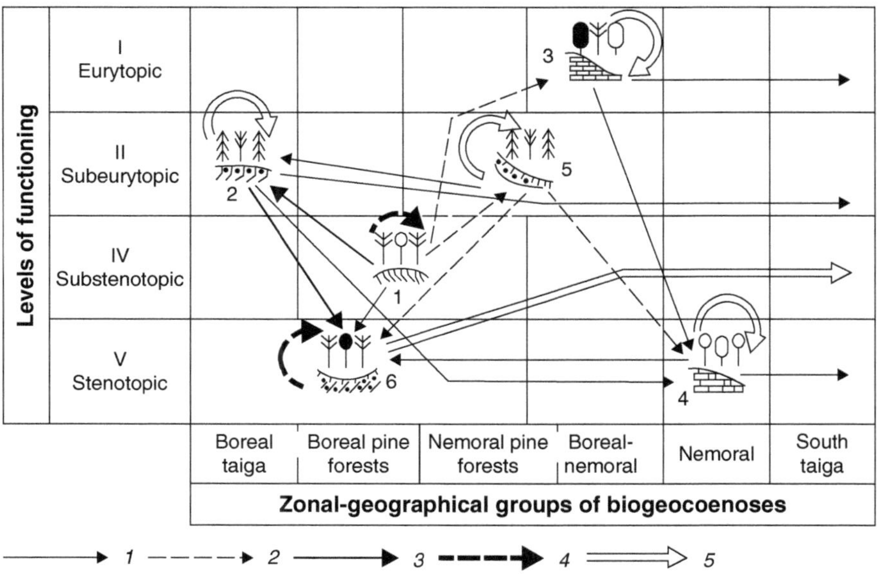


**FIGURE 12.5.** Ecoregion of By-Oka-Terrace Reserve. The cold–arid climatic trend. The oriented graphs of functional landscape–ecological transitions, according to the forecast climate model E GISS, 2050, between groups of facies (biogeocoenoses) with different zonal affiliations and at different levels of basic functioning. Transition probabilities: (*1*) 0.10 or less; (*2*) 0.11–0.20; (*3*) 0.21–0.30; (*4*) 0.31–0.40; (*5*) 0.61–0.70. The south taiga group of biogeocoenoses is an off-sample (extra-zonal).

form in the area that is currently occupied by the sub-zone of broadleaf forests and the southern part of the sub-taiga zone of the Middle Volga Region. Apparently, this will be the eastern variant of the overall "savannization" of broadleaf and mixed forests in the Russian Plain during global climate warming.

## 12.8 MAIN INFERENCES

The results of our studies are one of the main components of the *prognostic topological–ecological concepts* that we are developing. The experience of research into the mechanisms of reflection of global biospheric processes by the structural and functional characteristics of natural ecosystems of the topological dimension demonstrates that modern biogeocoenological methods are efficient for solving the systemic problems of ecological–geographical prediction.

Under such conditions, transgressive development of mixed-forest phytocoenoses should be expected in the future on the main catchment area of the Volga River basin not only in the south of the boreal belt but also in its "typicality core": the middle taiga. Even for the nearest decades, one should also assume a certain "savannization" of mesophilic broadleaf forests and their joining with the total mosaic complex of light forests, meadows, and steppes of typical forest steppe. We also anticipate further transformation of the complete canopy of mixed and even coniferous forests into light forest park-type stands disconnected by sites of meadow–steppe communities (the analogs of modern glades). The latter, as is known, are formed in the course of long-term human influence on coniferous forests, i.e., are a consequence of the anthropogenic aridization of forest lands. Finally, one can speak with confidence about the enhancement of the mobility of geobotanic areas and the instability of zonal boundaries.

# Phytocoenotic and Soil Signs of Current Global Warming

# 13

## 13.1 MAIN FEATURES OF THE CLIMATIC TREND

Global warming began, according to earlier estimates (Houghton et al. 1996), in the mid-1980s. Now its onset is attributed to the period 1970–1975, which is confirmed by the results of averaging meteorological observations from the global network of 3,288 ground stations – T3288 (Ran'kova et al. 2024). Since that time, the increase in the average annual surface temperature has been quite stable, and its 11-year moving averages show a clearly expressed linear trend (Figure 13.1). This trend is most pronounced in the Northern Hemisphere – at a rate of about 0.88°C per decade. At the same time, on the European continent, the rate of annual temperature increase averages 0.498°C per decade, and in the summer season, it reaches 0.531°C per decade. The year 2023 turned out to be the warmest in the entire history of global meteorological observations, starting from 1850. Its average temperature was 1.48°C higher than the temperature of the pre-industrial period (Ran'kova et al. 2024).

The results of climate time series analysis show that the mean annual value of temperature increase on the continents in the 20th century was within 0.7°–1.6°C (Klige 2000; Kondrat'ev and Demirchyan 2000). As early as 1980, in the Northern Hemisphere, the mean July temperature at 55°–60° N changed by 0.4°–0.5° relative to the normal value for 1891–1979 (Krauklis 1991). In the past 100 years, the mean annual air temperature on the territory of European Russia, including the Volga River basin, has increased by 0.9°–1.1° on average (Izrael 2001; Perevedentsev et al. 2001). After the known "Arctic warming" of the 1930s and 1940s, followed by a wave of cooling in the 1950s and 1960s (see Figure 12.1), a new stable tendency of global climate warming has manifested itself since the early 1970s of the 20th century (Budyko and Hecht 1991; Golitsyn and McBain 1992) and has continued up to now.

DOI: 10.1201/9781003733119-13

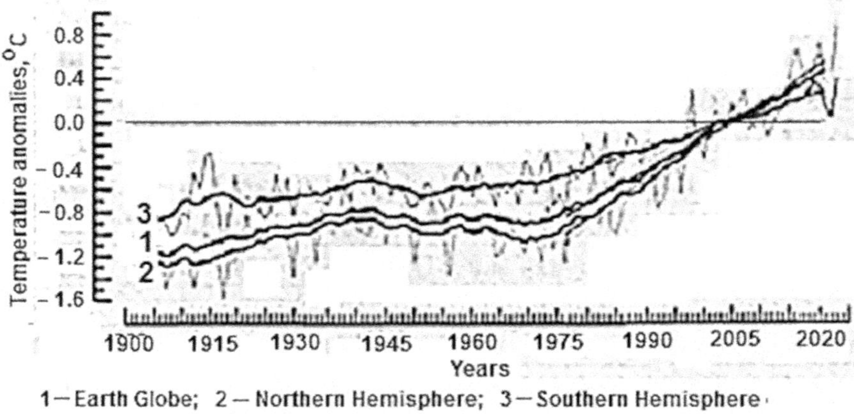


**FIGURE 13.1** Time series of 11-year moving averages of mean annual temperature anomalies over the Earth's territory. The data and the technique of special averaging of the IGCE were used (Ran'kova et al. 2024).

In the taiga and forest–steppe regions of the neighboring West Siberian Plain, the rate of total annual warming observed at meteorological stations in the last third of the 20th century was 0.04°C–0.08°C/yr, while the increase in the mean surface temperature after 1900 in the Altai-Sayan region, Central Europe, Central America, and other regions of the world was no less than 0.6°C–0.8°C, according to geothermal data. The mean annual temperature had been rising at a rate of 0.02°C–0.08°C/yr along the whole network of forest–steppe and steppe regions of southern Siberia during the 60-year period of meteorological observations (1938–1998), intensifying the general aridization of exogenous processes under the conditions of unstable fluctuations of atmospheric precipitation (Bazhenova and Martyanova 2000).

Thus, the global character of warming in the last century is undoubted, and it is a fluctuating and accelerating process. For example, during 1995–1999, the deviation of the mean annual temperature from the 20-year norm of the "pre-industrial period" (1886–1905) on the European territory of Russia was 0.7°–0.8° to 1.9°C (Izrael 2001). Paleo-botanic data for the Russian Plain (Dinesman 1976; Serebryannaya 1982) also demonstrate that the sub-Atlantic cold and humid trend has become slightly weaker and even tends to change to a thermo-arid one.

Assuming that the warming that started in the 20th century is for the most part and even predominantly anthropogenic and caused by the increasing emission of techno-genic "greenhouse" gases into the atmosphere, it should not be forgotten that it coincided with the beginning of a "conservatively stable", according to Shnitnikov (1957), warm and dry phase of the latest 1800–1900-year climatic rhythm and, therefore, may be a result of the positive interference of the anthropogenic and natural factors of climate changes.

Atmospheric humidification could be characterized by negative interference: a general increase in precipitation should be expected under global anthropogenic warming

(Budyko et al. 1992), whereas, in the modern phase of natural super-century climatic rhythm, the temperature rise must be accompanied by general, though slower, aridization of the climate. Both factors, neutralizing each other, are able to significantly weaken the modern century-long trend of atmospheric humidification. Indeed, based on factual data (Izrael 2001; Perevedentsev et al. 2001), annual precipitation in the territory of the Volga River basin, like in European Russia as a whole, has hardly changed in the past 100 years (there was only a very weak and statistically unreliable tendency to increase). Nevertheless, in the last third of the 20th century, there was still a marked increase in precipitation. Such was, e.g., the 24-year period of 1967–1990, when annual precipitation on the European territory of Russia increased by 30% (Efimova et al. 1990).

## 13.2 GENERAL CHANGES IN SOIL–PLANT COVER

The important question now arises of whether there are any signs of modern aridization and steppification of the boreal and nemoral forests and what forms of phytocoenotic structures and soil characteristics they assume.

The current century-long warming in the territory of the Volga River basin could not help but cause, first and foremost, the northward expansion of the areas of grass steppe (i.e., the most dynamic) ecosystems. Experiments with mathematical models of biosphere processes show (Moiseev et al. 1985) that forest ecosystems can much more efficiently neutralize the thermal signal compared to herbaceous ecosystems under conditions of increased temperature or atmospheric $CO_2$, which induce a simultaneous increase in both primary bio-productivity and the rate of decomposition of dead organic matter. Herbaceous ecosystems prove to be more sensitive to natural and/or anthropogenic climate changes. This difference manifests itself already in the first decades and is maintained throughout several centuries, suggesting the first-priority response of exactly grass steppe, not forest ecosystems, to stable climate changes.

As we have already shown (Kolomyts 2005), the stored summer soil moisture, which is the main hydrothermal factor directly influencing vegetation and soils, almost equally depends on variations in both the mean July temperature and annual precipitation. This means that the thermal and pluvial signals are rather interchangeable and can exert the same ecological effect. Consequently, the general warming that occurred in the 20th century must have influenced the state of zonal geo(eco)systems even in the absence of any substantial changes in precipitation. According to our estimates based on empirical data and statistical information analysis (Kolomyts 2024), the increase in the mean July air temperature in the south of the forest steppe zone of the Volga River basin by 0.5°, i.e., by half of the total annual warming in the 20th century, is equivalent to a 30-mm increment in annual evaporation capacity. The latter causes a decrease in summer productive moisture reserves in the 0–50 cm soil layer of herbaceous ecosystems by 5–7 mm, which is equivalent to a reduction in annual precipitation by more than 50 mm. We have established that the taxonomic (average-weighted) norm of annual precipitation is 507–509 mm for the southern forest steppe (both East-European and By-Ural) and 438 mm for the northern trans-Volga steppe. Hence, it may be supposed that the above decrease in edaphic humidification must inevitably cause the replacement of forest steppe plants by northern steppe vegetation and the replacement of the latter by dry steppe plants. At the same time, dark gray forest soils and common chernozems have no time for significant changes during this short period and remain in the state of northern relicts (Table 13.1).

These soil relicts could emerge in a comparatively short period of time (within a century or even several decades) because they were determined by the active northward transgression of grass steppe ecosystems (especially grasslands), which should cause (Walter 1976) intensive stand breakup and suppression of forest reproduction.

The described situation seems to be typical not only of the south of the middle belt of the Russian Plain. In the low- and middle-mountain regions of the South Ural, I.M. Krasheninnikov (1951) found, as early as the mid-1930s of the 20th century, under the canopy of broadleaf and even fir–spruce forests, the "originally alien" xerophilic herbaceous species from the larch–pine–birch forest steppe of the foothills of South Ural, which he characterized as a heritage of cold–arid "ancient Pleistocene forest steppe". In turn, forest steppe communities proved to be a field of transgression of dry steppe elements that are parts of the rocky North Caucasus steppes. West Siberia is also characterized by a modern expansion of meadow steppe into forest and of semi-desert into meadow steppe (Tsyganov 1950).

The incremental warming of the 20th century also affected some properties of soil moment, according to Stankov (1951), particularly the fractional composition of humus in horizon $A_1$. In contrast to vegetation, soil underwent these changes not in the southern but in the northern area of the boreal ecotone: in mixed forests and, to a lesser extent, in the southern and middle taiga. The dominant values of the humic to fulvic acid ratio ($C_{ha}$: $C_{fa}$) in these soils (1.00–1.50) proved to be slightly above their zonal norm (0.75–1.25). This shift was especially considerable in sod-podzol and transitional podzol–brown forest soils of pine and complex spruce forests of the Middle Kama region, where the predominant humus type changed from humate–fulvate to fulvate–humate (Table 13.2).

The predominance of humic acids over fulvic acids is also observed to increase in the light gray forest soils of oak–lime forests and in the leached chernozems of typical forest steppe (both coniferous and broadleaf). Such a noticeable increase in the depth of organic matter humification in these soils was due not only to the temperature increase but even more to the respective extension of the period of biological activity of soils. As is known (Orlov et al. 1996), the close dependence of the $C_{ha}$:$C_{fa}$ ratio on the latter factor is one of the main biogeochemical "rules" of humus formation.

Certain phytocoenotic changes are also observed in boreal forests. Stationary geobotanic studies in the Moscow region in the sub-taiga zone over a 20–25 year period (Maslov and Peterson 1993) have shown the following regularities of the succession trend: (1) substitution of spruce for pine and/or oak, with the loss of aspen and a general decrease in the degree of forest density for the stands, and (2) nemoralization of herbage, with a reduction of the number of boreal species in the ground cover. The former process may be ascribed to the increase in atmospheric humidification in the preceding centuries (the heritage of super-century climate fluctuations, see Kolomyts 2003), while

**TABLE 13.1** Distribution of ecological niches in plant formation groups of the forest–steppe and steppe zones of the Volga River basin by soil groups

| *SOILS* | *VEGETATION* | | | | | |
|---|---|---|---|---|---|---|
| | *TYPICAL FOREST–STEPPE* | | *SOUTHERN FOREST–STEPPE* | | *NORTHERN STEPPE* | |
| | *A* | *B* | *C* | *D* | *E* | |
| Broadleaf forests | $GF_{light}$ | | | •↑ | | |
| | $GF_{own}$ | • | | | | |
| | $GF_{dark}$ | • | • | | •↑ | |
| Typical and southern forest–steppe | $Ch_{pod}$ | × | • | • | | •↑ |
| | $Ch_{med}$ | • | × | • | • | |
| north steppe | $Ch_{typ}$ | | • | | • | |
| | $Ch_{ordin}$ | | | | ×↓ | × |
| Southern steppe | $Ch_{south}$ | | •↓ | ×↓ | ×↓ | ×↓ |

*Note:* The symbol "×" denotes an ecological optimum, and the symbol "•" denotes a "fuzzy" part of the niche. The symbol "↓" indicates that this soil group is a southern relict in relation to the plant formation, and the symbol "↑" denotes a northern relict. Formation groups: A – forb–grass pine forests with steppe shrubs; B – meadow steppes and steppe meadows in combination with oak forests; C – rich forb–feather–grass steppes of the Black Sea type; D – the same in trans-Volga–Kazakhstan; E – trans-Volga fescue–feather–grass steppes in combination with wormwood on solonetz. For the designations of soil groups, see the Note to Table 13.2.

**TABLE 13.2** Areas of dominance of humus types by the ratio of $C_{HA}$:$C_{FA}$ in various soil groups of the Volga River basin

| | | SOIL GROUPS | | | | | | | | | | | | | |
|---|---|---|---|---|---|---|---|---|---|---|---|---|---|---|---|
| | | NORTHERN AND MIDDLE TAIGA | | SOUTHERN TAIGA AND MIXED FORESTS | | | | BROADLEAF FORESTS | | | TYPICAL AND SOUTH FOREST–STEPPE | | NORTHERN STEPPE | | SOUTHERN STEPPE |
| HUMUS TYPE | CORRELATION $C_{HA}$: $C_{FA}$ | PODZOL (P) | | SODDY-PODZOL (SP) | | | | GRAY FOREST (GF) | | | CHERNOZEM (CH) | | | | |
| | | GUM | OWN | GLEY | SMALL | FERRU | 2GUM | LIGHT | OWN | DARK | PODZ | MED | TYP | ORDIN | SOUTH |
| Humate–fulvate | 0.50–0.75 | ⊗ | ⊗ | | | | | | | | | | | | |
| | 0.75–1.00 | ◆ | | ⊗ | ○ | ⊙ | ⊙ | ○ | | | | | | | |
| Fulvate–humate | 1.00–1.25 | | ◆ | | ✕ | | ○ | ⊗ | ⊙ | | ◆ | ◆ | | | |
| | 1.25–1.50 | | | ◆ | | ✕ | | ✕ | ⊗ | ⊗ | | | | | |
| | 1.50–1.75 | | | | ◆ | ◆ | ✕ | | | ○ | ⊗ | ○ | | | ○ |
| Humate | 1.75–2.00 | | | | | | | | | | ⊗ | ⊙ | ⊙ | ◆ | ◆ |
| | 2.00–2.25 | | | | | | | | | | | ✕ | ✕ | ✕ | |
| | 2.25–2.50 | | | | | | | | | | | | ⊗ | ⊗ | |

*Note:* The signs ⊗ . ⊙ and ○ indicate the zonal norm of the $C_{HA}$ : $C_{FA}$ ratio in the A1 horizon for a given soil group according to Kovda (1973), Ponomareva and Plotnikova (1980), Glazovskaya (1981), and Orlov et al. (1996).

Podzol soil groups: **gum** – illuvial-iron and illuvial-humus podzols, and peat and podzolic-gley; **own** – properly podzolic soils (finely, shallowly, and deeply podzolized).

Soddy-podzol soil groups: **gley** – soddy-podzolic gley, shallowly podzolized; **small** – the same. finely podzolized; **ferru** – soddy-polzolic illuvial-iron; **2hum** – soddy-podzolic with a second humus horizon, clarified and residually carbonate.

Gray forest soil groups: **light** – light gray forest; **own** – gray forest; **dark** – dark gray forest.

Chernozem soil groups: **podz** – podzolized and leached chernozems; **med** – meadow chernozems and meadow–chernozem leached soils; **typ** – typical and non-carbonate chernozems; **ordin** – ordinary chernozems; **south** – southern chernozems and carbonate and dark chestnut soils.

Additional ecological dominants in each vector-column of the niche are defined by the symbol ◆, and "fuzzy" links of the niche by the symbol ◆. The remaining designations are the same as in Table 13.1.

the latter seems to be caused by the modern temperature increase (first of all during the vegetation period) and, consequently, demonstrates an example of one of the initial phytocoenotic responses to global warming.

This example shows the obvious signs of enhanced contrast in the vertical component of three-dimensional local phytocoenotic ecotones: reinforcement of the positions of boreal-type vegetation in the arboreal layer and sub-boreal vegetation in the ground cover. Such mutually antithetical transgression of boreal and sub-boreal species in the mixed forest zone (boreal along the upper bio-horizons in the north and sub-boreal along the lower horizons in the south) seems to be a result of two climatic trends of different frequency: a super-century from 1800 to 1900 (Shnitnikov 1957) and a century-long trend from 1980 to 1990, respectively.

Thus, the modern century-long thermal signal has had a well-defined effect both on the structure of herbaceous communities (not only in steppe and forest steppe but also in forest zones) and on the properties of soil moment (mainly in boreal forests, which are most sensitive to temperature changes).

## 13.3 PHYTOCOENOTIC CHANGES AT THE MAIN LANDSCAPE BORDER OF THE RUSSIAN PLAIN

Let us refer to the results of geobotanic studies in Nizhegorodskoye Povolzhye (Lukina 1974). As is shown in Kharitonychev (1978), this territory is characterized by the most marked natural latitudinal–zonal ecotonization of the entire Russian Plain: high diversity of natural zones of the boreal and sub-boreal belts and considerable closeness of zonal–sub-zonal boundaries to each other. Figure 13.2 shows the meridian profile of the vertical structure of the phytocoenotic environment encompassing the northern part of the By-Volga Upland and the southern area of the low-lying Vetluga–trans-Volga Region, i.e., the parts of territories on both sides of the Main Landscape Border of the Russian Plain, according to the terminology of Mil'kov (1981). The profile characterizes the distribution of three zonal types of vegetation: taiga, nemoral forest, and steppe. As we can see, the above types regularly replace each other not only in the horizontal direction (from south to north) but also vertically from the upper biogeocoenotic horizons, in terms of Byallovich (1960), to the lower ones. This is a result of the sequential wedging from below of more southern into more northern types, resulting in the appearance of a kind of super-zonal stratification of phytocoenoses in the south taiga sub-zone, in the mixed-forest zone, and in the nemoral sub-zone.

According to the descriptions made in the 1940s and 1950s by Stankov (1951), the upper tree layer of well-drained watershed complex fir–spruce (coniferous) forests of the Povetluzhye is referred to as the boreal (taiga) type of natural environment. At the same time, oak–forest elements (lime, maple, hazel, honeysuckle, buckthorn, etc.) are widespread in the second layer and in the undergrowth. The ground cover also acquires much of the nemoral–forest appearance, where boreal species of dwarf shrubs and

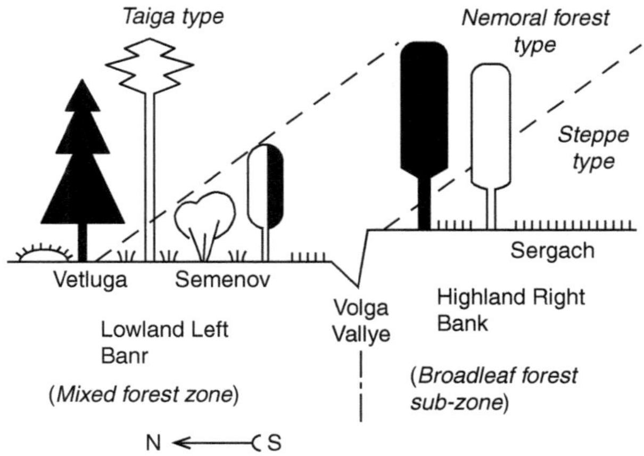


**FIGURE 13.1** Three-dimensional phytogeographic ecotone on the Main Landscape Boundary of the Russian Plain (within the Nizhny Novgorod Volga region). Explanations are provided in the text.

herbs (mountain cranberry, ferns, shamrock (oxalis), Lithuanian mannagrass, etc.) grow alongside widespread nemoral species (hair-like sedge, woodruff, glague (*Aegopodium*), asarabacca (*coltsfoot*), high aconite, etc.). Stankov believes that such vertical structure of ramens results from the struggle between plant groups dramatically different in zonal respect, their thrusting into each other. Southern elements are also widespread in the herbaceous layer of pine forests of the trans-Volga sub-taiga. Such are, for example, lily-of-the-valley pine forests (Lukina 1974).

In turn, representatives of typical forest steppe intrude into the zone of broadleaf forests from the south; they have already "forced" their way into the Volga River valley and spread over the lower reaches of the Kerzhenets and Vetluga. As a result, oak and pine forests of the By-Volga Upland have acquired multi-layeredness and species abundance (diversity), with evident steppification of the lower biogeocoenotic horizons in the southern belt of the sub-zone (Lukina 1974). The undergrowth often contains steppe shrubs (Russian broom (*Chamaecyticus ruthenicus*), dyer's-broom (*allelulia*), and creeping (*juniper*)), and the ground cover contains forest steppe and steppe grasses (vetch, *Corydalis marschalliana*, narrow-leaf lungwort, Mediterranean chrysanthemum, cinquefoil, mountain sedge (*Carex scopulorum*), geranium, etc.). The broad transgression of steppe plants resulted in the appearance of large "islands" of typical forest steppe in the broadleaf forest sub-zone (e.g., Arzamas and Sergach forest steppes).

The described phytocoenological profile, having the appearance of puff pastry, is one more example of a three-dimensional biogeographical ecotone, which has come into existence due to the transgression of only grass–shrub forest layers under the influence of the modern climatic trend. The ecotone has been formed as a result of the penetration of steppe plant elements into broadleaf forests along the lower bio-horizons and, in turn, the penetration of representatives of the nemoral flora into mixed forests and even south taiga. Such unidirectional wedging of horizons causes strong mobility of their lateral boundaries (Byallovich 1960), thereby initiating the reorganization of the vertical phytocoenotic structures over vast areas.

The above example characterizes the initial stage of structural transformations of natural ecosystems under the influence of the thermo-arid climatic trend, and this period seems to have been completed in the territory of the Volga River basin. As has been said above, soil-forming processes have undergone respective changes as well.

## 13.4 PHYTOECOLOGICAL EFFECT OF ENHANCED NON-STATIONARITY OF THE CLIMATE

We should also note the considerable enhancement of interseasonal climatic contrasts and the higher recurrence of extreme hydrothermal conditions during the past 100 years. Statistical analysis of time series showed (Perevedentsev et al. 1985) a general tendency toward greater abnormality of the temperature field, which developed in the territory of the European part of Russia, West Siberia, and Kazakhstan during 1891–1961 and has been especially noticeable since the 1920s. The materials presented in also showed an increase in the amplitude of interannual fluctuations of summer temperatures and precipitation in the territory of European Russia during 1891–1980 (quasi-biennial oscillation with a period of 15–20 years), indicating unambiguously increasing climatic extremity. The winter regime of the typical and southern forest steppe of the Russian Plain has also undergone noticeable changes. In the last quarter of the 20th century, the recurrence of winters with little and unstable snow increased here from 28% to 40% (Nefedyeva and Yashina 1985).

Thus, there was an intensification of the climatic imbalance of the territory, providing development of exactly steppe ecosystems (Mordkovich 1982) and having an extremely unfavorable effect on the state of oak forests not only near the southern boundaries of the area of their distribution but, as it seems, over the whole area, as indicated by G.F. Morozov (1949). The latter is confirmed by the widely known facts about the waves of large-scale drying of oak forests on the Russian Plain building up throughout the 20th century (Nikolaevsky and Nikolaevskaya 1993), which is associated first of all with the higher recurrence of strong droughts in the spring–summer period. The extreme values of thermal conditions in May in 90% of cases cause abnormalities in the course of oak growth even near the northern boundaries of its area (Pushin 1997).

Numerous materials from different authors (Molchanov 1961; etc.) show that the large-scale oak drying observed since the early 1940s of the 20th century, with maximum intensity in the 1950s and 1960s, has been altogether noninfectious. The fungal diseases of oak that ultimately result in stands drying and dying are a secondary event. Their main causes are changes in the state of ecotopes induced by external climatic fluctuations; therefore, the still continuing large-scale drying of oak forests noted, e.g., in the Middle Povolzhye (Nikolaevsky and Nikolaevskaya 1993), is quite probably a phytocoenotic feature of the modern global climate changes.

The mechanism of such connection is as follows: oak and lime at the periphery of their areas are very sensitive to the extreme states of the atmosphere: oak to winters with strong frost and little snow, especially in the initial period of winter (Kharitonychev 1968), and lime to dry and hot summers (Kurnaev 1980). In the forest steppe, oak also suffers a lot from droughts. Late frosts are also important; arboreal species become more sensitive to them under general warming (Morozov 1949). At the same time, one of the paramount phenomenal indicators of global climatic changes of the 20th century is considered to be the increase in high-frequency instability, the growth in the amplitude of temperature fluctuations and atmospheric humidification from 1 year to another, with a higher frequency of occurrence of extreme weather: abnormally cold and abnormally warm, as well as excessively humid and extremely dry (Klige 2000; Kondrat'ev and Demirchyan 2000). In particular, rather stable 10–30-year cycles of pronounced climatic fluctuations with abnormalities of the mean annual air temperature of 2° and more have been revealed for the territory of steppe, forest–steppe, and mixed-forest zones of the Russian Plain beginning from the 1920s of the 20th century (Vereshchagin et al. 1997). There have also been regularly repeating (approximately every two decades) 7–8-year periods of quasi-biennial oscillation with outbreaks of regional climate extremity (Krenke et al. 1995). The trend of intensified climate aridity in the warm period of the year was most significant, first of all due to abnormally hot summer seasons. For example, the June and July air temperatures in 1981, 1988, 1991, and 1998 regularly increased up to 36°–39° in the Central-Chernozem region and up to 42° in the Lower Povolzhye (Izrael 2001), exceeding the long-term temperature maximums for these regions. As is known, strong drought sets in when the mean temperature goes beyond the long-term norm by 3°–4°.

The first cause of intensification of climatic aridity and extremity in the Volga River basin may be considered, according to the data of Perevedentsev et al. (1985) and Popova (2001), to be the considerable general increase in the recurrence of the eastern (E) form of atmospheric circulation (by the Vangengeim–Dzerdzeevsky–Girs classification) during 1899–1997, with maintenance of a rather high frequency of meridian (C) circulation and reduced recurrence of western (W) circulation. According to other data (Krauklis 1991), from 1890 to 1980, there was a considerable increase in the total recurrence of E+C forms of atmospheric circulation and a weakening of the W form. The main features of such a synoptic situation are (1) continuous alternation of warm and humid air inflows from the Mediterranean, resulting in abundant rains and snowfalls in the fall–winter–spring period, with deep intrusions of dry and cold Arctic air (Popova 2001) causing light frosts; and (2) the inflow of abnormally warm and dry air from the intra-continental regions of Eurasia, causing early snow thawing in winter and prolonged strong droughts in summer. At the same time, the activation of meridian circulation increases winter snowiness, while the lower zonal circulation increases summer atmospheric aridity (Krenke et al. 1995).

The winter temperature conditions of the 20th century altogether resembled the time of the medieval climatic optimum (12th century) but with a higher recurrence of extremely cold winters rather than warm winters, while the abnormally cold winters were characterized by both little and much snow. Summer seasons were the coldest on average throughout the past 800 years. Nevertheless, every 10 years there were up to three extremely warm summer seasons, and the recurrence of extremely humid and extremely arid spring–summer periods was approximately equal (Krenke et al. 1995).

The described peculiarities of the climate of the middle area of the Russian Plain in the 20th century should have had a very unfavorable effect on the state of oak forests and promoted the northward expansion of steppe phytocoenoses, primarily along the prominent, driest elements of meso-relief. Vegetation of exactly eluvial and trans-eluvial locations is a primary indicator of background climatic changes. This tendency is apparently bound to be maintained in the near future because a new significant increase in the mean annual air temperature is predicted by 2005–2010 (Vereshchagin et al. 1997).

## 13.5 MAIN INFERENCES

Our data show that modern global warming is, to a certain extent, reflected in the structural and functional changes in natural ecosystems in the southern boreal belt and in the region of its transition to the sub-boreal belt. These changes are determined by a rather stable thermo-arid bioclimatic trend, which started to develop as early as the last century and is being predicted for the entire first half of the 21st century.

For plain territories of the temperate zone, it has been empirically established that the main channel of connections between regional and local ecosystems and the background climatic system passes through summer soil moisture content, which is a reliable geophysical indicator of the state of ecosystems, on the one hand, and the most powerful local ecological factor, on the other hand. Probabilistic scenarios of the expected changes in summer soil moisture content until the mid-22nd century and the respective structural and functional changes in forest ecosystems for different zonal–regional conditions of the Volga River basin are described.

The predicted scenarios of intensive aridization of soil–plant cover represent a general process of "savannization" in the sub-zone of broadleaf and mixed forests of the Russian Plain.

# Mechanisms of Stability of Forest Ecosystems

# 14

## 14.1 PROBLEM STATUS

The concept of two alternative types of ecosystem stability (Holling 1973; Odum 1983; Svetlosanov 1990; Thompson et al. 2009) has been established by the present time. These are (1) resilience, which characterizes the capability of an ecosystem to return to its initial status after a disturbance and to regenerate its main structure and functioning; and (2) resistance, which determines the capability of an ecosystem to resist transformation into another system. This property is related to ecosystem stability, which shows the capability of a system to maintain a dynamic balance for a particular time period by means of absorbing external effects.

It is also proposed to specify the engineering resilience and ecological resilience (Holling 1973; Peterson et al. 1998; Gunderson 2000). Engineering resilience refers to the existence of only one stable status (Gunderson 2000). Ecological resilience is defined as the capability of a system to absorb effects to a threshold level, at which it is transformed into a qualitatively new system. For example, during a drought, a resilient forest ecosystem corresponding to the status of engineering resilience would recover from this stress without significant changes in species composition. An ecosystem with the status of ecological resilience may have more than one stable status, and ecosystem resilience (capability for regeneration) characterizes the capability of a forest to withstand a long-term drought. During this period, the forest may undergo succession, transforming into several different stable ecosystems with a new species composition, with the final stage represented by a meadow. Forests at many of these successive stages are as valuable for forestry as the initial ecosystem, and the community may be recognized as a particular forest type. This is the concept of an unbalanced dynamic (Thompson et al. 2009).

The problem of forest stability in our country has been studied for a long time. Early in the 20th century, G.F. Morozov (1949) set an objective for forestry specialists to use forests so rationally that they could preserve their biological stability and productivity


DOI: 10.1201/9781003733119-14

for the quick natural regeneration of indigenous forest types, which, in his opinion, have the highest stability. In geo-ecology, the stability or homeostatic rate (Armand 1983) is still evaluated by qualitative gradations such as slight, medium, and high, and rigorous classification criteria are sometimes absent.

Nevertheless, formal characteristics were already used in the evaluation of the stability of zonal–regional geosystems of large areas (Buks 1977). Five stability levels of landscape–ecological complexes were specified with respect to landscape initial biological productivity and the provision of heat. Nevertheless, the stability type and the kinds of external effects are unclear; therefore, the specified stability levels are abstract.

A detailed qualitative analysis of forest stability under the effect of climate change and economic activity is given by Rozhkov and Kozak (1989). Nevertheless, the reactions of forest ecosystems to external impacts are only generally evaluated on the basis of qualitative parameters of the substrate, and the conclusions are often very subjective.

General ideas on the stability theory of geo-ecosystems are presented in the work by Armand (1983) and in monographs (Kupriyanova 1989; Glazovsky and Armand 1992). The authors describe the stability mechanisms of various natural components using qualitative and quantitative parameters that characterize (mainly indirectly) the stability of a particular geo-component system. A principle of levels of biotic stability mechanisms of natural complexes is proposed, and the stability of zonal ecosystems under the impact of a number of natural and anthropogenic factors is evaluated semi-quantitatively by scores (Glazovsky and Armand 1992).

System stability may also be characterized by its dynamic and hierarchic structure based on the oscillation concept (Puzachenko 1985), taking into account the ratio between its own fluctuation frequency (the periods of relaxation) and the frequencies of the external effects that cause forced fluctuations of the system.

In national landscape geography, there are issues devoted to the use of the concepts of reliability theory in the calculation of the partial stability rate of a geosystem (Grodzinsky 1987). Stability is evaluated by the behavior of particular natural components and is related to the notion of failure. It signifies that the component loses its homeostasis and begins to change within particular limits. The evaluation of these limits is, unfortunately, difficult. Hence, the determination of the threshold values of the structural and functional parameters of particular geo-components used for calculating the failure probability may be subjective.

In foreign works on forest ecology, ecosystem stability is usually qualitatively described using quantitative indicators that characterize stability indirectly. The growth rate, productivity, and compactness of the tree stand, species and functional diversity, plant damage, and other parameters are used for stability evaluation (Holling 1973; Chapin et al. 1997; Diaz and Cabido 2001; Thompson et al. 2009). A number of works by foreign researchers are devoted to the ratio between various types of forest stability: initial resistance and subsequent resilience. For example, research on the response of trees of 12 various species in New Zealand to unfavorable external impacts has shown that fast-growing species are characterized by low resistance but high resilience (Bee et al. 2007). That is to say, fast-growing trees are damaged first but then quickly restored.

The stability of forest ecosystems is sometimes correlated with topography. For example, in a study of the functional resilience of a subtropical forest on Okinawa

Island, it was revealed that it depended on species diversity, which in turn is maintained by the diversity of features of topography (Kubota et al. 2004).

The specification of so-called topographic types of vegetation repeats the classification of geomorphological complexes of forests in the Far East (Kolesnikov 1956). Physical nonuniformity of the environment may be more important for the evaluation of the resilience of a plant community compared to species diversity (Zaret 1982).

It is also assumed that there is a basic ratio between biological diversity, productivity, and stability of forests, and that this ratio is of practical importance for the adaptive monitoring of forests during climate changes (Begon et al. 1986; Thompson et al. 2009). This concept has already been formulated, and the use of the measure of negative entropy diversity by K. Shannon to evaluate ecosystem stability has been proposed (Margalef 1958). There is evidence that forest ecosystems with greater biological diversity are more productive under similar conditions when compared to ecosystems with a small number of plant species. The latter are prone to catastrophic degradation, including diseases and invasions.

More productive ecosystems are more resilient and thus regenerate more quickly after a disturbance (Stone et al. 1996). Some scientists point to a correlation between the functional diversity of a forest ecosystem and its productivity (Chapin et al. 1997; Diaz and Cabido 2001), and many plant species in forests are unnecessary for the purpose of total productivity. According to the insurance hypothesis (Naeem 1998; Yachi and Loreau 1999), the presence of excessive plant species is an important feature of most forest ecosystems; it favors their stability and protects the forest from the disappearance of species.

In general, the more complicated the community (the greater the species diversity), the higher its stability (Ricklefs 1976). This theoretical statement was not only adopted in general ecology (Odum 1983), but it also found its place in landscape ecology (Sochava 1978). Nevertheless, it is often rejected (Begon et al. 1986) and is still discussed (Whittaker 1975; Loreau et al. 2002). First, the notions of complexity and stability are very indefinite (Svirezhev 1983). Second, the Shannon entropy diversity measure (its analog is represented by the well-known stability measure by MacArtur (1955) cannot be applied to complicated natural ecosystems with rather well-developed internal relations (Svirezhev and Logofet 1978; Svetlosanov 2009).

There are comprehensively elaborated mathematical models of ecosystem stability in general ecology. The models describe a narrow range of relatively simple ecological systems and are usually deterministic. They analyze particular populations or their combination (a biotic community) (Smith 1974; Logofet and Svirezhev 1983; Svetlosanov 1990; etc.). The stability of these ecosystems is calculated as the ratio between the deviation of the disturbing effect and the response of the chosen parameters to this effect, or as the ratio between the homeostasis value and the parameter of distribution nonuniformity of a dependent variable (Fedorov 1974; Fedorov 1975). In other works, the stability of ecosystems is evaluated by balance equations of their equilibrium conditions, which are described by flows of energy and substances in the food chains (Svirezhev and Elizarov 1972; Logofet and Svirezhev 1983; Svetlosanov 2009). The stability is calculated by linear differential equations. Great attention is paid to the stability of the hierarchy of food chains (of the predator–prey type in particular).

Mathematical stability, according to the approach by A.A. Lyapunov, is also taken into account, though it has been revealed (Svetlosanov 1990) that it is hardly applicable for a stability evaluation of natural ecosystems.

Unfortunately, approaches involving mathematical modeling of the stability of ideal ecosystems are not widely applied in landscape ecology, where the investigation objects are considerably more complicated, poly-component, and discrete, and are composed of heterogenic elements with different characteristic times. The ecosystems of the landscape level are distinguished by the probabilistic nature of internal and external interactions (Armand et al. 1969) and nonlinear behavior related to a regulated network of positive and negative feedbacks (Zeidis et al. 2001; Svetlosanov 2009). The stability and functioning of these systems are primarily determined by geochemical cycles (annual and long-term turnover of organic matter), not by the number and dynamics of plant and animal species (Svirezhev 1983).

These natural systems with a so-called poor structure are better described by discrete mathematics based on the probability–statistical approach, which was proven by our landscape–ecological investigations (Sharaya 2009; Kolomyts et al. 2012). Geo-ecosystem stability should be evaluated with the use of these approaches and based on the most representative parameters of biological turnover.

We have searched for unified parameters (indices) of the potential stability of elementary biochorologic units – biogeocenoses (Timofeev-Resovskii and Tyuryukanov 1966) – as integral formations according to the hypothesis by Sukachev (1972).

## 14.2 ANALYTICAL MODELING OF THE STABILITY OF FOREST BIOGEOCOENOSES

Two stability categories of natural ecosystems: structural and functional – are usually studied in accordance with the two aspects (the structure and functioning) of their organization. Stability is usually treated as the constant system structure (or a slight fluctuation, which does not exceed a threshold point) kept by variations in the functional parameters (Smith 1974; Armand and Dolgushin 1983; Kupriyanova 1989; Svetlosanov 2009). The concepts of homeostasis and geo-ecosystem homeostatic reserve, as well as the norms of anthropogenic impact on the system, are related to this approach.

Nevertheless, the quantitative evaluation of the structural stability of multicomponent systems with different characteristic times is a difficult problem (Odum 1983). It is much easier to determine the functional stability, which is vital for ecosystems. This raises the efficiency of energy flow and nutrient turnover, improving the capability of the community to withstand different disturbing factors (Ricklefs 1976). This approach is based on the interpretation of stability as the capacity of a system to change its structural parameters in order to preserve the initial functioning level (mode) (Kerzhentsev 2006, 2010). The ecosystem structure is formed according to its main functions in the biosphere: the maintenance of a closed substance turnover, which provides stability for the ecosystem itself (Gerasimov 1985; Tishkov 2005), and regulation of the biota in the environment according to Le Chatelier's principle (Gorshkov 1994). A close relationship between soil and phytocoenotic components in the indigenous biogeocenoses is formed in the course of long-term evolution and results in optimal substance migration in the biosphere (Dobrovolsky and Nikitin 2006).

Therefore, we will discuss the functional stability (not structural stability) of natural ecosystems as their most fundamental parameter. This approach allows for the calculation of ecosystem stability by discrete parameters of biological turnover.

## 14.2.1 Labile Phytocoenotic Resistance

Biological turnover is a complicated multi-cycle process that includes numerous elementary soil processes and cycles of organic matter production, decomposition, mineralization, and humification with different characteristic times (Kononova 1963; Duchaufour 1965; Glazovskaya 1981). In boreal forests, where the soddy-podzolic process and percolation water regime of soils predominate, the lower biological turnover is slightly related to humification processes of the mort-mass (Odum 1971). It is represented by a kind of "autonomic turnover of elements between living organisms and dying remains entering the soil surface in forests" (Ponomareva and Plotnikova 1980, p. 115). This is a partial phytocoenotic cycle, which may be treated as a short-term (and hence accelerated) component of general biological turnover.

This statement is confirmed by empirical data on coniferous, mixed, and even broadleaf forests (Kolomyts 2005). We have not revealed any significant partial dependence of humus mass on the annual production of the forest community and on the mass of forest litter or roots. A reliable multiple correlation is also absent. A sufficiently close relationship between humus mass and the mass of roots and forest litter is only revealed for zonal forest and steppe ecotones, with a predominance of evaporative soil regimes and the soddy-meadow process. Nevertheless, the drop in humus mass in these systems as a result of global warming will be related only to the drop in the mass of forest litter and roots by 10%–25%. The correlation between soil organic matter and yearly production of green mass is also very slight. Humus participation in the short cycle of biological turnover is sufficiently small, even in these sub-boreal zonal–climatic conditions.

Therefore, the phytocoenotic (including forest litter) cycle of organic mass turnover for boreal and sub-boreal forest formations, in general, may be specified as an autonomous and rather independent component of the total soil–biological turnover. Its most dynamic part is represented by the yearly cycle of regeneration and decomposition of living above-ground phytomass, which is assigned to quick cycles (Duchaufour 1965).

It corresponds with the first level of the partial functional stability of forest communities, which we term *labile phytocoenotic stability.* Labile stability may be characterized by two complex discrete parameters of metabolism (Bazilevich and Rodin 1971; Krut'ko et al. 1982): the coefficient of the annual turnover of the above-ground phytomass ($CR = PV/BL$) and the forest litter–debris index ($CY = PV/LM$), i.e., the coefficient of annual destruction (see Section 3.4). In the abovementioned equations, PV is the annual production of green mass, BL is the total living above-ground phytomass, and ML is the mass of the forest litter. Both parameters characterize the initial transportation of organic matter and energy in the entire trophic chain and thus characterize the impact of plant mass on the stable functioning of the ecosystem.

The spatial variations of a particular function of the status of the ecosystems within a statistical sample were calculated to evaluate labile stability (Kolomyts 2005). Let us denote the indices of resistance and resilience (labile stabilities) for the level of phytocoenosis (the first level) as $I_{resist}(1)$ and $I_{resili}(1)$. The resilience index for a forest biogeocenosis is calculated as a measure of the Euclidean distance from the optimum functional status by the equation:

$$I_{i\ resili}(1) = 1 - \left[\sqrt{(\Delta CR)^2 + (\Delta CY)^2}\right] \Big/ \sqrt{2}, \qquad (14.1)$$

where $\Delta CR_i = (CR_{max} - CR_i)/(CR_{max} - CR_{min})$; $\Delta CY_i = (CY_i - CY_{min})/CY_{max} - CY_{min})$; and $i$ is the number of the investigation plot. The maximums and minimums of CR and CY are taken over $i$. Therefore, the resilience index is evaluated in dimensionless units. If $CR_i$ approaches $CR_{max}$ and $CY_i$ approaches 0, then $I_{iresile}(1)$ approaches 1. It can be seen that both prognostic parameters are considered equivalent and are accepted with weights equal to 1.

The index of the potential resistance ($I_{resist}(1)$) of a geo(eco)system was calculated in a similar way. In this case, the minimum value of CR and the maximum value of CY were taken as the optimum. The resistance index will approach one when $CR_i$ approaches $CR_{min}$ and $CY_i$ approaches $CY_{max}$.

The ecosystem stability indices characterize its capability to withstand integral disturbing effects and show its distance from critical status. The minimum and maximum CR and CY are taken from a particular statistical sample; hence, it is obvious that each of the indices characterizes a relative and not absolute stability of the ecosystem within the area of sampling. The indices show which part of the maximum possible stability is typical for a particular natural complex. $I_{resili}(1)$ reflects the well-known law of requisite variety (Ashby 1956), which states that the diversity of its states or reactions (adaptations) must be no less than the diversity of external impacts for stable system preservation. The stability (invariance) of the ecosystem is related to two factors: its multiple relations and the numerous means of compensating for external impacts (Puzachenko 1983).

Let us examine the role of each of the two prognostic parameters in both types of forest ecosystem stability. For this, we have used data on eight experimental plots (377 points) characterizing the Oka-Volga River basin between the southern part of the sub-taiga zone and the southern forest steppe sub-zone (Kolomyts 2022). The resilience is characterized by a very close parabolic correlation with the annual production of the above-ground phytomass:

$$I_{resili}(1) = -0.173 + 20.95\,CR - 98.08(CR)^2;\ R^2 = 0.966\ \left(P < 10^{-6}\right). \qquad (14.2)$$

The resilience rise is greatest (0.05 → 0.63) in the interval of small values of CR (0.01 → 0.05), and then it decreases. The highest $I_{resili}(1)$, equal to 0.90–0.98, is seen at CR=0.09–0.10. The partial correlation between $I_{resili}(1)$ and the mass of forest litter

(ML) is insignificant. The resistance correlates pronouncedly, though not very closely, to the forest litter mass:

$$I_{\text{resist}}(1) = 0.374 + 0.002161\,\text{ML};\; R = 0.785;\; R^2 = 0.615. \tag{14.3}$$

The sensitivity of forest biogeocoenoses to external impacts becomes lower, in contrast to the forest litter mass: when ML rises from 50 to 200 t/ha, the index $I_{\text{resist}}(1)$ increases from 0.45–0.50 to 0.80. In the area of small forest litter mass (lower than 70 t/ha), the field of values of $I_{\text{resist}}(1)$ is diffused from 0.25–0.30 to 0.40–0.50. Thus, the first reaction of forest biogeocoenosis to external effects depends mainly on the reserve formed by the forest litter mass. At subsequent stages, the adaptive regeneration potential is almost completely determined by the annual production of the above-ground biomass. The CR parameter plays a negative role in the first case and becomes a factor of functioning regeneration in the second.

When water and temperature conditions become stress factors, the regime of mineral plant nutrition is usually disturbed. On the basis of this regularity, we may determine the period of resistant labile stability of forest biogeocoenosis during climate successions (prior to irreversible disturbance of its functioning). It is equal to CY, expressed in the number of years required for complete decomposition of forest litter at a given volume of annual debris. It can be seen that this time period is equal to the first several years after the beginning of the external impact if biological turnover intensity is high or moderate (Bazilevich and Rodin 1971; Kovda 1971). In some cases, the length of the period may exceed10 years and even reach 30 years in the stagnation type of biological turnover.

The resilience period of forest biogeocoenosis necessary for regeneration of the initial level of the closed short biological turnover may be evaluated by the following approach. The empirical dependence of the utilization coefficient of the dead phytomass (CU) on ML and BD (of the total dead above-ground mass) shows that if mort-mass accumulation in a coniferous forest is equal to 40–60 t/ha, no more than 40% of it is utilized at the following trophic levels, while a constant amount of dead mass in the forest at the rate of 10–15 t/ha is sufficient to maintain a continuous flow of organic matter in this trophic chain, as well as its utilization up to 98% (Kolomyts et al. 2012). A simple calculation based on these data shows that resilience activity over a 1.5–2-year-long period is sufficient to restore the initial functioning level at a high annual production of green mass (>8–10 t/ha) typical for most productive coniferous and foliage forests. For forest communities with a relatively low production of green mass (lower than 3–5 t/ha), this period lasts 4–5 years. The functional regeneration potential of the forest community is obviously more effective compared to the initial forest capability to withstand external impact, all other factors being equal.

### 14.2.2 Inertial Soil–Biotic Stability

Humus mass (HU) is absent among the predictors in the calculation of the labile stability of forest ecosystems, though its significant role in the stabilization of natural ecosystems is well known. Calculation models with HU describe the second stability level of forest biogeocoenoses, which includes the larger and slower metabolic cycle. It incorporates a stability component related to processes of the formation and dynamics of soil organic matter. We refer to this level as inertial soil–biotic stability and denote its resistance and resilience by $I_{resist}(2)$ and $I_{resili}(2)$, respectively. This level characterizes the stability of soil organogenesis as a gradual cycle (the term by Duchaufour (1965).

The complete cycle covers the period from the first centuries to several millennia (Glazovskaya 1981). Nevertheless, a much more dynamic component may be specified in it. The short phytocoenotic cycle is not only a part of the gradual soil–biotic cycle; it also greatly affects its results. Laboratory experiments have shown that no less than 50%–75% of the initial mort-mass is mineralized upon humidification of plant remains (grass, leaves, and needles), and nitrogen and phosphorus compounds are immediately consumed by the root system of plants (Kononova 1963).

Let us replace CY with ML and introduce HU as a parameter of the constructional metabolism part, i.e., we enlarge the participation of the detritus branch in ecosystem stability. If CR is equal to 1, the normalized equations for the calculation of the inertial indices ($I_{resist}(2)$ and $I_{resili}(2)$) may be formed with new parameters and with weight coefficients $a$ and $b$. Normalization is made by the equation $CR_{ni}=(CR_i - CR_{min})/$ $CR_{max} - CR_{min})$ for ML and HU:

$$I_{resili}(2) = 1 - \sqrt{\frac{(1-CR)^2 + a \cdot ML^2 + b \cdot (1-HU)^2}{1+a+b}}; \qquad (14.4)$$

$$I_{resist}(2) = 1 - \sqrt{\frac{CR^2 + a \cdot (1-ML)^2 + b \cdot HU^2}{1+a+b}}. \qquad (14.5)$$

If the weight coefficients $a$ and $b$ are equal to 1, which signifies the equal role of all of the studied factors in forest biogeocoenosis stability, the calculated data on $I_{resile}(2)$ and $I_{resist}(2)$ for the eight test plots of the Volga River basin differ considerably from the data calculated by Equation (14.2) for $I_{resili}(1)$ and $I_{resist}(1)$ (Kolomyts 2022). The differences between ecoregions in the indices of both types of forest inertial stability are characterized by a sufficiently pronounced zonal type. At the boundary of the forest–steppe and steppe zones (the Zhiguli test site), the mass of forest litter plays the leading role in resistant stability, and humus mass determines resilient stability with an equally slight effect of CR. In the area from the southern forest steppe to the northern boundary of the forest–steppe zone (the Kud'ma and Shchelokovsky Farmstead test sites), the predominating positive effect of forest litter on the distribution pattern of the resistance of forest biogeocoenoses is preserved in parallel with a drop in the rate of the annual phytomass turnover. At the boundary of the forest–steppe and sub-taiga zones (the Green Town and Vyksa test sites), which represents the well-known main landscape border of the Russian Plain (Mil'kov 1981), the role of metabolic parameters in the formation of both types of stability is sharply changed. The role of the forest litter mass drops abruptly here from $c\pm 50\%–70\%$ to$\pm 15\%–35\%$. Humus mass becomes the predominant factor for resilience.

The two indices of resilience – $I_{resili}(1)$ calculated by Equation (14.2) and $I_{resili}(2)$ calculated from Equation (14.4) – are compared in Figure 14.1. In general, there is a close correlation between the parameters, but they are only equal to each other at the central values (about 0.50). When resilience is lower than 0.40, $I_{resili}(1)$ is by 0.05–0.10 higher than $I_{resili}(2)$. At high stability (>0.75–0.80), the ratio between the parameters becomes inverse. Humus reserves favor an increase in the regeneration potential only of forest ecosystems with high *CR*. When *CR* is small, a transformation of a part of dead organic matter into humus (Table 14.1) causes a drop in ecosystem resilience.

Resistance is characterized by a more diffused parabolic correlation, which is explained by the large number of significant predictors:

$$I_{resist}(2) = 2.27 I_{resist}(1) - 1.309\left[I_{resist}(1)\right]^2 - 0.384;\ R = 0.711;\ R^2 = 0.505. \qquad (14.6)$$

When $I_{resist}(1)$ does not exceed 0.5, its correlation with $I_{resist}(2)$ is almost linear, and the phytocoenotic resistance is somewhat higher than the soil–biotic one. The latter is in addition characterized by a large scatter of points near this value. With a rise in $I_{resist}(1)$ to 0.7–0.8, $I_{resist}(2)$ is stabilized at the level of 0.40–0.65.

Therefore, the stability of the annual regeneration of living above-ground phytomass (i.e., the labile phytocoenotic stability) may be tentatively used as an indicator of the general stability of forest ecosystems, including the inertial stability, which is related to the considerably longer soil–biotic metabolism cycle. This conclusion seems important from the point of view of methodology. Labile stability is measured on the basis of simpler and stricter ratios between the parameters of biological turnover when compared to the inertial stability, for which diverse characteristic times of metabolic predictors must be used, and humus separation into labile and conservative fractions is inevitable (Gerasimov 1986). It is difficult to solve this problem under field conditions.

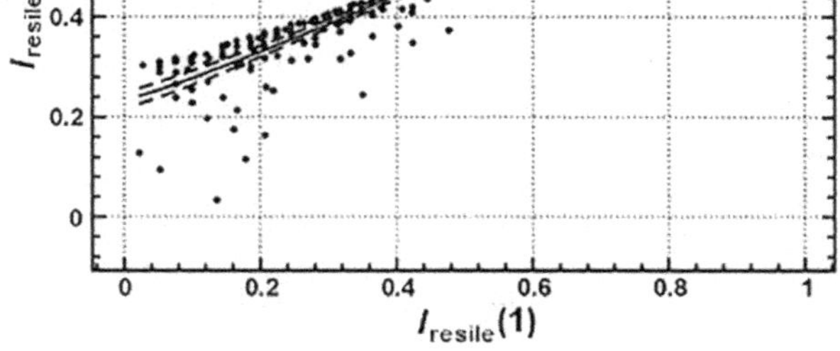


**FIGURE 14.1** Ratio between the resilience values $I_{resile}(1)$ and $I_{resile}(2)$ obtained by Equations (14.2) and (14.4), respectively. *R* is the correlation coefficient, and $R^2$ is the determination coefficient.

**TABLE 14.1** Positive or negative participation of normalized predictors in the indices of the inertial stability of forest ecosystems in the Upper and Middle Volga River Region

| *REGIONAL ECOSYSTEM (EXPERIMENTAL TEST SITE)* | $I_{RESIST}(2)$ | | | $I_{RESILE}(2)$ | | |
|---|---|---|---|---|---|---|
| | *CR* | *ML* | *HU* | *CR* | *ML* | *HU* |
| Southern boundary of southern forest steppe; pine, oak, and lime forests (the Zhiguli low-mountain massif) | –0.289 | +0.565 | –0.146 | +0.346 | –0.461 | +0.193 |
| The Volga River upland, typical forest steppe; pine, and lime–oak forests (the By-Sura River Region) | –0.107 | +0.798 | –0.095 | +0.149 | –0.617 | +0.234 |
| The Volga River Region, the Ni-zhny Novgorod domain, northern forest steppe; oak–lime, and pine–broadleaf forests (Kud'ma) | –0.132 | +0.507 | –0.136 | +0.131 | –0.449 | +0.420 |
| The Volga River Region within the Nizhny Novgorod domain, northern forest steppe; oak, and lime forests (the Shchelokovsky Farmstead upland massif) | –0.168 | +0.711 | –0.121 | +0.124 | –0.698 | +0.178 |
| The Volga River region within the Nizhny Novgorod domain, northern forest steppe; mixed, and coniferous forests (the Green Town habitat) | –0.158 | +0.194 | –0.648 | +0.222 | –0.050 | +0.728 |
| Lowland in the Lower Oka River Region, southern boundary of the sub-taiga zone; spruce–pine outwash woodlands (Vyksa) | –0.139 | +0.187 | –0.674 | +0.245 | –0.118 | +0.637 |
| Lowland in the Middle Oka River Region, the southern boundary of the sub-taiga zone; mixed, coniferous, and broadleaf forests (the By-Oka-Terrace Reserve) | –0.334 | +0.365 | –0.301 | +0.391 | –0.174 | +0.435 |

(*Continued*)

**TABLE 14.1 (*Continued*)** Positive or negative participation of normalized predictors in the indices of the inertial stability of forest ecosystems in the Upper and Middle Volga River Region

| *REGIONAL ECOSYSTEM (EXPERIMENTAL TEST SITE)* | $I_{RESIST}(2)$ | | | $I_{RESILE}(2)$ | | |
|---|---|---|---|---|---|---|
| | *CR* | *ML* | *HU* | *CR* | *ML* | *HU* |
| The lowland in the Volga River region within the Nizhny Novgo-rod domain, southern part of the sub-taiga zone, pine, and spruce forests (Kerzhenets) | −0.177 | +0.484 | −0.339 | +0.252 | −0.263 | +0.485 |
| The main catchment area of the total Volga River basin (statistical sample of all the test sites) | −16.2 | +55.8 | −18.6 | +61.6 | −36.1 | +2.3 |

*Note:* $I_{resist}(2)$ is resistance, $I_{resile}(2)$ is resilience, CR, ML, and HU are predictors (the abbreviation is explained in the text).

It is obvious that the sign of the weight values remains unchanged upon the introduction of the regression coefficients, but the mechanisms of forest community functioning, which are responsible for its resistance and resilience, become much clearer. First, the stability indices, including humus mass, differ considerably from the indices calculated by the initial Equation (14.1) and the similar equation for $I_{resist}$. In our investigations, this is especially the case for the By-Oka-Terrace reserve ecoregion, where the humus mass is relatively small. Mechanisms of resistance and resilience cannot work without the participation of humus mass. Its role in forest biogeocoenosis stabilization becomes greater, in contrast to its content, and corresponds to the known law of the minimum of limiting factors by Liebig (Odum 1971).

The high resistance of a forest community is obviously maintained to a greater extent by a weakening of autotrophic biogenesis (–CR) and a slowing down of destruction processes (+ML). The regeneration potential is mainly realized by a rise in the activity of the detritus component of metabolism, which is evidenced by the close adverse correlation between $I_{resili}(2)$ and the forest litter mass. It is also enhanced by a rise in autotrophic biogenesis (+CP). Its role in eco-genetic successions may be as important as that of active detrito-genesis, which decreases ML, or it may be secondary.

Therefore, the leading functioning processes in the forest biogeocoenosis change as it undergoes unfavorable external effects: auto-morphic biogenesis is replaced by forest litter decomposition. The two opposite mechanisms of the manifestation of the well-known buffering properties of forest litter begin to work (Kovda 1971). At the initial stage of resistance development, the rise in ML shows that the category of the ecosystem is exchanged for a more stagnant one (Bazilevich and Rodin 1971; Kovda 1971). At the second functional regeneration stage, the processes of resilience develop, the forest litter mass drops, and this signifies that the ecosystem becomes more active.

# 14.3 STABILITY OF FOREST BIOGEOCOENOSES IN CARTOGRAPHIC TERMS

New geomorphometric approaches were used in the mapping of forest ecosystem stability. Geomorphometry is a science dealing with topography modeling that has been developed over the last 30 years (Pike 2002; Shary et al. 2002; Shary 2006). The interpolation and extrapolation of data measured or calculated for representative test plots are more reliable when based on multi-parameter landforms. This considerably extends the boundaries of the studied area. In the methods of quantitative topography description known to date, only six base morphometric parameters were used. We have applied a system of 18 parameters elaborated by P.A. Shary and implemented in the GIS Eco computer program (Shary 2006). New approaches for the construction and classification of topographic elements and improved algorithms for the calculation of morphometric characteristics are utilized in the system. We used NASA's satellite topography data represented as a matrix of the altitudes of the Earth's surface with a resolution of 90 m to compile large-scale maps of stability indices. The data were obtained as part of the international project of the Shuttle Radar Topographic Mission performed by the United States, Italy, and Germany.

We constructed matrix maps that comprehensively describe the studied properties of an ecosystem by quantitative parameters. The matrix map is a digital matrix represented by a field of values of a particular ecosystem parameter. Each element (pixel) of the matrix has a numerical value accessible for other programs working with grids. The image of the matrix map was formed using a GIS software package with adequate explication. Data measured and calculated for the sampling points were interpolated and extrapolated based on a matrix of those environmental parameters for which the statistical relations revealed by multiple regression were the closest.

Let us analyze the example of mapping the indices of potential stability of forest biogeocoenoses in (1) the Samarskaya Luka (SL), i.e., Zhiguli low-mountain massif, and (2) the By-Oka-Terrace Reserve (BOTR) ecoregions. The first test plot is allocated to the forest–steppe zonal ecotone. The water and temperature conditions here are extreme for the development of pine–broadleaf forest (Kolomyts 2005), and therefore it is very sensitive to climate variations. With the construction of maps on a scale of 1:50,000 (Figures 14.2 and 14.3), we used the weight coefficients ($a$ and $b$) obtained from Equations (14.4) and (14.5) for the SL ecoregion: $a=1.93$, $b=0.12$ and $a=0.42$, $b=0.40$; we used $a=16.1$, $b=1.07$ and $a=0.31$, $b=1.97$ for the BOTR ecoregion.

The stability indices for each observation site were calculated from Equations (14.4) and (14.5). The topography parameters and soil moisture and temperature were used as environmental variables. The mean resistance index of the low-mountain forest communities of the SL test site (Figure 14.2), which is characterized by a semi-humid climate, is greater (0.614) when compared to the resilience index (0.478). This is evidence of a particular homeostasis reserve in the southern forest communities at the boundary of

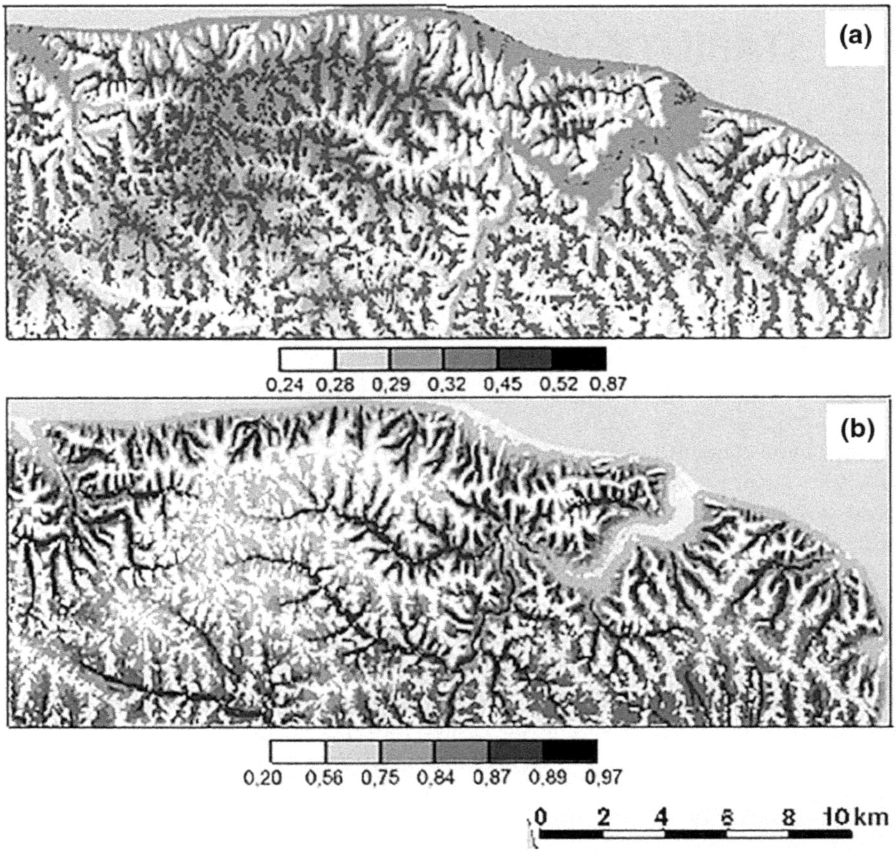


**FIGURE 14.2** Map of indices of resilient (a) and resistant (b) inertial stability for the Samarskaya Luka test plot. The Gauss–Krueger projection is zone 9.

the forest zone, where zonal climatic conditions critical for forest development result in the balance of relatively low rates of synthesis and destruction.

The mixed forests of the BOTR region (Figure 14.3) are developed near the southern boundary of the forests of this type under a moderately humid climate and are characterized by a similar mean resistance (0.624) and resilience (0.656). The minimized coefficient of variation is relatively low: 25.96% for $I_{resist}(2)$ and 23.15% for $I_{resile}(2)$. This is evidence that metabolic processes in this region are sufficiently well balanced; thus, the forest communities must be characterized not only by high sensitivity to climate variations but also by a well-developed functional regeneration potential.

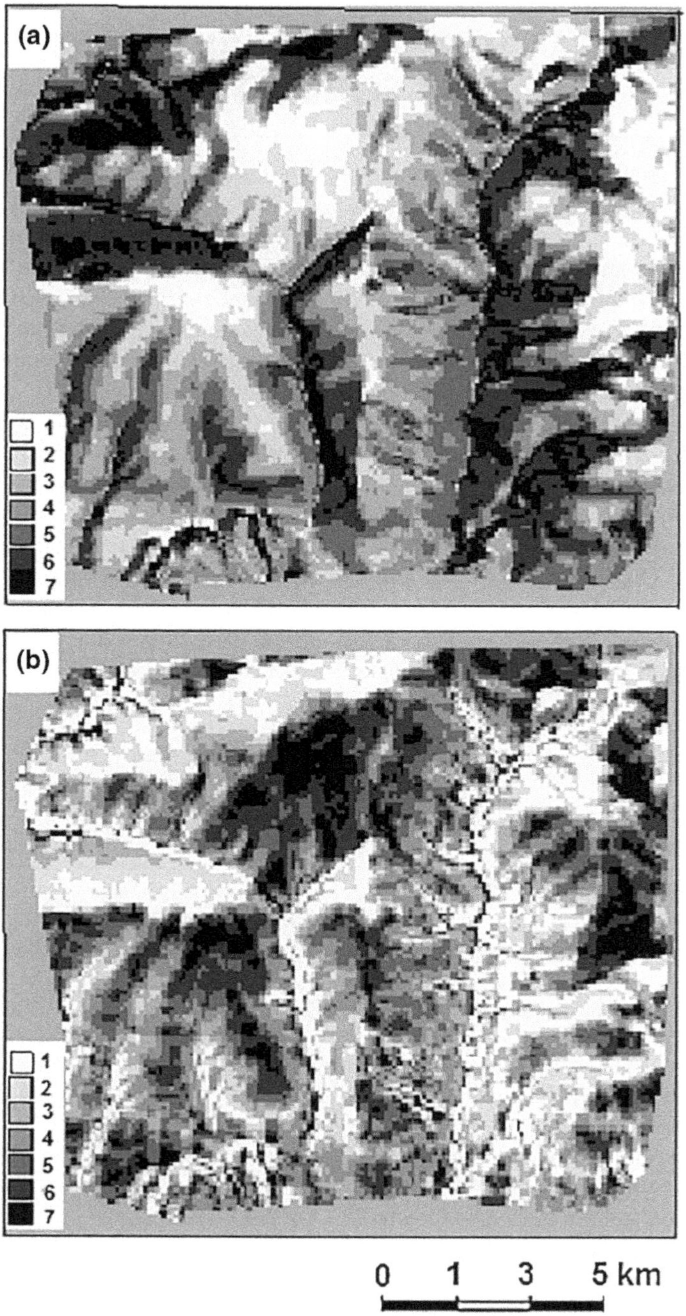


**FIGURE 14.3** The map of indices of resilience (a) and resistance (b) for the By-Oka-Terrace Reserve calculated by Equations (12.4) and (12.5). The Gauss–Krueger projection is zone 9. The biogeocenotic structure of the reserve can be seen in Figure 5.6.

## 14.4 MODELS OF STABILITY FOR FOREST FORMATIONS

Among the metabolic predicting parameters, the mass of forest litter and the coefficient of the annual turnover of the above-ground mass are the most significant for resistance and resilience, respectively (Table 14.1, lower line). Their positive differentiating effect is no less than 55%–60% over the entire main catchment area of the Volga River basin, i.e., on the sub-continental level of geographical zoning (Bazilevich et al. 1986; Kolomyts 2005). This made the analysis considerably simpler, and the functional stability of forest formations in this area was generally evaluated by their phytocoenotic labile stability (Table 14.2).

For the transition from local to regional mapping level (the scale from 1:2,500,000 to 4,000,000) of forest stability in the Oka-Volga River basin, we transited from the local to the regional level of prognostic modeling with the use of *inductive–hierarchic extrapolation*, a method that we developed on the basis of the empirically established phenomenon of the poly-zonal nature of local ecosystems as a response to global climate changes (Kolomyts 2022). The patterns of the modification of the zonal/regional climatic background by local geomorphological and hydro-edaphic factors serve as the basis for analyzing *regional systems of local zonal structure*. They consist of vector series of upland biogeocoenoses reflecting the zonal/regional background of the area

**TABLE 14.2** Taxonomic (weighted mean) norms of resistant ($I_{resist}(1)$) and resilient ($I_{resile}(1)$) labile stability of plant formations in the Oka-Volga River basin

| *CLASSES (SUB-CLASSES), TYPES (A–G), AND GROUPS OF PLANT FORMATIONS* | $I_{RESIST}(1)$ | $I_{RESILE}(1)$ |
|---|---|---|
| **Northern-European spruce forests** | | |
| A. Spruce southern taiga and broadleaf–spruce (sub-taiga) forests | | |
| Southern taiga true moss/grass/dwarf shrub spruce forests | 0.568 | 0.390 |
| Southern taiga birch and aspen forests and forest lands | 0.310 | 0.567 |
| Multilayer, nemoral–grass spruce forests with pine | 0.606 | 0.284 |
| Broadleaf–spruce nemoral–grass forests | 0.338 | 0.460 |
| Boggy tall grass birch–black alder forests with spruce | 0.270 | 0.627 |
| Nemoral–grass sub-taiga birch and aspen forests | 0.247 | 0.671 |
| Sub-taiga fragmentary forest lands | 0.268 | 0.652 |
| **Pine and broadleaf–pine forests and their derived communities** | | |
| **Northern-European pine forests** | | |
| B. Middle- and southern taiga pine forests | | |
| Dwarf shrub/true moss pine forests with spruce and birch | 0.446 | 0.398 |
| Dwarf shrub *Polytrichum–Sphagnum* pine forests with birch | 0.388 | 0.587 |
| True moss and *Sphagnum* pine forests with raised bogs | 0.468 | 0.485 |
| True moss–*Sphagnum* pine forests with spruce and eutrophic bogs | 0.384 | 0.601 |
| Grass/dwarf shrub birch forests and forest lands | 0.268 | 0.322 |

(*Continued*)

**TABLE 14.2 (*Continued*)** Taxonomic (weighted mean) norms of resistant ($I_{resist}(1)$) and resilient ($I_{resile}(1)$) labile stability of plant formations in the Oka-Volga River basin

| *CLASSES (SUB-CLASSES), TYPES (A–G), AND GROUPS OF PLANT FORMATIONS* | $I_{resist}(1)$ | $I_{resile}(1)$ |
|---|---|---|
| C. Broadleaf–pine (sub-taiga) forests | | |
| Dwarf shrub–grass pine forests with oak in the shrub layer | 0.514 | 0.434 |
| Bilberry forbs–grass broadleaf–pine forests | 0.419 | 0.457 |
| Moss–dwarf shrub–grass and lichen pine forests | 0.475 | 0.445 |
| Dwarf shrub–grass pine forests with eutrophic bogs | 0.300 | 0.547 |
| Birch and aspen forests with broadleaf trees in places | 0.395 | 0.536 |
| Fragmentary sub-taiga forest lands | 0.274 | 0.565 |
| Broadleaf forests and their derived communities: European broadleaf forests | | |
| D. Eastern-European broadleaf forests | | |
| Northern oak and lime–oak forests with spruce | 0.420 | 0.500 |
| Southern steppe–oak and lime–oak forests | 0.245 | 0.609 |
| Light oak and lime–oak forests of the Volga River Region | 0.311 | 0.541 |
| Lime–oak forests of the Cis-Volga and Trans-Volga Regions | 0.398 | 0.567 |
| Tall grass–lime–oak forests of the Ural piedmonts | 0.408 | 0.519 |
| Mesophyte forbs–grass lime forests with oak and elm | 0.320 | 0.517 |
| Aspen and birch forests with broadleaf trees | 0.283 | 0.540 |
| **Near-steppe and steppe broadleaf and pine forests** | | |
| E. Outlier and ravine forests in the typical (meadow) and southern forest steppe | | |
| Broadleaf–pine forests with steppe bushes | 0.558 | 0.464 |
| Pine–birch steppe forests and forest lands | 0.540 | 0.295 |
| Steppe thin oak forests of Central Russia | 0.436 | 0.563 |
| Oak forests with pine of the cis-Ural Syrt Region | 0.264 | 0.529 |
| Outlier and ravine forests among agricultural land | 0.310 | 0.371 |
| F. Southern taiga, sub-taiga, and nemoral forest–bog combinations | | |
| Swamped coniferous forests | 0.371 | 0.551 |
| Forest bogs | 0.323 | 0.544 |
| Swamped nemoral forests | 0.248 | 0.656 |
| G. Nemoral floodplains | 0.340 | 0.559 |

*Note:* Here and in Table 11.5, group numbers of plant formations are given according to Gribova et al. (1980).

and extra-zonal topological-level ecosystems as representatives of other, often remote, zonal types of the geographic environment.

Each group of plant formations specified on a small-scale geobotanical map is identified with a particular group of biogeocoenoses of their upland–extra-zonal series. The map of meso-habitats is simultaneously compiled. In this investigation, we have used our ecological map of landscape groups (see Chapter 5). Each area of the particular formation represented by a poly-vector set of meso-catenas is divided into regional landscape–geochemical types of habitats from eluvial and trans-eluvial to accumulative and super-aqual. New methods of geomorphologic statistics are used (Shary et al. 2002; Sharaya 2009). An intermediate map of habitat types is compiled for each group of formations.

A meso-catena is taken as a homomorphic image of a micro-catena, according to the concept of landscape interrelations (Polynov 1956). An intermediate map of the types of meso-localities is drawn for the entire forested part of the region. Assuming meso-catenae to be a homomorphic image of micro-catena, we assign biogeocoenoses of each previously identified group to meso-geotypes of the corresponding geobotanical range. On the basis of this assumption, the biogeocoenoses of all the specified groups for each of the experimental plots are assigned to habitats of the corresponding geobotanical area. The obtained regional mosaic of phyto-catenas is then filled with the particular metabolic parameters of biogeocoenoses, which makes it possible to specify the zonal–regional types of the geographical environment. In our investigation, forest stability indices were taken.

The distribution pattern of the labile and resistant stability indices of forest formations is shown in Figures 14.4 and 14.5. The lower stability of both types is typical for the

**FIGURE 14.4** The map of indices of resistant labile stability of the groups of forest formations in the Volga River basin on a scale of 1:4,100,000 (calculated and compiled by L.S. Sharaya).

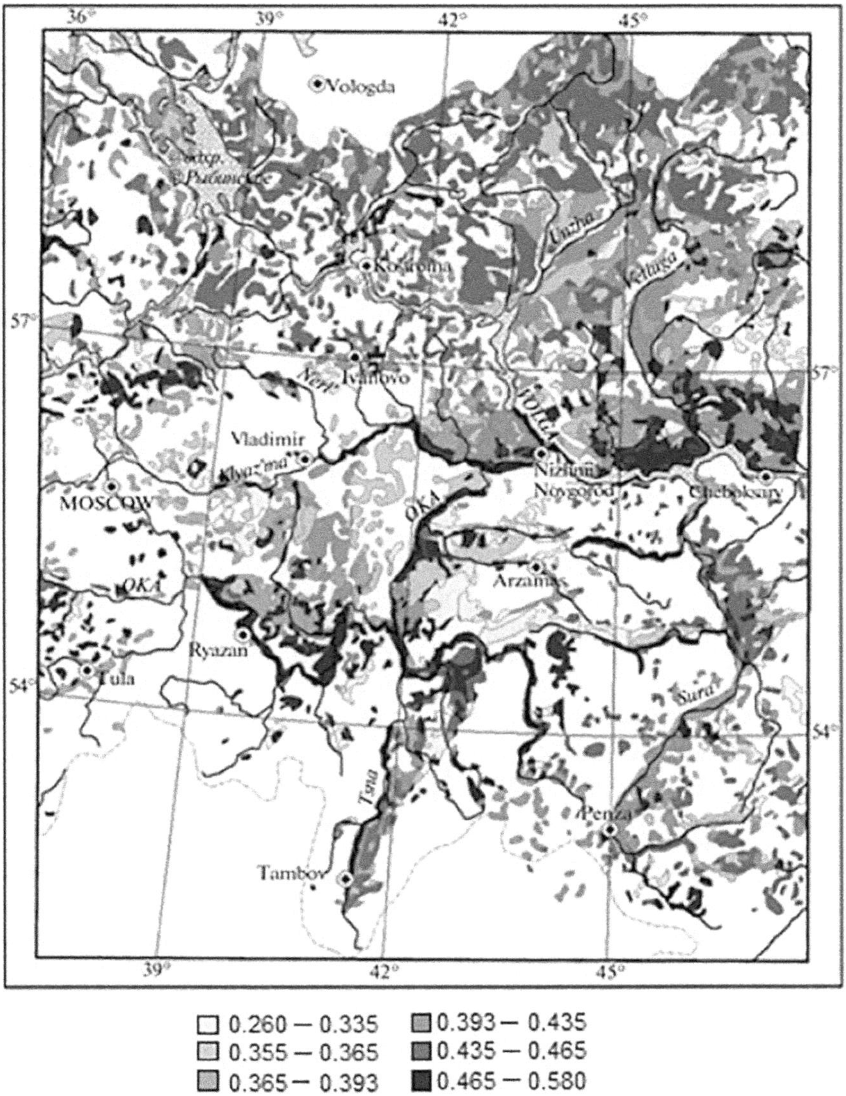


**FIGURE 14.5** The map of indices of resilient labile stability of the groups of forest formations in the Volga River basin on a scale of 1:4,100,000 (calculated and compiled by L.S. Sharaya).

indigenous medium and southern taiga spruce and broadleaf–spruce sub-taiga forests allocated to the extremely northern part of the Volga River basin ($I_{resist}(1)=0.37–0.40$, and $I_{resile}(1)=0.35–0.37$).

Their derived small-leaf forests are characterized by somewhat higher regeneration potential ($I_{resili}(1)=0.41–0.45$). Higher sensitivity to climate variations ($I_{resile}(1)=0.43–0.48$) but less developed mechanisms of functional regeneration ($I_{resile}(1)=0.36–0.40$) are revealed for pine forests of medium and southern taiga located in the interfluves on the left banks of the Volga, Unzha, and Vetluga Rivers. The contrast is more pronounced in the broadleaf–pine forests of the sub-taiga zone in the lowlands of the trans-Volga Region and the lower reaches of the Oka River. They are very sensitive to external effects ($I_{resist}(1)=0.30–0.36$) but are capable of relatively quick functional regeneration ($I_{resile}(1)\geq 0.46–0.50$). A similar ratio between resistance and resilience is typical for small massifs of broadleaf forests in the basins of the Moksha and Sura Rivers and the middle reaches of the Oka River.

## 14.5 MAIN INFERENCES

The methods of calculating and large-scale mapping of the two main types (resilience and resistance) of potential forest ecosystem stability were elaborated. They permit the evaluation of the stability of biogeocoenoses as integral elementary chorological units rather than by some structural or functional characteristics. The elaborated approaches may be inherently assigned to landscape–ecological methods. They are effective because they are based on discrete parameters of biological turnover. These parameters are responsible for the main mechanisms of natural ecosystem stability and provide a large amount of empirical data for stability calculation.

Using the Euclidean distance, we calculated the mobile and inertial stability of forest biogeocoenoses. The obtained data reveal the metabolic diversity of the two main stability types: resistance and resilience. They were expressed as corresponding indices that represent the complicated parameters of the functional status of biogeocoenoses. The relative role of each metabolic factor in a particular index was quantitatively evaluated, and the calculated models were verified. Based on the obtained data, matrix maps of the biogeocoenose stability indices were calculated using new geomorphometric approaches. The stability indices were initially calculated for statistical samples based on matrices of environmental factors for which the statistical correlations revealed by multiple regression were the closest. The calculated stability values were then interpolated and extrapolated for the compilation of maps of natural ecosystem stability.

The analytic and cartographic modeling of forest biogeocoenosis stability in the southern part of the forest belt in the Middle Volga Region allows for the following conclusions. At the initial stage of an unfavorable external effect, the reaction of those forest biogeocoenoses that tend to survive consists of a decrease in the intensity of the annual turnover of the above-ground phytomass and/or the enhancement of mineral plant nutrition by substances from the humus soil horizon. After that, the regeneration potential is realized by the activation of destruction processes of dead organic matter in the forest litter and then by an increase in the annual production of green mass, i.e., by a general enhancement of biological turnover.

Most forest formations in the Oka-Volga River basin are characterized by a high sensitivity to external effects. At the same time, the mechanisms responsible for their resilience are also sufficiently developed. This signifies that the dynamics of the functional and structural parameters of the regional forest ecosystems under the effect of exogenic factors reflect a change in their strategy. The initial adaptive strategy gives way to the strategy of stable development under new environmental conditions. It is aimed at returning the forest ecosystem to its initial status or the formation of a new, stable functional status after the disappearance or weakening of the effect (climate stabilization, reduction of environmental contamination, a drop in recreation or pasturing load, and other conditions).

# Role of Boreal Forests in Regulation of the Carbon Cycle and Mitigation of Present Global Warming

# 15

## 15.1 ADAPTATION AND MITIGATION IS THE FINAL STAGE OF GEO-ECOLOGICAL MONITORING

The problem of the interaction of terrestrial ecosystems and the climate system through the carbon cycle, along with the assessment of the greenhouse effect of the atmosphere, has already been covered quite widely in both domestic and foreign literature (Kobak 1988; Isaev and Korovin 1999; Stinson et al. 2011; Neigh et al. 2013; Le Quere et al. 2015). For example, a method is described for compiling a hybrid map for Russia using satellite and ground data, including both land cover classes and some dominant tree species. On this basis, and using ground-based measurements of organic carbon stocks in the 0–100 cm soil layer, a map of carbon stocks for Russia with a resolution of 1 km was constructed. Using satellite laser scanning and auxiliary data, regional maps of the above-ground carbon stock were constructed in Central Siberia (Nelso et al. 2009), in boreal forests (Neigh et al. 2013), and in forests of China (Chi et al. 2015).


DOI: 10.1201/9781003733119-15

Significant progress has been made in studying the response of the global carbon cycle to changes in the climate system. A picture is provided of the general patterns of the global carbon cycle and the greenhouse effect of the atmosphere in the Holocene and in the modern period in connection with temperature changes (Kondrat'ev and Demirchyan 2000). It was also possible to construct a forecast of the global carbon content in the ocean, atmosphere, and vegetation until 2100, as well as to estimate the increase in the $CO_2$ content in the atmosphere and its effect on air temperature (Quegan et al. 2011).

The knowledge of large-scale biospheric processes is closely associated with the solution of the problem of conservation and reproduction of forest resources over vast territories under the conditions of global climate change. Forest cover is one of the decisive factors for the stability of the continental biosphere (Morozov 1949; Sukachev 1972; Moiseev et al. 1985). Forests occupy more than 49% of the land area of Russia, which corresponds to 20% of the total area of the planet's forest cover (Isaev 2012). The knowledge of large-scale biospheric processes is closely associated with the solution of the problem of conservation and reproduction of forest resources over vast territories under the conditions of global climate change.

Global predictive estimates of the role of forest cover in the regulation of the greenhouse effect of the atmosphere are presented in small-scale scenarios of the carbon budget of the circumpolar boreal forests of Eurasia and North America based on the correlations of their biomes with climatic areas of temperature and precipitation (Gauthier et al. 2015; Shvidenko et al. 2017). The expected impacts of climate change on Russian forests and mechanisms for forest adaptation are considered here to use the latter as a means of mitigating climate signals, which is included in the concept of transition to sustainable forest management in Russia developed by the authors. This concept is still, as the authors themselves write, characterized by 'theoretical and methodological premises.'

Our research outlines the experience of further development of the well-known concept of biotic regulation of the carbon cycle in the biosphere by studying the local mechanisms of the small biological cycle, which ensure the stability of the natural environment in accordance with Le Chatelier's principle (Gorshkov 1994). The landscape–ecological approach to modeling the functional parameters of local ecosystems is based on the biochorological concept: "The material and energy cycles of biogeocoenoses are interconnected and form a giant cycle of the biosphere" (Timofeev-Resovsky 1970, p. 90).

The practical purpose of climate-genic monitoring is to justify the transition to sustainable forest management, i.e., towards adaptive forestry (Shvidenko et al. 2014). We are talking about the creation of such adapted forest communities that provide optimal regulation of the carbon cycle and thus ensure their sustainability in a changing climate. This is precisely the task of unlocking the ecological potential of forests as a necessary condition for the transition to adaptive forestry.

Some general proposals for the environmental support of adaptive forestry can be found in the concept of transitioning to sustainable forest management in Russia, developed in the works of Shvidenko et al. (2014); Shvidenko et al. (2017). This concept involves managing and using forests in such a way as to maintain their biodiversity, sustainability, and productivity, as well as their ability to perform ecological functions, including mitigating current global warming. These four indicators should be considered at the third and final stage of geo-ecological monitoring under this project. According to the aforementioned authors, the system of consistent actions for sustainable forest management should include (1) analysis of the feedback between vegetation and climate; (2) calculations and descriptions of regional indicators of the functional sustainability of forests; (3) creation of a new unified system of forest management and operational monitoring on a zonal–typological basis. The third point in this project can be covered only in terms of setting the task.

Also noteworthy are the methods for solving problems outlined in the Forest Management Program of the Split Rock Lighthouse National Park in Wisconsin, USA (Ontl et al. 2020). This program includes activities aimed at both maintaining existing carbon (defensive actions) and increasing the capacity of forests to sequester carbon in the future (offensive actions). According to this program, one can focus on solving the following tasks to ensure adaptive forestry: (1) maintaining or increasing the area of forest lands and restoring forests after disturbances; (2) maintaining the fundamental ecological functions of the forest by promoting its species and structural diversity; (3) identification of areas with the highest carbon value in the landscape; (4) changing the species composition and structure of the forest to maximize carbon stocks.

## 15.2 THE CONCEPT OF ECOLOGICAL RESOURCES OF FORESTS IN THE REGULATION OF THE CARBON CYCLE

To solve the problems of “adaptation and mitigation” at the third and final stage of geo-ecological forest monitoring, the concept of carbon balances and functional sustainability of forest ecosystems under global climate change is proposed. This concept considers the coupled absorbing and adaptive capacity of forest, mainly boreal, biomes of European Russia under the conditions of current global warming. The scientific search was carried out in accordance with the provisions of the Paris Agreement on climate change (Paris Agreement 2015).

The ideological basis of our research into the above problem was the new aspects of an environmentally oriented paradigm in the doctrine of forests put forward in the works of Gorshkov (1994), Utkin (1995), and Losev (2001). In addition to the known provisions of this paradigm (e.g., the water protection and sanitary protection role of forests), the new aspects contain conceptual propositions on the ecological resources of forest cover, such as its ability to adsorb greenhouse gases in addition to carbon dioxide under climate change. This regulation is aimed at returning the environment to the state optimal for forest ecosystems and contributes to the maintenance of relative stability of its production under varying climate conditions, which provides stability to the mechanisms of carbon cycle regulation as a key element of the biological cycle. The task of “maintenance of … the reproductive capability of forests … to protect forest resources

... at the local, national and global levels" is a component of the *strategy of sustainable forest management* (Shvidenko et al. 2014, p. 14). One of the most important trends of this strategy is to use forests for climate change mitigation (Shvidenko et al. 2017).

One of the important global goals of the Paris Agreement is planning and implementing actions in the field of adaptation – "strengthening adaptive capacities, increasing resilience and reducing the vulnerability of socio-economic and ecological systems to climate change, to promote sustainable development and ensure an adequate adaptive response in the context of the above temperature target" (Article 7 of the Agreement). This is essentially about the need to switch to adaptive forestry, taking into account the special sensitivity of forests to "rapid" climate changes (Shvidenko et al. 2014), to ensure the adaptation of forests to global changes and their use to mitigate unwanted climate changes (Gauthier et al. 2015).

Thus, the concept of ecological resources of forest ecosystems should be expanded to include their ability to adapt to a changing climate. The adaptive properties of forests were quantified by indicators of their functional stability based on the parameters of the biological cycle (Kolomyts et al. 2012). We will consider the dual problem of ecological resources of forest ecosystems: on the one hand, their "work" on additional adsorption/emission of $CO_2$ during global climate changes, and on the other, the degree of sustainability of their functioning under the conditions of these changes.

The total area of relatively undisturbed forests of Russia that can perform the function of stabilizers and regulators of the environment to the maximum extent possible (Utkin 1995) is estimated to be from 3.45 to 4.65 million km$^2$, i.e., up to one-third of all virgin forests in the world (Losev 2001). At the same time, almost half of these are boreal forests, providing more than 90% of the carbon sink of all boreal forests worldwide (Shvidenko et al. 2017). In bioclimatic interpretation, these are the climax or related quasi-climax (coniferous, mixed, and broadleaf) forest communities representing the final stage of endo-ecogenetic (restorative) successions, according to Sukachev (1972). The latter consists of a series of derivative (secondary) small-leaf formations – mostly birch and aspen forests. The final succession stage brings the structure and functions of forest communities into conformity with the given zonal–regional climate conditions. Forest biomass is stabilized, and the closed biological cycle is restored. In mature ecosystems, the carbon increase due to photosynthesis outpaces its losses from heterotrophic respiration, thereby increasing their productivity. The depots of live phytomass and phyto-detritus maintain their stability, thereby stabilizing carbon deposits and blocking their emission (Utkin 1995). Hence, the maximum efficiency of soil–biotic mechanisms of stabilization and regulation of the environment is provided. This suggests that the primary climax and related forest formations, in contrast to derivative formations, must have the maximum ecological resources. This hypothesis should be verified.

The given research provides a comparative quantitative assessment of ecological resources in forest formations by specific and total values of their *carbon balances under the predicted scenarios of climate change*. The ecological resources of two forest categories were assessed: primary and derivative. The main predictive climate scenarios were taken from E GISS (Schmidt et al. 2006), belonging to the family of general circulation models: AOGCMs (Houghton et al. 1996). This model gives the limits of climate changes corresponding to the purposes of the Paris Agreement (2015) (see above). In the boreal belt of the Volga River basin, a decrease in the mean July temperature of –(0.5 to 1.9)°C is expected by 2050, with an increase in the atmospheric humidity factor $F_{hum}(1)$ from 1.15–1.52 to 1.27–1.79 (humidification), and an increase by 0.5 to $0.7^0$ with a decrease in $F_{hum}(1)$ to 1.00–1.36 (aridization) is expected by 2200. Both prediction periods were considered irrespective of their timing, i.e., they have been used as models to answer the "what if ..." question.

However, the currently existing warming trend can lead to a $4^0$% increase in the mean global temperature by 2100 (Le Quere et al. 2015; see also Figure 12.1). At the same time, regional warming over the territory of Russia can be about 6°–11° (Shvidenko et al. 2017), which aligns with the climate prediction by the known extreme model HadCM3, version A2 (Pope et al. 2000), with a decrease in $F_{hum}(1)$ from 1.29–1.37 to 0.78–0.75 in the boreal belt, i.e., to the level of southern forest–steppe (Kolomyts et al. 2012). The results of our landscape–ecological studies based on this model describe a quite probable pattern of dramatic aridization of the natural environment in this part of the East-European subcontinent over the next 100–200 years.

# 15.3 CARBON BALANCE OF FOREST ECOSYSTEMS FOR DIFFERENT CLIMATE TRENDS

## 15.3.1 The General Trends and Objectives of Research

The research strategy consists of identifying the local mechanisms of the biotic regulation of the carbon cycle based on the predicted changes in the discrete parameters of the biological cycle in forest biogeocoenoses for specific variants of climate prediction. The objects of the prognostic analysis should be forest ecosystems with a protective status or a slight degree of disturbance, on the one hand, but with highly contrasting local geomorphological and hydrothermal conditions, on the other hand. It is necessary to quantitatively estimate the responses of different carbon pools in forest biogeocoenoses to global climate changes and predict their functional parameters pertaining to the biotic regulation of atmospheric $CO_2$ content either according to or contrary to Le Chatelier's principle. This will make it possible to determine the most probable role of boreal and sub-boreal forests in reducing the greenhouse effect of the atmosphere during global climate warming and to estimate to what degree forests of the temperate zone can actually maintain the stability of the biosphere under conditions of climate change.

Having solved this problem, we will be able to determine the spectrum of the main forest formations of boreal and nemoral forests, some of which improve the stability of the terrestrial ecosystem cover and others, conversely, enhance its destabilization under the predicted thermo-arid climatic trend in the given region. This objective is one of the key issues of modern global ecology that remains an open question thus far, which, in particular, causes heated debates about the Kyoto Protocol (Grace 2004).

Prognostic ecological maps should reflect the local diversity of changes in the biological cycle and carbon pools. This diversity is accounted for by the spatial variation of the hydrothermal and edaphic conditions of geo(eco)systems. Special attention should be paid to the mechanisms and consequences of the biotic regulation of the carbon cycle in forest and steppe zonal ecotones, where forest communities are always under hydrothermal conditions that are close to critical ones and, hence, are the most sensitive to climate changes.

In global ecology, it has already been understood that the spatial variation of carbon fluxes, mainly related to the distribution of forests, should be taken into consideration (Grace 2004). The changes are expressed as peaks and troughs of $CO_2$ emissions in mature forest ecosystems. This results in flows of biogenic elements between local ecosystems, which are part of the biotic regulation of the environment (Gorshkov et al. 1999). We suggest methodological approaches that agree with these theoretical views.

## 15.3.2 Methods of Prognostic Analysis

The results of large-scale landscape–ecological surveys in experimental test grounds, characterizing different zonal–regional conditions, are used for preliminary estimation of the pattern of climate-related changes in the discrete parameters of the biological cycle in forest ecosystems at the local level. In our studies on the middle and upper Volga regions, each of these grounds consisted of 40–50 test plots (specific mapping units of facies). The state of biogeocoenoses was described using 80 structural and functional parameters (Kolomyts 2022). The resultant comprehensive discrete data allowed us to implement the biochorological approach, enunciated by Timofeev-Ressovsky back in the 1960s, as a tool for prognostic landscape ecological study and to develop analytical and cartographic empirical statistical models of the changes in the biological cycle and carbon balance with spatially distributed parameters (Sharaya 2009; Kolomyts et al. 2012). The prediction was based on (1) our methods of empirical statistical modeling of topological-level ecosystems and discrete metabolic parameters and (2) hydrothermal ordination of these parameters in the space of local landscape series (catenae) for different zonal conditions. For estimating changes in the carbon contents of individual biotic components and forest biogeocenoses in general, we used the traditional forest management method (Tselniker 2006), based on the estimation of the dynamics of live phytomass and necromass (and the humus mass), which yields the best results in $CO_2$ balance calculations over long periods of time. The change in the carbon mass flux $\Delta C(Fa)$ over the plant cover, i.e., the *carbon balance* of the soil–plant system for a specified prognostic time, was calculated as

$$\Delta C(\mathrm{Fa}) = \Delta C(\mathrm{WD}) + \Delta C(\mathrm{ML}) + \Delta C(\mathrm{HU}) - \Delta C(\mathrm{BS}) - \Delta C(\mathrm{BV}) - \Delta C(\mathrm{BR}) \quad (15.1)$$

where WD and BV are live skeletal (wood) and green biomasses, BR is the root biomass, ML and HU are the weights of the forest litter and humus; $\Delta C$ is the change in carbon weight. The $\Delta(Fa)$ parameter was calculated for every group of biogeocoenoses in all regional ecosystems and for all prognostic times. According to the cybernetic concept

(Ashby 1956), the phytomasses were regarded as "output" functional characteristics of the system, without analysis of metabolic processes within the "black box". This inevitably made the resultant prognostic estimates less accurate; however, it also allowed us to present the functional states of biogeocoenoses as integrated formations in the vein of Sukachev's (1972) theory, thereby obtaining a spatially differentiated result of all biogeochemical processes involved in $CO_2$ gas exchange. Thus, we attempted to solve the problem in terms of geographic ecology, which is the essential difference of our approach from those used in other published studies on this subject (both in Russia and worldwide).

The transition from the phytomass itself to the carbon content in it was carried out using carbon coefficients known in the literature (Kobak 1988; Orlov et al. 1996; Isaev and Korovin 1999; Osipov 2004; etc.). The values of some of these coefficients (e.g., for BV, ML, and HU) were differentiated for each experimental polygon and for each facies group at the polygon, depending on the zonal and local habitat conditions (Table 15.1).

Each member of the right part of Equation (15.1) may have both positive and negative values. At positive values, the first three items give an increase in $CO_2$ emission from soil–plant cover into the atmosphere, while the other three give a decrease in this flow. In the latter case, the dead mass pool acts as an additional carbon source, while the live phytomass plays the role of its sink (deposit). With negative values of the above parameters, the picture is quite the opposite: in the reduced branch of biological turnover, carbon dioxide release into the atmosphere decreases, while autotrophic biogenesis becomes less intensive and consumes less $CO_2$, thus compensating for the resulting deficiency of carbon dioxide in the atmosphere.

As a result, the total balance of changes in carbon exchange between biogeocoenoses and the atmosphere $\Delta C(Fa)$ is formed, which must show whether this group of forest biogeocoenoses is a source of additional emissions. In the former case, there is a negative feedback directed at realization of Le Chatelier's principle for stabilization or even weakening of the primary thermo-arid climatic signal; in the latter case, there is a positive feedback, which leads to intensification of the greenhouse effect in the atmosphere and, consequently, to warming itself.

Such *carbon balance* of local ecosystems for all five predicted periods is presented in Figure 15.1. The zero balance shows the maintenance of the base level of influence, for which the following condition is accepted: the direct and reverse carbon flows between biogeocoenoses and the atmosphere are balanced. The negative and positive carbon balances show that the biotic regulation of the carbon cycle is directed at additional destabilization of the biosphere in the first case and at maintenance of its stability in the changing climate in the second case.

The results of the local prognostic analysis are presented in both graphic–analytical models (Kolomyts et al. 2012) and large-scale maps (with the scale varying from 1:50,000 to 1:100,000) of the changes in ecosystems at the topological level and the components of the carbon cycle for individual ecoregions (Kolomyts 2005; Sharaya 2009).

The quantitative approaches to prognostic ecological and geobotanical mapping recently developed abroad (MacMillan et al. 2009) are intended for drawing large-scale maps. However, these approaches have never been used for creating prediction maps

**TABLE 15.1** Carbon coefficients (*Cc*) for live and dead phytomass in different groups forest biogeocenoses at the experimental sites of the Volga River basin

| *GROUPS OF BIO-GEOCOENOSES (TABLE 5.3)* | *EXPERIMENTAL TEST SITES* | | | | | | | |
|---|---|---|---|---|---|---|---|---|
| | *ZHIGULI* | *BY-SURA* | *GREEN TOWN* | *SHCHELO-KOVSKY FARMSTEAD* | *VYKSA* | *KERZHE-NETS* | *KUD'MA* | *BY-OKA-TERRACE RESERVE* |
| **Green mass, BV** | | | | | | | | |
| 1 | 0.59 | 0.52 | 0.52 | 0.59 | 0.52 | 0.51 | 0.52 | 0.52 |
| 2 | 0.57 | 0.52 | 0.52 | 0.55 | 0.51 | 0.51 | 0.52 | 0.52 |
| 3 | 0.54 | 0.51 | 0.52 | 0.54 | 0.52 | 0.52 | 0.52 | 0.52 |
| 4 | 0.54 | 0.52 | 0.52 | 0.55 | 0.51 | 0.52 | 0.52 | 0.52 |
| 5 | 0.54 | 0.52 | 0.52 | 0.54 | 0.50 | 0.50 | 0.52 | 0.52 |
| 6 | 0.55 | 0.52 | 0.52 | 0.55 | 0.50 | 0.50 | 0.52 | 0.51 |
| **Forest litter mass, ML** | | | | | | | | |
| 1 | 0.52 | 0.49 | 0.49 | 0.70 | 0.47 | 0.49 | 0.49 | 0.49 |
| 2 | 0.51 | 0.50 | 0.50 | 0.50 | 0.45 | 0.48 | 0.49 | 0.49 |
| 3 | 0.50 | 0.49 | 0.49 | 0.55 | 0.46 | 0.49 | 0.49 | 0.49 |
| 4 | 0.50 | 0.49 | 0.48 | 0.50 | 0.46 | 0.48 | 0.49 | 0.49 |
| 5 | 0.50 | 0.49 | 0.49 | 0.45 | 0.045 | 0.47 | 0.49 | 0.48 |
| 6 | 0.51 | 0.48 | 0.49 | 0.52 | 0.46 | 0.47 | 0.49 | 0.49 |

*Note:* For the rest of the phytomass in general for all test sites and groups of biogeocoenoses, the parameter *Cc* is equal to: for living and dead above-ground skeletal, as well as the root mass of trees and shrubs (BS, BR, WD) – 0.50; for the mass of humus *H* – 0.58.

on a time scale. Moreover, we do not know of any experience in using these methods for mapping the functional parameters of forest ecosystems within the framework of a global climatic scenario. In landscape ecological analysis, we use approaches based on *new geomorphometric methods* (Shary et al. 2002) for creating spatiotemporal maps.

Geomorphometry, as the science of quantitative description of terrain, which has been substantially developed over the past 30 years, provides methods for obtaining the pattern of the spatial distribution of soil–biotic and hydrothermal parameters of geo(eco)systems.

The multifold edificatory properties of the terrain make it possible to interpolate and extrapolate data measured (or calculated) in representative test grounds (which was justified during the analysis); this substantially extends the analyzed area. Earlier methods of quantitative description of terrain are restricted to the use of six basic morphometric parameters. We have introduced a system of 18 parameters, developed new geomorphological classification frameworks, and derived better algorithms for calculating morphometric characteristics (Shary et al. 2002).

The maps of the basic and predicted metabolic parameters based on statistically significant relationships between the functional characteristics of the ecosystems and the terrain allow the changes in the metabolic parameters to be quantitatively estimated, taking into account their spatial differentiation and the relative contributions of the

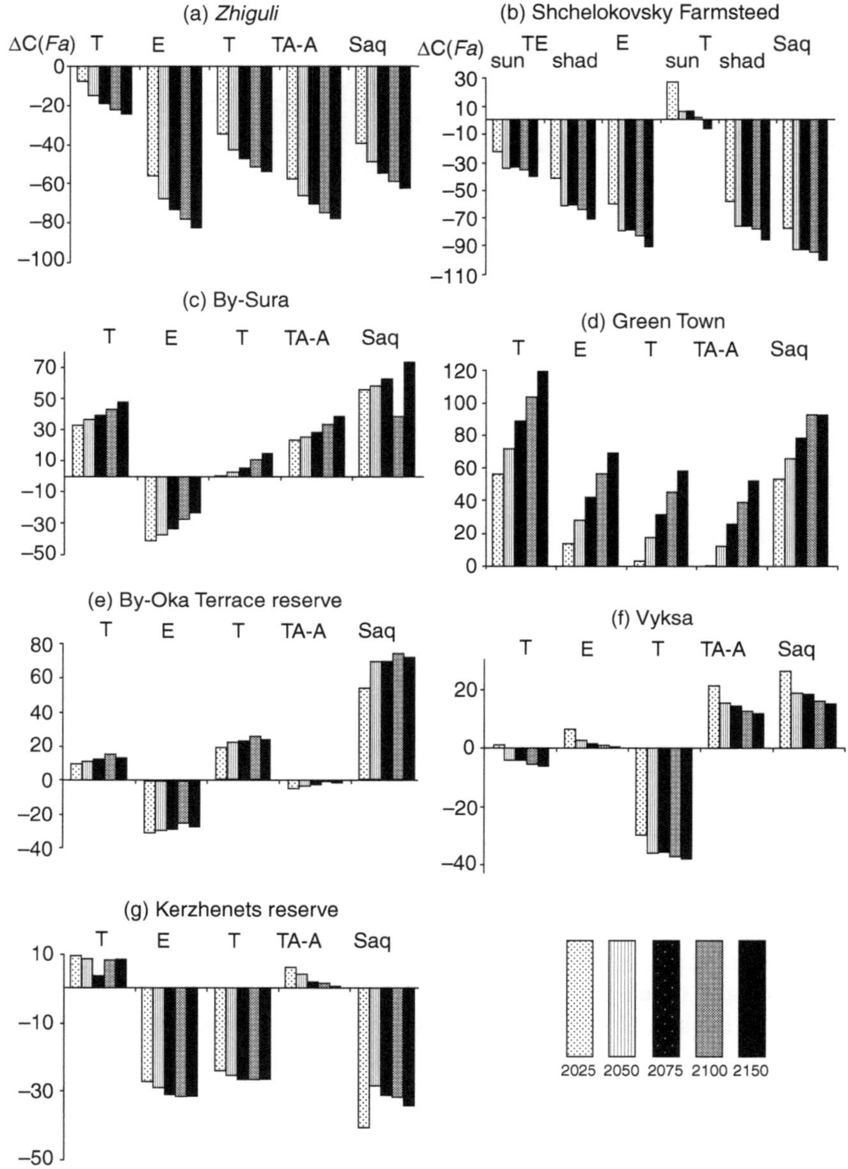


**FIGURE 15.1** Landscape couplings of predicted *C* balance dynamics in forest biogeocoenoses under different zonal–regional conditions of the Middle and Upper By-Volga Regions (see Section 5.3), according to the scenarios of the HadCM3 model. Δ*C*(*Fa*) – carbon balance (t/ha); TE, E, T,... – local sites (see in the text); solar exposed slopes: sun – sunny, shad – shadow and neutral.

main forest associations and formations of the southern forest zone to the changes in the carbon cycle. To map the initial (basic) carbon contents of different pools of forest biogeocoenoses in a given ecoregion, we used NASA satellite data on the terrain, namely, the matrices of elevations of the Earth's surface with a resolution of 90 m obtained during the international (United States–Italy–Germany) Shuttle Radar Topography Mission implemented in the year 2000.

We transitioned from the local to the regional level of prognostic modeling using the inductive–hierarchic extrapolation method, which was described in the previous chapter. The resultant vegetation–catena mosaic was filled with the basic or prognostic metabolic parameters of biogeocoenoses, which are now used as local representatives of different zonal–regional types or sub-types of the geographic environment. In this case, such parameters included the total carbon base stocks in the groups of forest formations and the total carbon balance for the predicted periods. Finally, the weighted average values $\Sigma C(\text{bas})$, $\Delta C(Fa)$ were calculated according to the area share in the given areal range of particular groups of formations based on information analysis of each group of plant formations in the basin. The methodology described here is a fundamentally new approach to the regional cartogramming of the basic carbon content and predicted carbon balance in forest ecosystems based on the results of field landscape–ecological surveys.

Regional maps were composed using the NASA Shuttle Radar Topography Mission elevation data. The Oka River basin was taken as a model territory at the regional level. The matrix for this basin was transformed into the Kavraysky VII INT projection for European Russia with a 900 m resolution. Beforehand, we applied "GIS Eco" software (Shary 2001) to perform the statistical analysis of spatial differentiation of changes in total C content for predictive periods for the plant formations and their combinations.

## 15.3.3 Analytical Predictive Modeling

The local mechanisms of biotic regulation of the carbon cycle were investigated based on predicted changes in the biological cycle of forest biogeocoenoses under climate prediction scenarios up to 2150–2200, according to the global models HadCM3 and E GISS. The second model shows a more complex picture: the cold arid trend until the middle of the 21st century is followed by weak climate warming and humidification (Table 15.2). Our calculations have demonstrated that the sensitivity of parameters of the biological cycle to external impact is not adequate to the scope of changes that occur in the course of this impact. The main results of empirical–statistical prediction are given in Table 15.2.

1. The decomposition part of the biological cycle is the most sensitive to climatic impacts. Nevertheless, during sufficiently long periods of time, *the maximal (by absolute values) changes occur not in branches but in autotrophic biogenesis, indicated by the general net production of forest communities.* Moreover, skeletal tree–shrub phytomass (*BS*), mainly wood gain, constitutes a significant portion of the shifts in productivity and carbon

content (Figure 15.2). Such regularity is outlined in both climate prediction models and is typical for all the forest ecosystems under consideration (Tables 15.2 and 15.3). In 2100, the changes in total carbon content in its extreme values will range from –78÷100 t/ha at the zonal forest–steppe ecotones (test grounds Zhiguli and Shchelokovsky Farmstead) to +115÷120 t/ha in coniferous/broadleaf forests near the northern boundaries of the nemoral forest sub-zone (Green Town). The changes in carbon content in *BS* will be, respectively, 25÷40 t/ha and +85÷105 t/ha. This means that in the coming century, the content of carbon conserved in skeletal phytomass may change by ±25÷50% or more of the base value on average. This fact indicates that the phytomass of forest ecosystems of the Middle Volga River may significantly change, affecting the regulating role of the productivity of mixed and nemoral forests in their carbon exchange with the atmosphere.

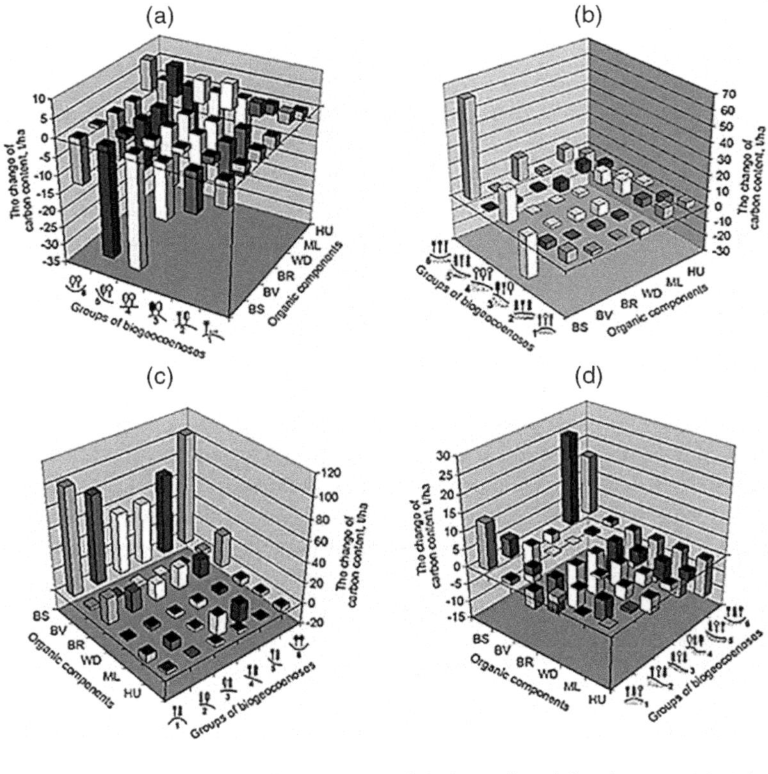


**FIGURE 15.2** The changes in carbon content (t/ha) predicted for the year 2150 in separate biogeocoenotic pools for different zonal–regional conditions of the Volga River basin, according to the HadCM3 model. Test grounds: (a) Zhiguli; (b) By-Oka-Terrace reserve; (c) Green Town; (d) Kerzhenets. For conventional meanings of biogeocoenoses groups, see Table 5.3.

**TABLE 15.2** The predicted climatic parameters for different periods by three types of hydrothermal trends, according to the HadCM3 (version A2) and E GISS models

*HADCM3 MODEL*

| *TEST GROUND* | *BASIC* $C_{hum}$ | *2075 (THERMO-ARID TREND)* $\Delta t_{Jan}$ | $\Delta t_{Jul}$ | $\Delta r_{year}$ | $C_{hum}$ | *2150 (THERMO-ARID TREND)* $\Delta t_{Jan}$ | $\Delta t_{Jul}$ | $\Delta r_{year}$ | $C_{hum}$ |
|---|---|---|---|---|---|---|---|---|---|
| Zhyguli | 0.95 | 2.7 | 3.2 | –38 | 0.71 | 5.9 | 7.1 | –50 | 0.36 |
| By-Sura | 1.07 | 0.8 | 3.1 | –16 | 0.76 | 7.2 | 7.0 | –34 | 0.45 |
| BOTR | 1.52 | 3.2 | 4.0 | –96 | 1.12 | 6.5 | 5.1 | –31 | 0.74 |

**E GISS model**

| *TEST GROUND* | *BASIC* $C_{hum}$ | *2050 (COLD–HUMID TREND)* $\Delta t_{Jan}$ | $\Delta t_{Jul}$ | $\Delta r_{year}$ | $C_{hum}$ | *2200 (THERMO-HUMID TREND)* $\Delta t_{Jan}$ | $\Delta t_{Jul}$ | $\Delta r_{year}$ | $C_{hum}$ |
|---|---|---|---|---|---|---|---|---|---|
| Zhyguli | 0.95 | –1.0 | –0.5 | –19 | 0.99 | 1.9 | 1.2 | 13 | 0.80 |
| By-Sura | 1.07 | –1.0 | –1.0 | 15 | 1.25 | 1.7 | 1.3 | 16 | 0.92 |
| BOTR | 1.52 | –1.3 | –1.9 | 61 | 2.02 | 2.6 | 0.9 | 58 | 1.47 |

*Note:* The base period falls within the interval of weather observations from 1885 to 1985. Climatic parameters: $\Delta t_{Jan}$, $\Delta t_{Jul}$, and $\Delta r_{year}$ – the deviations of the mean January and mean July temperatures, as well as annual precipitation, from their base values; $C_{hum}$ –Vystosky–Ivanov's annual atmospheric humidity factor.

**TABLE 15.3** Predicted partial and balance changes in the carbon content (t/ha) relative to the base period in flat-interfluve forest biogeocoenoses in different zonal–regional conditions of the Middle and Upper Volga Region, according to the HadCM3 model

| *PARAMETERS* | *BASIC VALUES* | *CHANGES IN PREDICTED PERIODS* 2025 | 2050 | 2075 | 2100 | 2150 |
|---|---|---|---|---|---|---|
| **a. Zonal ecotone of forest and steppe (Zhyguli low-mountain range – National Nature Park "Samarskaya Luka")** | | | | | | |
| BS | 101.75 | –23.05 | –27.32 | –31.55 | –35.78 | –39.66 |
| BV | 4.16 | –1.60 | –1.77 | –1.93 | –2.10 | –2.26 |
| BR | 30.25 | –11.90 | –14.08 | –16.13 | –18.06 | –19.72 |
| WD | 10.24 | –6.98 | –7.53 | –8.02 | –8.45 | –8.80 |
| ML | 5.35 | 3.56 | 4.90 | 6.34 | 7.78 | 8.99 |
| HU | 113.30 | –10.54 | –15.29 | –18.62 | –20.68 | –21.66 |
| Balance | **265.05** | **– 50.51** | **–61.09** | **–69.91** | **–77.29** | **–83.11** |
| **b. Right-bank of Sura River, mixed, and broadleaf forest – National Nature Park "Chavash Forest" (test ground By-Sura)** | | | | | | |
| BS | 123.42 | –18.99 | –14.77 | –10.84 | –3.31 | 2.90 |
| BV | 5.17 | –0.60 | –0.61 | –0.84 | –0.93 | –1.13 |
| BR | 40.40 | –8.88 | –8.04 | –7.82 | –6.46 | –5.64 |
| WD | 4.90 | –1.06 | –1.23 | –1.46 | –1.71 | –1.91 |
| ML | 5.88 | –1.28 | –1.60 | –2.03 | –2.45 | –2.79 |
| HU | 31.64 | –9.76 | –11.09 | –11.45 | –13.53 | –14.76 |
| Balance | **211.41** | **–40.56** | **–37.34** | **–34.42** | **–28.39** | **–23.33** |

*(Continued)*

**TABLE 15.3 (*Continued*)** Predicted partial and balance changes in the carbon content (t/ha) relative to the base period in flat-interfluve forest biogeocoenoses in different zonal–regional conditions of the Middle and Upper Volga Region, according to the HadCM3 model

| *PARAMETERS* | *BASIC VALUES* | *CHANGES IN PREDICTED PERIODS* | | | | |
|---|---|---|---|---|---|---|
| | | *2025* | *2050* | *2075* | *2100* | *2150* |
| **c. High-plain Volga Region at the northern boundary of the nemoral forest zone. Mixed forests in the southern vicinity of the city of Nizhny Novgorod (stow "Green Town").** | | | | | | |
| BS | 134.14 | 13.73 | 26.61 | 38.16 | 49.82 | 60.19 |
| BV | 6.59 | 1.90 | 2.32 | 2.65 | 3.01 | 3.30 |
| BR | 39.48 | 4.44 | 7.99 | 11.09 | 14.28 | 17.06 |
| WD | 4.06 | –3.46 | –3.63 | –3.73 | –3.80 | –3.84 |
| ML | 5.48 | –0.13 | –1.00 | –1.60 | –2.05 | –2.38 |
| HU | 11.16 | –1.71 | –2.28 | –2.65 | –3.11 | –3.43 |
| Balance | **201.36** | **14.77** | **30.0** | **43.91** | **58.15** | **70.9** |
| **d. Southern strip of the mixed-forest zone. Left bank of the middle Oka River. Moraine–erosion plain. Pine–lime–oak forests (By-Oka-terrace biosphere reserve).** | | | | | | |
| BS | 310.18 | –27.14 | –27.85 | –27.76 | –27.99 | –27.91 |
| BV | 11.54 | –1.98 | –1.88 | –1.89 | –1.87 | –1.88 |
| BR | 84.81 | –4.11 | –4.30 | –4.28 | –4.34 | –4.32 |
| WD | 31.96 | –7.44 | –7.19 | –7.21 | –7.15 | –7.17 |
| ML | 14.96 | 6.58 | 9.44 | 9.58 | 12.69 | 11.14 |
| HU | 22.05 | 2.84 | 2.52 | 2.55 | 2.48 | 2.51 |
| Balance | **475.5** | **–31.24** | **–29.25** | **–29.0** | **–26.17** | **–27.61** |
| **e. South strip of sub-taiga zone on the over-Volga Region, moraine–outwash forested lowland. Fir–pine forests – biosphere reserve "Kerzhenets".** | | | | | | |
| BS | 84.7 | –13.00 | –12.64 | –11.86 | –11.85 | –1.15 |
| BV | 4.41 | 1.21 | 1.19 | 0.82 | 0.86 | 0.60 |
| BR | 17.08 | –9.44 | –9.81 | –11.68 | –11.60 | –12.59 |
| WD | 7.85 | 0.40 | –0.26 | –0.31 | –0.40 | –0.64 |
| ML | 10.15 | –3.91 | –4.16 | –4.71 | –4.73 | –3.45 |
| HU | 13.69 | –2.45 | –3.41 | –3.50 | –3.66 | –4.09 |
| Balance | **137.88** | **–27.19** | **–29.09** | **–33.24** | **–31.38** | **–31.32** |

*Note:* BS – living skeletal phytomass; BV – verdure mass; BR – root mass; WD – dead skeletal phytomass; ML – forest litter mass; HU – humus mass.
The bold values are the Balance (i.e. Sum by vertical).

2. Global warming will induce, in some cases, an abrupt drop and, in other cases, a significant increase in forest productivity. These effects are expected to have a direct response in the carbon balance and to be the key control over the biotic decline of the production potential of forest communities that is

being observed. Carbon deposits in skeletal phytomass will decrease most quickly in flat-interfluve mesophilic elm–lime and oak forests of erosion–loamy uplands (see Figures 14.1a–c, and 14.2a and b). Their role as a sink for atmospheric carbon will be reduced.

3. The productivity of marginal forest communities at the zonal forest–steppe ecotone will decrease to the greatest extent. In the skeletal tree–shrub pool of flat-interfluve broadleaf forests of the Zhiguli mountain range, the loss of carbon content in 2050 will vary from 2–6 to 26 t/ha and from 9–14 to 34 t/ha by E GISS and HadCM3, respectively. By the end of the 21st century, this deficiency will reach 32 and 39 t/ha, respectively. It is obvious that the climate prediction models do not differ too much.

   Simultaneously, the rate of decay of skeletal dead mass and humus will significantly increase, though the forest litter mass will grow due to the increasing deficit of precipitation (see Table 15.2a). As a whole, the detritus branch here will contribute to further accumulation of $CO_2$ in the atmosphere and, in combination with the production branch, will induce the progressing disturbance of Le Chatelier's principle at the southern boundary of the forest belt, resulting in the intensification of global warming.

   In the fluvio-glacial plains of the nemoral forest sub-zone, the moderate hydromorphic birch–spruce–pine forests on the transit sites are able to markedly reduce the rates of carbon conservation in skeletal phytomass as well (see Figure 15.1f). Rather small changes in productivity are expected in xeromorphic biogeocoenoses, particularly forest–meadow–steppe communities of steep south-facing slopes and pine forests with steppe species at well-drained watersheds. Hence, their contribution to carbon deposition will be insignificant.

4. On the contrary, a significant increase in productivity is anticipated in high-plain mixed and broadleaf forests of the nemoral forest sub-zone and in the southern sub-taiga zone. Mixed and dark coniferous forest biogeocoenoses will significantly increase both their primary productivity and reserves of living organic matter. Accordingly, conservation of atmospheric carbon in perennial skeletal phytomass of these topo-ecosystems will increase and, as a consequence, the significance of this phytocoenotic pool as a carbon.

   The sink will increase as well (see Figures 15.1d and 15.2c). Positive biotic regulation of the carbon cycle will be widely spread here. This process will be particularly significant in TE xeromorphic and super-aqual (Saq) hydromorphic boreal forests, which will play the role of the major carbon sink. In 2050–2075, the value of additional carbon deposited in wood will range from 9–11 to 38–57 t/ha according to GISS and from 36–53 to 77–83 t/ha according to HadCM3; by the end of the 21st century, its values will be 38–87 and 75–105 t/ha, respectively. Simultaneously, an increased rate of decomposition of forest litter and humus is expected. The effect of forest productivity growth, which is positive for the carbon balance, will be reduced to a certain extent by the resulting $CO_2$ emissions into the atmosphere.

In sub-taiga outwash forested lowlands, Le Chatelier's principle will be realized only due to the functional shifts of semi-swamped coniferous forests at the lower elements of the catena, where the improvement of hydrothermic conditions in the soil will induce a significant increase in productivity (see Figure 15.1f). At the same time, in flat-interfluve and neighboring transit spruce–pine and spruce–oak–lime communities, autotrophic biogenesis will weaken against the significant activation of decomposition of above-ground dead mass and humus, resulting in a general tendency toward negative biotic regulation of the carbon cycle.

Far more unambiguous and intensive changes may occur in the detritus branch of the biological cycle (ΔML) in the outwash forested lowlands of the more northern sub-taiga region (see Figure 15.2d). Soil temperature increases and desertification abruptly activate the degradation of dead organic matter, with a higher degree of its utilization and additional release of $CO_2$ into the atmosphere.

At a moderate cold–humid signal, the negative values of $\Delta C(Fa)$, up to –(45–60) tons/ha, will be prevalent in upland pine/lime/oak forests of the Middle By-Oka region (the By-Oka-Terrace reserve) and eluvial lime forests of loamy sediments of the By-Sura region. The thermo-humid signal will cause a rather wide range of the carbon balance: from +(65–90) t/ha in TE spruce/pine and pine/broadleaf forests of the By-Sura outwash–moraine plain up to –55ч65 t/ha in flat-interfluve and TA broadleaf forests of the Zhiguli mountain range (see Table 15.4).

## 15.3.4 Mapping Predictive Scenarios of the Carbon Balance

The base and predictive matrix maps of forest ecosystem characteristics derived from such models (Figures 15.3 and 15.4) are a helpful tool not only for a better understanding of elementary and complex intra-system interactions but also for decision-making in forestry. In the Zhiguli low-mountain range, the maximum $CO_2$ emission under global warming will be characteristic of gentle near-watershed poorly insolated slopes, with the minimum catchment area. On steep slopes, carbon emissions will be negligible. In the By-Oka-Terrace reserve, the maximum total carbon stocks in the base period are characteristic of low-inclined near-watershed regions with the minimum catchment area. In the predicted period, these sites, as well as the relatively steep slopes, especially south-facing ones, will be characterized by the maximum $CO_2$ emission. The forest stands in ravines and near-talweg locations will be the pools of maximum carbon deposition.

Let us now consider plant formations, considering that their carbon characteristics involve not only vegetation but soils as well. The maximum basic total carbon stocks (above 200 t/ha) were observed in the southern taiga and sub-taiga pine forests and the birch forests that replace them during succession. The broadleaf–pine sub-taiga forests,

**TABLE 15.4** The changes in the total carbon content (t/ha) in different forest biogeocoenoses and in their organic components in three natural areas of preferential protection in the Oka-Volga River basin for two prognostic periods: 2050 (cooling) and 2200 (warming) according to the E GISS climate model

| | | *NATIONAL NATURE PARK "SAMARSKAYA LUKA" (SOUTH FOREST–STEPPE* | | *NATIONAL NATURE PARK "CHAVASH FOREST" (TYPICAL FOREST–STEPPE)* | | *BY-OKA-TERRACE BIOSPHERE RESERVE (SOUTH SUB-TAIGA)* | |
|---|---|---|---|---|---|---|---|
| *CHARACTERISTICS* | | *2050 г.* | *2200 г.* | *2050 г.* | *2200 г.* | *2050 г.* | *2200 г.* |
| Groups of Biogeocoenoses (Types of Sites) | 1 (TE) | 9.55 [–] | –12.23 [–] | 77.33 [–] | 90.29 [+] | 20.39 [–] | 29.46 [+] |
| | 2 (TE, *E*) | –16.20 [–] | –38.72 [–] | 46.83 [+] | 63.60 [+] | –13.13 [+] | –0.44 [–] |
| | 3 (*E*, *t*) | 3.34 [–] | –27.76 [–] | **–61.03** [+] | **–32.51** [–] | **–44.80** [+] | **–27.82** [–] |
| | 4 (*E*–TA) | **–15.20** [+] | **–58.88** [+] | 12.34 [–] | 31.82 [+] | 13.85 [–] | 29.01 [+] |
| | 5 (TA, *A*) | –36.04 [+] | –65.03 [–] | 7.05 [–] | 27.24 [+] | 7.26 [–] | 11.35 [+] |
| | 6 (Saq, EA) | 1.22 [–] | –35.79 [–] | 35.72 [–] | 63.65 [+] | –26.02 [+] | 11.82 [+] |
| | *Average* | *–8.89* [+] | *–39.74* [–] | *19.71* [–] | *40.68* [+] | *–7.07* [+] | *8.90* [+] |
| Organic components | BS | –12.65 [+] | –22.93 [+] | 3.56 [–] | 21.69 [+] | –6.38 [+] | 6.17 [+] |
| | BV | –0.57 [+] | –1.14 [–] | – 0.14 [+] | 0.12 [+] | –0.63 [+] | –0.08 [–] |
| | BR | 1.98 [–] | –5.11 [–] | 2.92 [–] | 5.23 [+] | –1.81 [+] | 2.19 [+] |
| | ML | –1.54 [+] | 1.38 [+] | –1.54 [+] | –3.12 [–] | –5.02 [+] | 0.29 [+] |
| | WD | –2.83 [+] | –5.36 [–] | 24.72 [–] | 28.62 [+] | –0.68 [+] | 0.74 [+] |
| | HU | 0.42 [–] | –6.52 [–] | –9.81 [+] | –11.85 [–] | 7.45 [–] | –0.42 [–] |
| | *Sum* | *–8.89* [+] | *–39.74* [–] | *19.71* [+] | *40.68* [+] | *–7.07* [+] | *8.90* [+] |

*Note:* Types of sites by Glazovskaya (1964): *E* – eluvial; TE – trans-eluvial; *T* – transit; TA – trans-accumulative; *A* – accumulative; Saq – super-aqual; EA – eluvial–accumulative. Flat-interfluve biogeocoenoses are in bold. [+] and [–] denote the attenuation and intensification of the climatic trend, respectively. Other symbols can be seen in Table 15.3.

Italic values show the Sum more distinctly.

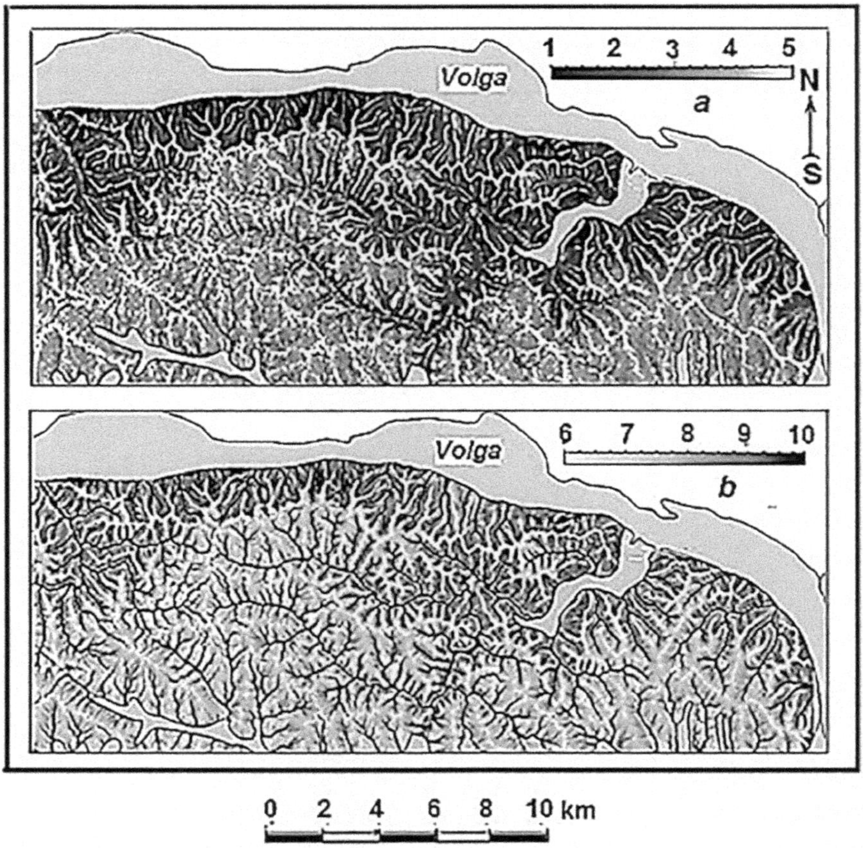


**FIGURE 15.3** Zhiguli low-mountain ecoregion – National Nature Park "Samarskaya Luka". Base total *C* content in forest ecosystems (a) and changes in total *C* for the predicted period of 2075 (b), in t/ha, according to the HadCM3 model. The maps were calculated and compiled by L.S. Sharaya. Total basic carbon stocks (t/ha): 1–100; 2–174; 3–200; 4–227; 5–420. Changes in carbon stocks (t/ha): 6÷(–297); 7÷(–101); 8÷(–63); 9÷(–25); 10÷(+15). Main predictors for the calculation of maps – insolation – coming to the slopes (illumination), steepness of slope, maximal catchment area. Step of lattice – 50 m. Mean change, calculated by matrix: for 2075, –65.5 t/ha; for 2150, –74.0 t/ha.

as well as the oak and lime forests of the nemoral forest sub-zone and their derivatives, contain much less carbon: 140–180 t/ha. The minimum total carbon stocks (100–130 t/ha) account for the middle and southern taiga spruce and broadleaf–spruce sub-taiga forests (Figure 15.5). The empirical data are generally in agreement with the review materials on forest phytomass regarding individual species and their carbon stocks in the respective regions of European Russia (Usoltsev 2003) but make them much more precise as they characterize the real forest formations and not idealized pure plantations. In the works of the cited author, it has been noted that the total carbon stocks in

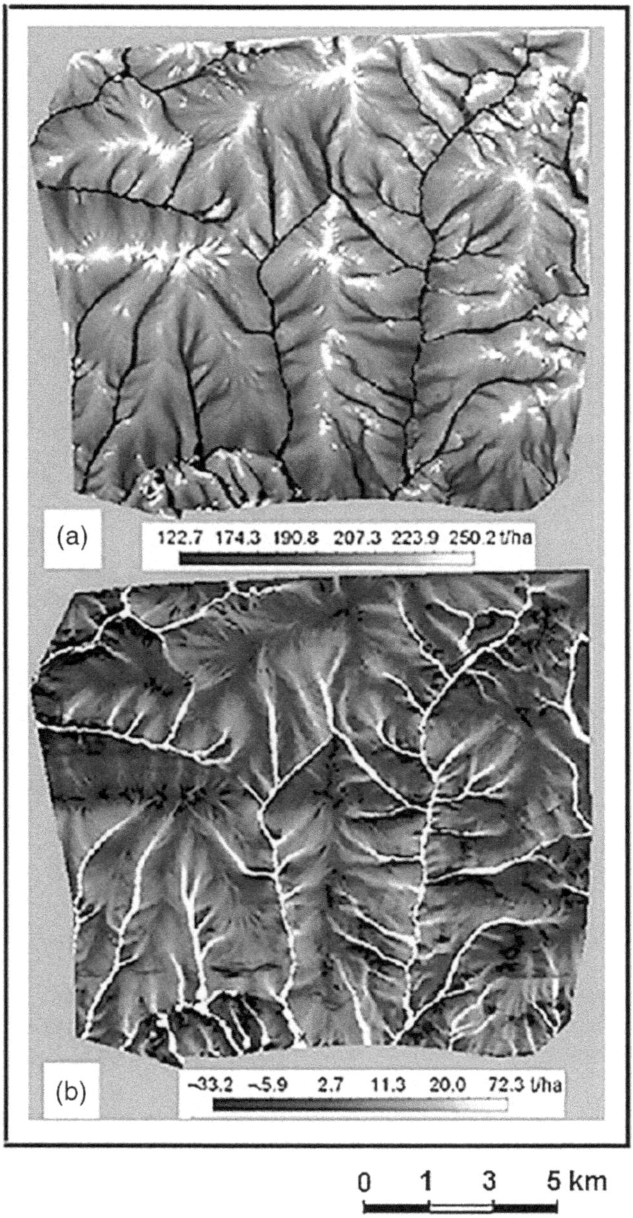


**FIGURE 15.4** Lowland Middle left-bank region By-Oka River basin – By-Oka-Terrace reserve. Changes of sum carbon content (carbon deposit, t/ha) to predictive dates 2075 (a) and 2150 (b), in t/ha, according to the HadCM3 model. The maps were calculated and compiled by L.S. Sharaya. Main predictors – illumination, maximal catchment area, and horizontal curvature showing flux convergence regions, with the growth of carbon deposits. Step of lattice – 20 m. Mean change, calculated by matrix: for 2075, +0.35 t/ha; for 2150, +0.67 t/ha.

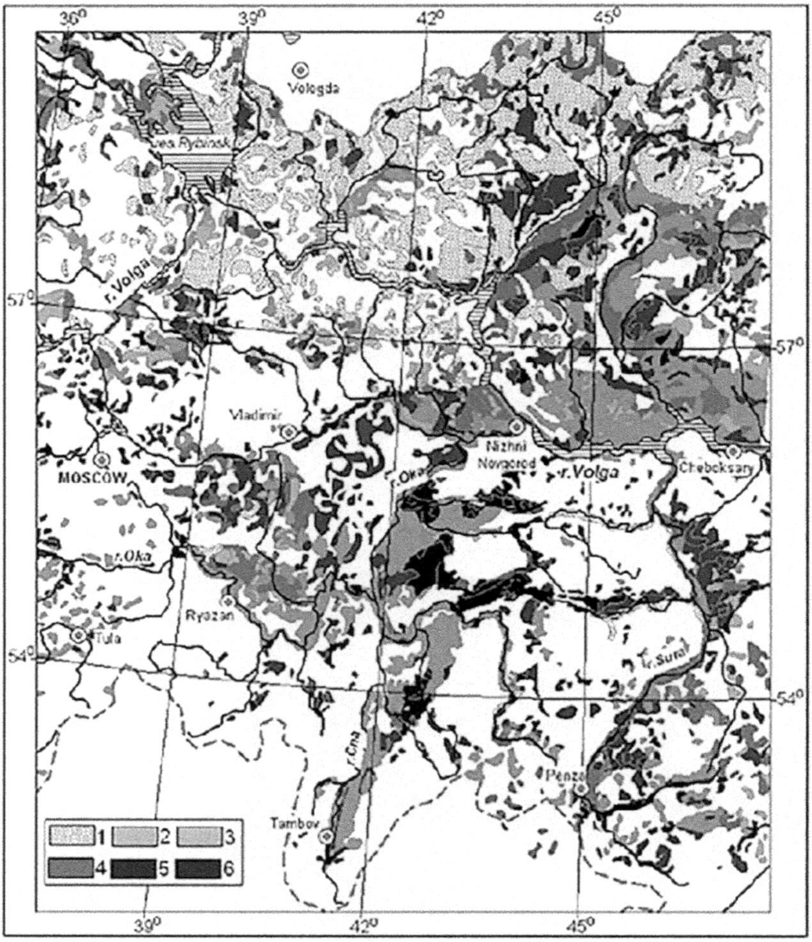


**FIGURE 15.5** Map of basic total carbon content in forest ecosystems of the Oka-Volga River basin. The maps were calculated and compiled by L.S. Sharaya. Total *C* content, in t/ha: 1–104÷116; 2–116÷130; 3–130÷150; 4–150÷170; 5–170÷206; 6–206÷280.

the territory of the Upper and Middle Volga Region account for the following: (1) in pine and spruce forests, 100–200 t/ha (up to 350 t/ha in the pine forests of the southeasternmost regions); (2) in oak and lime forests, 100–200 t/ha and 200–350 t/ha in the northwestern part of the region and in its southern and southeastern part, respectively; (3) in birch forests, from 100–200 t/ha in southern taiga and sub-taiga to 25–100 t/ha in the forest–steppe zone; and (4) in aspen forests, from 10–200 t/ha in the western sector to 200–350 t/ha in the By-Ural region.

The maps show evidence that substantial differences are inherent to spruce forests only. Our values for these formations approximate the data presented in Bobkova (1987) on total carbon stocks in the phytomass of middle taiga upland spruce forests in the north of European Russia: about 80–100 t/ha.

For the territory of the Oka-Volga River basin, the maps of the carbon balance of forest formations have been composed for the cold humid (2050) and thermo-humid (2200) climatic scenarios according to E GISS (Figures 15.6 and 15.7). According to the former scenario, the most effective positive regulation of the carbon cycle that would mitigate the process of cooling will be ensured by small-leaf forests that have replaced the primary middle and southern taiga and sub-taiga pine forests in the Unzha and Vetluga river basins and along the Oka-Moksha right bank, where their productivity will be maximally reduced ($\Delta C(Fa) = -(40 \div 75)$ tons/ha).

At the same time, the aforementioned indigenous communities, as well as the broadleaf–pine sub-taiga forests, will increase their productivity ($\Delta C(Fa)$ in the range from 25–40 to 100–150 t/ha), thereby exerting a negative effect on the carbon balance. The small sparse stands of broadleaf forests in the Sura and Moksha river basins will play a very weak positive regulatory role ($\Delta C(Fa) = -(5 \div 15)$ t/ha). In general, this scenario is characterized by the small-scale mosaic of values $\pm\Delta C(Fa)$.

The thermo-humid scenario presents a much more distinct pattern (see Figure 17.7). The middle and southern taiga spruce forests and the substituting birch forests over the entire northern and northeastern part of the Oka-Volga River basin will reduce productivity (due to increased atmospheric moistening) and thereby must contribute to global warming ($\Delta C(Fa)$ will account for $-(15 \div 30)$ to $-(45 \div 65)$ t/ha). The weak negative regulation of the carbon cycle ($\Delta C(Fa) = -(10 \div 35)$ t/ha) should also be expected from oak and lime forests of the forest–steppe zone. As we have already seen, this regulation must be much more pronounced under extreme thermo-aridization according to HadCM3. At the same time, the primary deciduous–pine forests and their derivatives (birch and aspen forests) in the broad belt encompassing the sub-taiga zone and the nemoral forest sub-zone will show a sufficiently pronounced positive regulation of the carbon cycle as they will increase their productivity ($\Delta C(Fa) = 25$–45 t/ha, here and there up to 60–85 t/ha).

## 15.4 ECOLOGICAL RESOURCES OF RESTORED PRIMARY FORESTS OF THE VOLGA RIVER BASIN

In Chapter 5, Figure 5.4, a basic raster map of zonal–provincial groups of plant formations of the Volga River basin is presented at a scale of 1:8,000,000. The areas of each of the phytocenological groups included indigenous forest communities (dark and light coniferous, mixed, broadleaf) – both modern and restored on the site of long-term derived (small-leaf) forests, as well as agro-landscape complexes (see Table 5.2). Thus, this map is a model of the potential bioclimatic system of the forests of the Volga River basin, according to cartographic data (Kotova 1987). On its basis, we provided predictive estimates of the carbon balance of forest cover, hypothetically represented only by

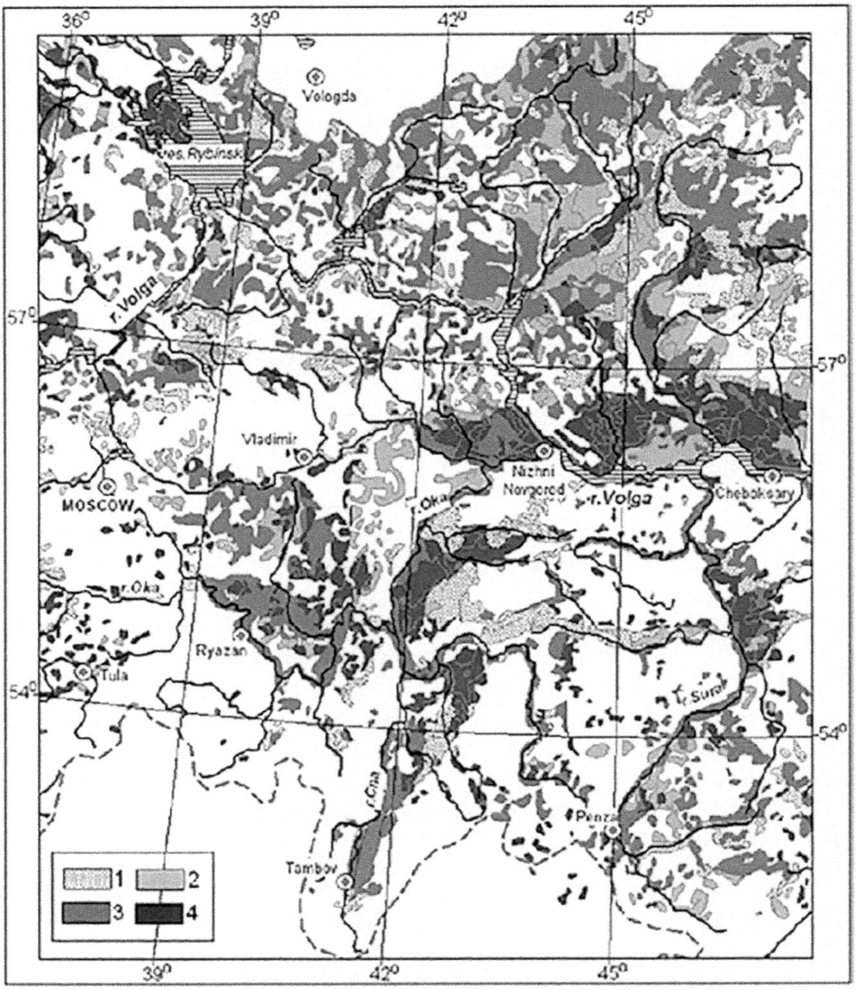


**FIGURE 15.6** Map of *C* balance (t/ha) in vegetation formations of the Oka-Volga River basin for the period 2050 by prognostic climate E GISS model (cool–humid trend). The maps were calculated and compiled by L.S. Sharaya. The changes in total *C* content, in t/ha: 1 – (–76)÷(–14); 2 – (–14)÷0; 3 – 0÷20; 4–20÷162.

primary formations, which is very important for comparison with the predicted carbon balance of real forests.

The map demonstrates the initial arrangement of the classes and groups of primary forest formations in the pre-industrial period, in the absence of intensive land and forest use. The basal total carbon content, as well as the specific and total carbon balances of these hypothetical forests under changing climate conditions, are presented in Table 15.5 and Figure 15.8. They inherently disclose the carbon cycle regulation by the

**FIGURE 15.7** Map of C balance (t/ha) in vegetation formations of the Oka-Volga River basin for the period 2200 by prognostic climate E GISS model (thermo-humid trend). The maps were calculated and compiled by L.S. Sharaya. The changes in total C content, in t/ha: 1 – (–66)÷(–10); 2 – (–10)÷0; 3 – 0÷30; 4–30÷89.

zonal–regional structure of the bio-climatic system of the Volga River basin for three mild (according to E GISS and HadCM3) scenarios of predicted climate changes: cooling and warming.

In general, the primary boreal and sub-boreal forests of the basin have two mutually opposite effects on the carbon balance between the Earth's surface and the atmosphere: (1) comparatively weak negative – under the cold–humid trend (additional $CO_2$ deposit and the respective intensification of cooling); (2) much stronger positive – under the thermo-arid trend (a three-fold increase in $\Delta C(Fa)$, which reduces

**TABLE 15.5** The total basal carbon reserves, as well as the total and specific carbon balances in restored primary (zonal-climax) forests of the Volga River basin (see Figure 5.4) for the predicted periods up to cooling (2050) and warming (2100 and 2200), according to the E GISS and HadCM3 climate models

| *GROUPS OF FORMATIONS (SEE TABLE 5.2)* | *TOTAL AREA SPARE, SQ. KM* | *TOTAL BASE CARBON STOCKS, MILLION TONS* | *TOTAL (SPECIFIC) CARBON BALANCE, IN MILLION TONS (T/HA) ACCORDING TO CLIMATIC SCENARIOS* | | |
|---|---|---|---|---|---|
| | | | *E GISS MODEL* | | *HADCM3 MODEL, EXTREME THERMO-ARID, 2100* |
| | | | *COLD–HUMID, 2050 г.* | *TEMPORAL THERMO-ARID, 2200* | |
| 1 | 45,927 | 508.596 | −70.932 [+] (−16.06) | +187.489 [+] (+42.45) | +91.205 [+] (+20.65) |
| 2 | 128,709 | 2368.889 | −122.565 [+] (−9.93) | −51.100 [−] (−4.14) | +2.222 [+] (+0.18) |
| 3 | 96,957 | 1975.984 | +36.362 [−] (+3.91) | +137.822 [+] (+14.82) | +103.785 [+] (+11.16) |
| 4 | 32,319 | 566.358 | −64.274 [+] (−9.45) | −26.800 [−] (−3.94) | +221.865 [+] (+32.62) |
| 5, 6 | 65,772 | 1246.708 | +9.426 [−] (+3.64) | +35.685 [+] (+13.78) | +40.372 [+] (+15.59) |
| 7 | 93,555 | 2445.425 | +111.725 [−] (+12.47) | +189.404 [+] (+21.14) | +147.563 [+] (+16.47) |
| 8, 9 | 66,339 | 1157.35 | +117.881 [−] (+18.57) | +173.537 [+] (+27.33) | +55.448 [+] +(9.05) |
| 11 | 133,245 | 2736.852 | +18.118 [−] (+1.37) | −102.622 [−] (−7.76) | +110.689 [+] +(8.37) |
| 10,12 | 148,554 | 2656.309 | +58.859 [−] (+14.96) | −56.891 [−] (−14.46) | − 13.613 [−] −(3.46) |
| 13 | 8,505 | 10.266 | +4.165 [−] (+13.09) | +2.416 [+] (+7.60) | +0.767 [+] +(2.41) |
| Swampy forests | 5,850 | 82.315 | +8.463 [−] (+16.84) | +7.172 [+] (+14.27) | +20.370 [+] +(40.53) |
| Forest swamps | 2400 | 33.014 | +2.142 [−] (+10.39) | +5.798 [+] (+28.12) | +4.182 [+] +(20.28) |
| Nemoral floodplains | 17361 | 255.762 | +0.223 [−] (+0.15) | +11.798 [+] (+7.91) | +12.469 [+] +(8.36) |
| **Sum (Average)** | **845493** | **16043.828** | **+109.586 [−] +(4.61)** | **+513.708 [+] +(11.31)** | **+797.319 [+] +(14.02)** |

*Note:* The forested area for groups of formations 10 and 12 is taken to be equal to 26%, and for the 13th group of formations - 11%. The [+] sign indicates the positive regulation of the carbon cycle under the given climatic trend, and the [-] sign indicates the negative regulation.

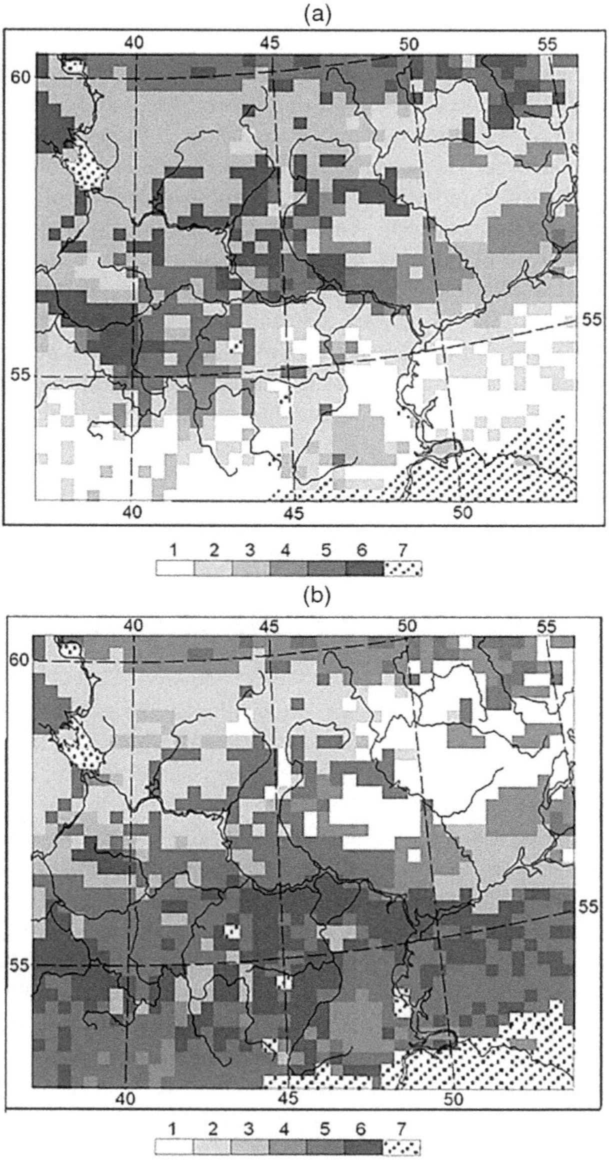


**FIGURE 15.8** The distribution of specific carbon balance in restored primary forest formations of the Volga River basin for the scenarios of: (a) regional warming (2200) and (b) regional cooling (2050), according to the E GISS model. The maps were calculated and compiled by L.S. Sharaya.

***Notes:*** Carbon balance (tons/ha) for a scenario: 1 – (–234.3)÷(–162.3); 2 – (–162.2)÷(–90.3); 3 – (–90.2)÷0; 4 – 0÷53.8; 5–53.9÷125.8; 6–125.9÷197.8. Carbon balance (tons/ha) for b scenario: 1 – (–223.6)÷(–127.8); 2 – (–12.7)÷(–62.8); 3 – (–62.7)÷0; 4 – 0÷73.8; 5–73.9÷241.7; 6–241.8÷242.5. 7 – lakes, reservoirs and areas with no forest vegetation.

the concentration of greenhouse gases in the atmosphere and thereby weakens anthropogenic warming).

The total $CO_2$ adsorption from the atmosphere is about 1% of the total basal carbon content in boreal and sub-boreal forests in the former case and 3.5% in the latter case. Let us consider in more detail the ecological resources of the hypothetical forest cover of the Volga River basin for the scenario of regional moderate warming (see Chapters 10 and 12). According to this scenario, the maximum reduction of greenhouse gases in the atmosphere and consequently the maximally weakened regional warming are expected for two groups of primary forest formations – middle-taiga fir–spruce forests and middle and south taiga pine forests: in each of them, the total carbon balance ($\Sigma\Delta C(F)$) approaches +190 million tons. The sub-taiga spruce and pine–broadleaf forests in the western sector of the basin have an almost equally high ecological resource, with ($\Sigma\Delta C(Fa)$) ≈ +140 to 170 million tons. The total carbon balance in the eastern sector (the Kama-Pechora-Western Ural) decreases almost four times.

## 15.5 ANTHROPOGENIC CHANGES IN THE ECOLOGICAL RESOURCES OF MODERN FOREST COVER

The basal and predicted carbon parameters were calculated for each of the 13 groups of forest formations of the Volga River basin (see Table 15.5), both for the restored primary forests and for the real forest cover, including primary and derivative communities. The analyzed territory was within the boundaries of 52–60° N and 36–53° E, i.e., it occupied about half of the area shown in Figure 5.3. The generally comparative pattern of carbon balances remains the same as described for the hypothetical forest cover of the basin. Under the thermo-arid scenario, the maximum specific ecological resources remain in the middle and south taiga and sub-taiga pine forests, as well as middle taiga fir/spruce forests, while the minimum resources are in the northern steppe broadleaf forests and south taiga dark coniferous communities. Total resources are also determined by the areas of forest formations; therefore, the general regulatory effect on the carbon cycle, e.g., of rarely occurring Kama and Western Pre-Ural coniferous/broadleaf forests, as well as oak/lime forests of northern forest–steppe, is insignificant.

Specific carbon balances for separate groups of forest formations range from 1%–5% to 11%–16% of their specific basal carbon content. The exception is the sub-taiga broadleaf–pine forests and especially forest bogs, which additionally absorb up to 17 t/ha and more than 28 t/ha of carbon under warming conditions, respectively (see Tables 15.5 and 15.6).

The comparison of specific and total carbon balances of the restored primary forests and the real forest cover (primary + derivative forests) shows that the replacement of primary forests by derivative ones leads to a total and considerable decrease in the ecological resources of the forest cover of the Volga River basin. This is manifested by the decrease in total positive regulation of the carbon cycle under warming conditions

**TABLE 15.6** Specific carbon balances of the restored primary (zonal-climax) and real (primary+derivative) forest formations of the Volga River basin for the forecast periods of 2050 (cooling) and 2200 (warming), according to E GISS models

| | *SPECIFIC BASE CARBON CONTENT, T/HA* | | *SPECIFIC CARBON BALANCE, T/HA* | | | |
|---|---|---|---|---|---|---|
| | | | *2050* | | *2200* | |
| *GROUPS OF FORMATIONS (SEE TABLE 5.2)* | *RESTORED PRIMARY FORESTS* | *PRIMARY + DERIVED FORESTS* | *RESTORED PRIMARY FORESTS* | *RESTORED PRIMARY FORESTS* | *RESTORED PRIMARY FORESTS* | *RESTORED PRIMARY FORESTS* |
| 1 | 110.74 | 80.33 | −16.06 [+] | −11.64 [+] | +42.45 [+] | +30.78 [+] |
| 2 | 184.05 | 133.50 | −9.93 [+] | −1.94 [+] | −4.14 [–] | −6.48 [–] |
| 3 | 203.80 | 233.58 | +3.91 [–] | +0.68 [–] | +14.82 [+] | +16.45 [+] |
| 4 | 175.24 | 156.25 | −9.45 [+] | −2.27 [+] | −3.94 [–] | −7.58 [–] |
| 5, 6 | 189.55 | 169.27 | +3.64 [–] | +0.49 [–] | +13.78 [+] | +11.92 [+] |
| 7 | 154.50 | 179.87 | +12.47 [–] | +10.89 [–] | +21.14 [+] | +10.21 [+] |
| 8, 9 | 174.46 | 180.26 | +18.57 [–] | +8.94 [–] | +27.33 [+] | +16.39 [+] |
| 11 | 205.40 | 182.65 | +1.37 [–] | +5.83 [–] | −7.76 [–] | −5.82 [–] |
| 10,12 | 178.81 | 167.44 | +14.96 [–] | +13.50 [–] | −14.46 [–] | −12.73 [–] |
| 13 | 15.12 | | +13.09 [–] | | +7.60 [+] | |
| Swampy forests | 140.71 | | +16.84 [–] | | +14.27 [+] | |
| Forest swamps | 137.56 | | +10.39 [–] | | +28.12 [+] | |
| Nemoral floodplains | 158.65 | | +0.15 [–] | | +7.91 [+] | |
| **Average** | **156.01** | **148.89** | **+4.61 [–]** | **5.00 [–]** | **+11.32 [+]** | **+8.54 [+]** |

and by the increase in its negative regulation under cooling conditions (see Tables 15.5 and 15.6, bottom lines). The distribution of birch and aspen forests in the middle and south taiga and sub-taiga pine forests, making a decisive contribution to additional $CO_2$ adsorption (under warming, more than 70% of the total volume of absorption), has particularly negative effects. Their average $\Delta C(Fa)$ values decrease almost by half under warming conditions: from 16.16 to 8.87 t/ha (Figure 15.9). The ecological resource of the middle taiga fir/spruce forests also decreases by 27%, while the negative carbon balance in the south taiga East-European and Kama-Western Pre-Ural dark coniferous forests increases by 40%–50%. In the dark coniferous south taiga and sub-taiga communities, as well as in the broadleaf and steppe pine–broadleaf forests, secondary small-leaf formations have much less influence.

The wide distribution of restorative successions in the forest belt of the Volga River basin, with the predominance of long-derivative small-leaf communities, decreases the total additional absorption capacity of forests by 2.8 times for the thermo-arid predictive scenario compared to the hypothetically restored primary forests. On the contrary, this capacity increases two-fold under the cold–humid trend. In the former case, there is a dramatic attenuation of positive regulation of the carbon cycle, while in the latter case there is slightly lesser but still significant intensification of negative regulation. Both cases provide direct evidence of reduced adsorption capacity of forest formations, with the presence of long-derivative formations together with primary communities. This is the general result of the anthropogenic decrease in the ecological resources of forest cover in the Volga River basin since the beginning of intensive land and forest use in this region.

## 15.6 VERIFICATION OF PREDICTIVE CALCULATIONS OF THE CARBON BALANCE

The described regularities of carbon cycle regulation by forests of the Volga River basin demonstrate a rather high efficiency of using the actualistic method for numerical prediction of the ecological resources of forest cover. However, verification of computed prediction models is a highly difficult problem, and the ways of its solution have not been found up to now. The main problem is that it is necessary to obtain a sufficiently long perennial series of direct (remote or field) measurements of the carbon content in numerous forest ecosystems, which represent the phytocoenological structure of a particular region. Such a database has already started to be formed.

In this case, the analytical and cartographic data on carbon deposition/emission in forests of the boreal belt of central Canada were used (Stinson et al. 2011). Such data were obtained by repeated remote sensing over the period of 1990–2000 (note that it was the initial period of global warming), as well as terrestrial measurement and modeling for the central plain part of Canada – the Saskatchewan and Manitoba provinces. This area of "Hudson taiga" has the predominance of high, thick coniferous stands of Canadian spruce, American larch, balsam fir, and Italian stone (black) pine.

Forests are rather intensively cut down and suffer from fires and entomopests; nevertheless, beginning in 1990, the areas of managed Canadian forests have remained relatively stable, while the average age of plantations is 92 years (Stinson et al. 2011). First of all, it was necessary to assess the comparability of the Volga River basin and Central Plain Canada with respect to regional climate predictions, as well as the carbon cycle parameters and the ecological consequences of forest management in predictive periods. These data are given in Table 15.7. In general, both large ecoregions are quite similar with respect to major climatic trends. They describe an unambiguous increase

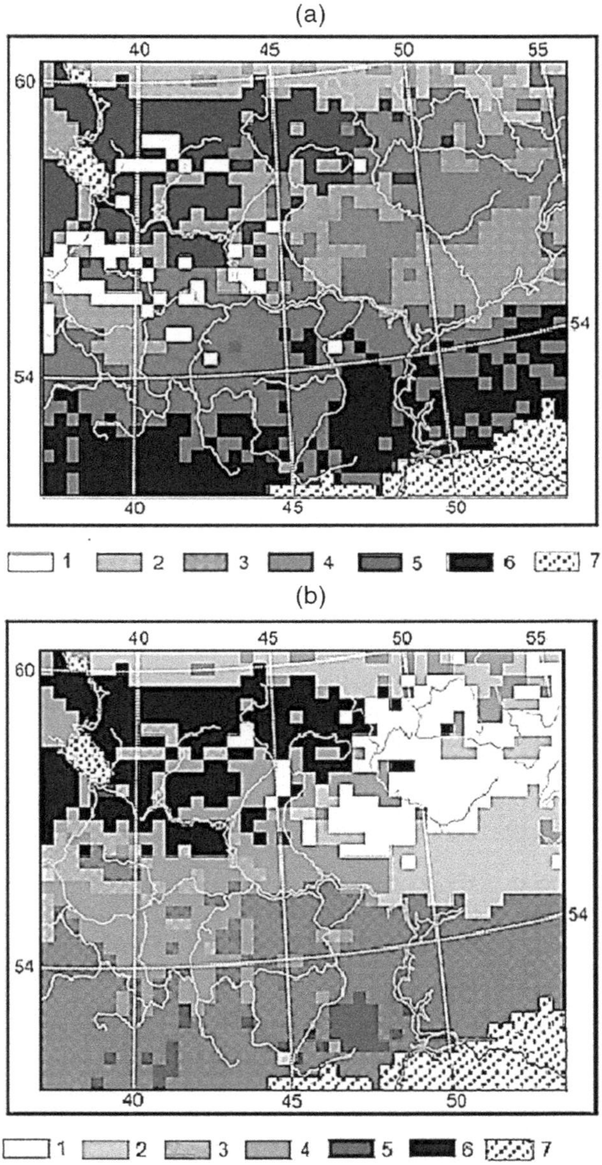


**FIGURE 15.9** Distribution of the specific carbon balance (t/ha) of restored primary forest formations (a) and real forest cover (b) of the Volga River basin for the extreme warming scenario, according to the HadCM3 model for 2100 (see online version for colors). The maps were calculated and compiled by L.S. Sharaya. Notes: Carbon balance (t/ha) for map *a*: 1–32.0÷28.0; 2–28.0÷17.0); 3–17.0 ÷13.0; 4–13.0÷8.0; 5–8.0÷3.0; 6 – 3.0÷(–3.5). 7 – lakes, reservoirs, and areas with no forest vegetation. Carbon balance (t/ha) for map *b*: 1–23.8÷19.0; 2–19.0÷14.0); 3–14.0÷9.0; 4–9.0÷6.0; 5–6.0÷(–6.0). 6 – (–6.0)÷(–10.3). 8 is the same as 7 for map *a*.

**TABLE 15.7** The factual and EGISS-predicted deviations from the basal period of major climatic parameters in different periods for the main catchment area of the Volga River basin and for the Central Canada

| *DEVIATIONS FROM BASE PERIOD** | *MIDDLE AND UPPER VOLGA REGION* | | | *CENTRAL CANADA* | | |
|---|---|---|---|---|---|---|
| | *1990* | *1998* | *2050* | *1990* | *1998* | *2050* |
| $t_{Jan}$ | 0.8–1.4 | 1.1–1.6 | 1.6–2.1 | 0.1–0.7 | 1.4–1.9 | 1.6–2.1 |
| $t_{July}$ | 0.0–0.4 | 1.0–1.6 | −0.5÷(+0.1) | 0.1– 0.4 | 0.3–0.8 | 0.7–1.1 |
| $r_{ann}$ | 110–292 | –(73÷292) | 73–328 | 73–292 | 36–256 | 73–328 |

*Note:* *Average temperatures: $t_{Jan}$ – January. $t_{July}$ – July; $r_{ann}$ – annual precipitation.

in winter and summer temperatures (except for the abnormal cooling in the Volga River basin in 2050), as well as an increase in annual precipitation (except for temporal aridization of all Europe at the turn of the century).

The modern carbon reserves in boreal forests of the Volga River basin are estimated to be 120–150 t/ha. In forests of the central plain part of Canada, they are a little lower: 80–110 t/ha. According to the data from the predictive simulation model IMAGE2 (Alcamo et al. 1994), the carbon flow from terrestrial ecosystems to the atmosphere in both large ecoregions was no more than 0.15–0.20 t/ha in 1990 due to deforestation, forest management, and age-related dynamics of forest restoration after fires and logging, but is expected to increase to 0.20–0.50 t/ha by 2050. It can be assumed that the Volga River basin and Central Canada are, and probably will remain, under similar conditions of forest management, which makes it easier to compare them with respect to climate-genic dynamics of the carbon cycle.

Let us show the yearly and average changes in the ecosystem carbon sink (ΔNEP-($CO_2$), t/ha·yr), as well as their total changes (in t/ha) in the period of 1990–2000, in the boreal forests of Central Canada, according to Stinson et al. (2011):

| | | | | | | | | | |
|---|---|---|---|---|---|---|---|---|---|
| Years | 1990 | 1991 | 1992 | 1993 | 1994 | 1995 | 1996 | 1997 | 1998 |
| ΔNEP($CO_2$) | 0.113 | 0.083 | 0.135 | 0.057 | 0.061 | −0.170 | 0.061 | 0.104 | −0.096 |
| Years | 1999 | 2000 | | | | | | | |
| ΔNEP($CO_2$) | 0.017 | 0.087 | | | | | | | |
| Average | −0.041 | | | | | | | | |
| Total | −0.452 | | | | | | | | |

Except for the reduced carbon sink in 1995 and 1998, when catastrophic forest fires spread over Central Canada, the mean value of ΔNEP($CO_2$) for these 11 years was 0.080 t/ha·yr. This value is close to our calculated data on the yearly carbon balance according to the extreme model HadCM3 – first of all, for the real forest cover (see Table 15.7). The E GISS model gives understated results.

The linear extrapolation of the dynamics of the carbon sink in Canadian forests will show the following total increase in their carbon reserves: about 0.40 t/ha for the predicted period up to 2050, 0.80 t/ha for the predicted period up to 2100, and 1.60 t/ha for the predicted period up to 2200, respectively. When comparing these values with our prediction data, we can see that they are lower by an order of magnitude. It is impossible to verify our predictive calculations for the long-term future (after 2050) based on factual data on Canadian forests in the interval of the first 10 years of global warming. Our calculation models of predicted changes in the carbon content in different pools of forest biogeocoenoses, according to the results of its basal hydrothermo-edaphic ordination, have the appearance of not only linear but also parabolic and exponential relationships (see Kolomyts et al. 2012, Table 12.1), which gives the total ascending trend of the $\Delta C(Fa)$ parameter not confirmed by the linear trend of its changes based on the empirical data from the initial period of global warming.

The predicted temperature trend is of great significance per se. Almost all prediction models of the AOGCMs family describe an exponential but not a linear increase in global temperature after 2000 (Albritton et al. 2003; Mokhov 2006), which must initially predetermine the increasing trend of the predicted positive carbon balance in the boreal forest cover.

## 15.7 TOWARD THE SOLUTION OF THE DUAL PROBLEM OF ADSORPTION AND ADAPTATION (IN ACCORDANCE WITH THE PARIS AGREEMENT)

As already stated, one of the most important pathways for implementing the objectives stipulated in the Paris Agreement on Climate Change (2015) is the solution of the dual problem: adsorption of atmospheric $CO_2$ by forest communities from the atmosphere under global warming and their adaptation to these climate changes, which must provide the efficiency of adsorption per se. Labile plastic–elastic resilience of forest ecosystems was considered an indicator of adaptation.

The numerical experiment on the assessment of the effect of labile elastic stability of forest formations and predicted climatic parameters on the carbon balance of forests, i.e., on their adsorption capacity, was carried out in collaboration with L.S. Sharaya in the territory of the Oka River basin (about 250,000 $km^2$). Climatic parameters were taken from the HadCM3 predictive scenarios for 2050. Plant formations with the respective indices of labile elastic resilience and changes in carbon content were distributed by meso-catenas. Thereby, an array of cartographic data was formed (more than 52,000 points). The spatial variability of the carbon balance of forests associated with their restoration potential in the basin was investigated by multiple regression methods. For this purpose, we also used the matrices of climatic parameters, 18 relief features (Shary et al. 2002), and some satellite data on the characteristics of forests (vegetation index NDVI, crown closure).

The resultant functional relationships between the carbon balance and adaptation for zonal/sub-zonal types/sub-types of forest formations (see Table 5.1) are given below (significance level $P < 10^{-6}$).

**A.** Spruce and broadleaf–spruce forests, middle- and south-taiga, and sub-taiga:

$$C(Fa) = 62.4I_{\text{resili}}\ (1) + 22.7r_{\text{year}} + 10.2t_{\text{July}} - 49.3;\ \text{Rs} = 0.50; \tag{15.1a}$$

**B.** Pine south-taiga forests:

$$C(Fa) = 38.6r_{\text{year}} + 22.4t_{\text{resili}}(1) - 24.8; Rs = 0.46; \tag{15.2}$$

**C.** Broadleaf–pine forests, sub-taiga:

$$C(Fa) = 217.4Iresili(1) + 78.0t_{\text{July}} + 55.3r_{\text{year}} - 165.6; Rs = 0.64; \tag{15.3}$$

**D1.** Northern broadleaf forests with an admixture of spruce:

$$C(Fa) = 6.23I_{\text{resili}}(1) - 4.650t_{\text{July}} + 4.30r_{\text{year}} - 5.63; Rs = 0.46; \tag{15.4}$$

**D2.** Southern steppified broadleaf forests:

$$C(Fa) = 32.9t_{\text{July}} + 24.1r_{\text{year}} - 54.3I_{\text{resili}}(1) - 19.6;\ Rs = 0.38 \tag{15.5}$$

All formations of the basin:

$$C(Fa) = 76.3I_{\text{resili}}(1) + 36.9t_{\text{July}} + 31.1r_{\text{year}} - 69.6;\ Rs = 0.13 \tag{15.6}$$

All formations of secondary forests:

$$C(Fa) = 123.5I_{\text{resili}}(1) + 22.8r_{\text{year}} + 18.7t_{\text{July}} - 64.6;\ Rs = 0.30 \tag{15.7}$$

All formations of primary forests:

$$C(Fa) = 68.6t_{\text{July}} + 37.1r_{\text{year}} - 3.4I_{\text{resili}}(1) - 51.0;\ Rs = 0.16 \tag{15.8}$$

Here, $\Delta C(Fa)$ is the change in the specific flow of $CO_2$ (t/ha·year) in the terrestrial surface–atmosphere system; $I_{\text{resili}}(1)$ is the index of labile resilience of a forest community; $t_{\text{July}}$ and $r_{\text{year}}$ are the normalized values (in fractions of 1) of the mean July temperature and the annual precipitation; $Rs$ is the Spearman's rank correlation coefficient. The generally low values of this coefficient are determined by the enormous number of statistical samples (see above) with a substantial "noise" of local geomorphological and edaphic factors forming the groups of mean values with opposite connections. Nevertheless, the general tendency of connections can be considered quite reliable for each equation, as is evidenced by a significant Pearson's correlation coefficient $P$.

Each zonal/sub-zonal type/sub-type includes primary and derived forest communities, as well as fragmented timberlands; i.e., it represents the state of real forest cover. In practice, it would be advisable to calculate parameter $I_{resili}(1)$ by the following empirical formulas (Kolomyts 2008):

$$I_{resili}(1) = 0.715 - 0.0555CY; \; R = 0.95; \; R^2 = 0.90; \tag{15.9}$$

$$CY = 0.0.993\, t_{Jun} - 0.5365\, t_{July} + 0.003508 r_{year}; \; R = 0.78; \; R^2 = 0.61; \tag{15.10}$$

$$CY = \exp\,(6.453 - 0.2993 t_{July}); \; R = 0.84; \; R^2 = 0.76. \tag{15.11}$$

Here, CY is the litter-fall index (see above); $t_{Jun}$ and $t_{July}$ are the mean January and July temperatures, respectively (°C); $r_{year}$ is the annual precipitation (mm); $R$ and $R^2$ are the correlation and determination coefficients, respectively.

For example, Table 15.8 shows the results of calculations (by Equations 15.1–15.8) of predicted specific and total values of the carbon balance of zonal/sub-zonal types/sub-types of forest formations of the Volga River basin concerning the basal and predicted values of $I_{resili}$, as well as by the $t_{July}$ and $r_{year}$ parameters prescribed by two different global climate prediction models. According to the moderate model GISS-93, the warming in the Middle Volga Region will be from 0.2°–0.6° in winter to 0.8°–1.1° in summer by the middle of the 21st century, which corresponds to the Paris Agreement Compatible Scenario (2015). The extreme model HadCM3 gives the annual temperature increase of 2.5°–4.0° for this period, which can occur at the current rates of global warming (Le Quere et al. 2015; Shvidenko et al. 2017).

The results of the new calculation of $\Delta C(Fa)$ are quite similar to the data obtained by other methods described above (see Tables 15.1–15.3), though there are some differences. According to the climate scenario for 2100 based on the HadCM3 prediction model, there is almost the same average specific carbon balance of forest cover in both cases (11.05 and 10.02 t/ha·yr, respectively). The maximum adsorption capacity is due to sub-taiga forest formations; however, it is typical of dark coniferous–broadleaf forests in the former case and shifts to the pine–broadleaf forests, with a dramatic increase in this maximum, in the latter case.

The total elastic–plastic resilience of forest formations must be increasing during the entire 100-year prediction period (to the greatest extent under extreme warming). Hence, one should also expect a significantly enhanced greenhouse gas adsorption capacity of boreal forests, i.e., an increase in their ecological resources (see Table 15.8). The comparison of values of the carbon balance of forest formations obtained for the initial (basal) and final (total) resilience indices shows an unambiguous pattern of considerably increasing adsorption capacity of boreal forests with the increase in their restoration potential.

However, these calculations do not take into account the changes in ecological resources of forest formations, which are caused by their functional and structural transformations throughout the given prediction period. These changes are ambiguous.

**TABLE 15.8** Specific and total values of carbon balances of forest formations in the sample area of the Volga basin, projected for 2100 with their base and final labile elastic stability and climatic scenarios, according to two global forecast models: moderate GISS-93 and extreme HadCM3

| *ZONAL TYPES AND SUB-ZONAL SUB-TYPES OF FOREST FORMATIONS (SEE TABLE 4.2)* | *WEIGHTED AVERAGE INDEX OF RESILIENT STABILITY, $I_{resill}(1)$* | | | *CLIMATIC PARAMETERS PROJECTED FOR 2100* | | | | *SPECIFIC CARBON BALANCE, T/HA* | | *WOODED AREA, KM²* | *TOTAL CARBON BALANCE, MILLION TONS* | |
|---|---|---|---|---|---|---|---|---|---|---|---|---|
| | | *PROJECTED* | | *GISS-93 MODEL* | | *HADCM3 MODEL* | | | | | | |
| | *BASE* | *GISS-93 MODEL* | *HADCM3 MODEL* | $T_{July}$, *⁰C* | $R_{year}$, *MM* | $T_{July}$, *⁰C* | $R_{year}$, *MM* | *MODEL GISS-93* | *MODEL HADCM3* | | *MODEL GISS-93* | *MODEL HADCM3* |
| **A.** Spruce and broadleaf–spruce forests, middle- and south-taiga, and sub-taiga | 0.554 | 0.630 | 0.670 | 20.1 | 950 | 22.2 | 725 | 9.70<br>14.44 | 8.87<br>16.04 | 75'706 | 73.435<br>109.319 | 67.151<br>121.432 |
| **B.** Pine south-taiga forests | 0.430 | 0.630 | 0.675 | 20.1 | 930 | 22.6 | 710 | 9.23<br>6.03 | 12.82<br>8.90 | 48'506 | 44.771<br>29.249 | 62.185<br>43.170 |
| **C.** Broadleaf–pine forests, sub-taiga | 0.531 | 0.664 | 0.684 | 21.6 | 830 | 23.0 | 705 | 18.70<br>57.83 | 44.51<br>82.76 | 37'841 | 70.763<br>218.835 | 160.430<br>313.172 |
| **D₁.** Northern broadleaf forests, with an admixture of spruce | 0.547 | 0.667 | 0.689 | 22.2 | 780 | 23.8 | 685 | – 5.58<br>– 2.32 | – 5.26<br>– 3.61 | 18'350 | –10.239<br>–4.257 | –9.652<br>–6.624 |
| **D₂.** Southern steppified broadleaf forests | 0.547 | 0.667 | 0.689 | 23.8 | 700 | 24.0 | 650 | –14.00<br>– 17.86 | – 10.83<br>– 18.57 | 7'544 | – 10.562<br>–13.434 | – 8.170<br>–14.009 |
| Waterlogged coniferous forests and forest swamps | 0.549 | 0.630 | 0.672 | 20.1 | 940 | 22.4 | 720 | 8.12<br>13.17 | 6.61<br>14.18 | 7'052 | 5.726<br>9.287 | 4.661<br>10.000 |
| Nemoral floodplains | 0.559 | 0.642 | 0.685 | 22.5 | 770 | 23.6 | 680 | 3.37<br>3.13 | 4.69<br>11.53 | 16'267 | 5.482<br>5.092 | 7.676<br>18.871 |
| **Weighted average/Sum** | **0.517** | **0.634** | **0.673** | **20.4** | **921** | **21.0** | **717** | **3.61**<br>**11.62** | **10.02**<br>**17.10** | **211'266** | **179.376**<br>**354.091** | **284.281**<br>**486.012** |

*Note:* $t_{July}$ is the average temperature in July; $r_{year}$ is the annual amount of precipitation. In the columns of the specific and total carbon balance, the first row characterizes the balance at the base value of the stability index of plant formations, and the second – at its final value (for the forecast period of 2100). Treeless raised and transitional bogs are not counted.

According to the climate scenarios of moderate warming (GISS-93 and E GISS), algebraic addition of the $\Delta C(Fa)$ values has been made for the zonal/sub-zonal types/sub-types of forest formations of the Volga River basin (see Tables 15.5 and 15.6). The following *indices of changes in the specific carbon balance of forests* [$\Delta C(Fa)$] have been obtained (t/ha): **A** (+4.998); **B** (–14.570); **C** (+27.773); **D$_1$** (+10.010). Generally, the ecological resources of the entire repertoire of boreal dark coniferous and dark coniferous–broadleaf, as well as sub-taiga broadleaf–pine forests, are preserved. The adsorption capacity of broadleaf forests also noticeably increases (mainly due to their transformation into boreal forests). At the same time, pure south-taiga pine forests, being transformed by more than 50% into mixed forest communities, substantially lose their ecological resources, though their $\Delta C(Fa)$ remains positive.

The presented data indicate, altogether even more clearly than it has been shown previously, the phenomenon of increasing adsorption capacity of forest cover in the Volga River basin with the intensification of the global warming signal and the respective increase in elastic–plastic resilience of ecosystems. This can also be verified by a comparison of $\Delta C(Fa)$ values from the GISS-93 and HadCM3 models (see Tables 15.7 and 15.8). The increase in forest restoration adaptation, being a direct ecological factor, makes a decisive contribution to the enhanced adsorption of greenhouse gases.

For all formations of the Oka River basin as a single group (Equation 15.6), it has been shown that their carbon balance has a positive relationship with the restoration potential and the major climatic parameters. At temperatures and precipitation close to the average values, carbon deposition becomes more intensive with the increasing index of elastic resilience, while $CO_2$ emission intensifies as the latter decreases.

The carbon balance close to the equilibrium state is associated with different values of elastic resilience in the territory of the basin. For example, it is $I_{resili}=0.418$ for the central areas of the Oka River basin, 0.465 for its northwestern area, and 0.385 for the drier and warmer southeastern area. If the restoration potential takes lower values, the carbon balance becomes negative. This means that forest formations more resistant to global warming have altogether higher adsorption capacity.

The carbon balance of primary forests as a single group in the Oka River basin is positively related to temperature and precipitation and negatively related to the elastic resilience of forest formations (Equation 15.10). Here, the elastic–plastic resilience of forests has a very low relative weight. This can be attributed to the absence of unambiguous dependence between the carbon balance of all primary forests and their restoration potential. A positive relationship has been revealed for spruce and broadleaf–spruce, broadleaf–pine, and northern broadleaf forests, while a negative relationship has been shown for pine south-taiga and broadleaf forests in more southern latitudes. In the latter cases, the carbon balance will be highly dependent on precipitation, with its decrease resulting in a negative carbon balance. It is also noteworthy that emissions in the primary and derived forests of the southern forest–steppe and northern steppe can be sustained by the relatively high restoration potential of these forests.

## 15.8 SOME INFERENCES

1. The specific and total carbon balances of forest formations in the Volga River basin were the basis for a quantitative assessment of their ecological resources, providing more favorable environmental conditions through carbon cycle regulation by forests. The generalized data on the ecological resources of the boreal and nemoral forest cover of European Russia and the Pre-Urals can be used for a quantitative assessment of the effects of forests in these regions on the fluxes of greenhouse gases in the Earth surface–atmosphere system.
2. It is predicted that different forest associations and formations at the same climatic signal will both additionally absorb greenhouse gases (the positive carbon balance of forest communities) and emit them to the atmosphere (the negative balance), exerting respective effects on the state of terrestrial ecosystems. Each ecoregion has shown particular zonal–regional types of regulation of $CO_2$ content in the atmosphere by forest cover during the development of climatic thermo-aridization. According to the area-related distribution of these types, the regional contribution of forest phytobiota to the stability of this part of the continental biosphere is valuable.
3. It has been empirically proven that the replacement of primary forests by derivative forests results in a very significant reduction of ecological resources in the boreal forest cover. The mean values of positive specific balances there under warming (even according to the moderate climate model E GISS) decrease two to three times. This represents the anthropogenic loss of ecological resources of the boreal forests of the Volga River basin since the beginning of intensive land and forest use there.
4. It has been shown that climate-genic transformation of forest ecosystems in the Volga River basin, even according to the moderate thermo-arid trend, results in a total reduction of positive carbon balance in most formations of boreal forests, whereas the initially negative balance of steppificated and insular forests in the sub-boreal belt increases to positive values. The stated experience of analytical modeling of changes in the carbon balances of forest formations of the Volga River basin in the system of their thermo-arid transformations, according to one of the global warming scenarios, sheds some light on the dual problem posed by the Paris Agreement (2015) concerning the necessity of a coupled analysis of the adsorption capacity of forest biomes and their adaptation to the changing climate.

# Empirical Simulation of Changes in Productivity and Carbon Stock in Forest Ecosystems under Climate Variations

# 16

## 16.1 OBJECTS AND TASKS OF THE STUDY

The analysis was carried out at the Vyksa test site, located in the northern part of the Oka-Moksha woodlands in the Middle Volga region (see Figure 5.2). The ecological region is situated at the northern limits of the broadleaf forest sub-zone, as supported by the main climatic characteristics (Table 16.1). However, the vegetative ground cover here is atypical for the broadleaf forest sub-zone. The lithological and hydrological factors, such as the presence of sandy glacial and ancient alluvial deposits and generally poor drainage, contribute to the prevalence of pine forest biogeocoenoses (see Table 5.3). There is a distinct phenomenon of interchangeability between climatic and lithogenic

DOI: 10.1201/9781003733119-16

ecological factors: in the northern subboreal climatic conditions, the sandy substrate simulates boreal conditions of the sub-taiga or even the southern taiga. However, the pine forests of the Oka-Moksha woodlands exhibit a more southern character compared to the pine forests of the sub-taiga zone. On the other hand, there is a certain shift in the types of pine forests towards higher hydromorphicity compared to their analogs from the more northern regions.

These ecological and phytocenotic features of the forests of the Oka-Moksha lowland zander allow us to consider the territory as a *zonal–regional biogeographic ecotone*, where a significant overlap of climatic niches of boreal and broadleaf forest communities occurs. According to Armand and Vedyushkin (1989), the trigger character of this transition zone determines the state of unstable equilibrium of its constituent biogeocoenotic systems, which should be sensitive to climatic fluctuations, primarily by their functional parameters, which are somehow related to primary bio-productivity.

The problem was, using this model territory as an example, to identify the origins and mechanisms of functional shifts of forest biogeocoenoses in different years, which differ quite clearly in their hydrometeorological regime. This was the case in 1989 and 1990, when large-scale landscape and ecological surveys were conducted at the Vyksa test site in the first decade of June (Kolomyts et al. 1993; Kolomyts 2008).

The first year (1989) was characterized by a general positive thermal anomaly and relative aridity (see Table 16.1). The previous exceptionally warm winter was accompanied by deep thaws, with partial snowmelt, which significantly reduced the spring moisturizing of the soil. The vegetation period was also abnormally warm, although not to the same extent. The annual precipitation was close to the long-term norm; nevertheless, the Vysotsky–Ivanov atmospheric humidity factor ($F_{hum}$) turned out to be significantly lower than one and 23% lower than the long-term average value. The territory of the test site was typical in the hydrothermal conditions for even the southern forest–steppe, which is characterized by the range of $F_{hum}$=0.76–1.03 (Kolomyts 2003). Thus, *a thermo-arid climatic signal* was clearly expressed for this hydrological year. At the same time, changes in soil moisture reserves at the beginning of the vegetation period of 1989 were caused almost exclusively by the shifts in the soil temperature regime.

On the contrary, there was a very sharp pluvial anomaly against the slightly warmer winter and moderate summer season in the second year (1990) (see Table 16.1). The annual precipitation exceeded the long-term norm by more than 65%, and the annual humidity coefficient increased by the same amount. This, combined with relatively low evaporation, could not but cause a total increase in the spring and summer moisture content of the soil. As you can see, this year *a humid climatic signal* has been clearly manifested, with almost no thermal component. The signal seemingly shifted the territory under study to the zonal boundary between the mixed forests and southern taiga, where for the Middle Volga region, the $F_{hum}$ norm is 1.62–1.88 (Kolomyts 2003).

There were six selected groups of biogeocenoses on the territory of the test site, which covered the main diversity of forest types, located across the spectrum of the local catena (see Figure 5.5). Empirical information was collected in full (see Chapter 5). Since repeated studies (1990) were carried out on new test sites, averaged data on analogous groups of forest biogeocenoses were used for subsequent comparative analysis.

**TABLE 16.1** Comparison of climatic parameters and productive moisture reserves in the soil of an open area (meadow) at Vyksa station during periods of landscape–ecological surveys at the experimental site with long-term standards for these characteristics (indicated in italics in brackets)

| *PERIOD OF METEOROLOGICAL OBSERVATIONS* | *TIME OF SURVEYS* | *AVERAGE TEMPERATURE, °C* | | *ANNUAL PRECIPITATION, MM* | *ATMOSPHERIC HUMIDITY FACTOR* |
|---|---|---|---|---|---|
| | | *JANUARY* | *JULY* | | |
| 1988–1989 | 01.06–11.06 | –4.6 *(–11.5)* | 20.6 *(18.9)* | 624[a] *(590)* | 0.89[a] *(1.15)* |
| 1989–1990 | 01.06–10.06 | –8.1 *(–11.5)* | 18.8 *(18.9)* | 894[a] *(590)* | 1.62[a] *(1.15)* |

| *PERIOD OF METEOROLOGICAL OBSERVATIONS* | *TIME OF SURVEYS* | *MOISTURE RESERVES (IN MM) DURING THE SURVEY PERIOD IN SOIL LAYERS, CM* | | | *CORRECTION COEFFICIENT* |
|---|---|---|---|---|---|
| | | *0–20* | *0–50* | *0–100* | |
| 1988–1989 | 01.06–11.06 | 22 *(24)* | 40 *(44)* | 76 *(90)* | 1.18 |
| 1989–1990 | 01.06–10.06s | 30 *(24)* | 63 *(44)* | 125 *(90)* | 0.74 |

[a] Parameters calculated for the hydrological year (October–September).

# 16.2 PROCEDURE OF THE ECOLOGICAL EXPERIMENT

Figure 16.1 gives us the diagram illustrating the sequence of empirical simulation modeling. The data from the Moscow region water balance station, as well as those from the agrometeorological stations of Nizhny Novgorod, Murom, and Ardatov, were used to interpret soil–geophysical data of landscape–ecological surveys. Preliminary thermo- and hydro-edaphic ordination of the metabolic characteristics of topo-ecosystems showed (Kolomyts 2008) that these characteristics depended maximally on two geophysical parameters: soil temperature at the depth of 50 cm ($t_{50}$) and reserves of summer productive moisture in the soil layer 0–50 cm (*W–50*). These correlations are not always high enough, but they are quite significant (Pearson's significance test $P \ll 0.05$). In the case of a weak relation, the latter can be interpreted as some general trend in changes of this metabolic parameter under the effect of a geophysical trend.

Since the long-term norms of soil humidity are known more for field (meadow) sites of agro-meteorological stations $(W\text{-}50)_{meadow}$, the sequence of calculations included

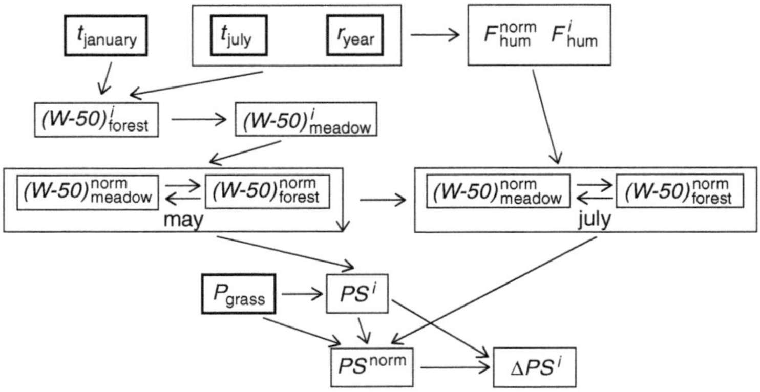


**FIGURE 16.1** Scheme illustrating the procedure for empirical imitation of environmental consequences of climate change. The thick frames contain the initial features for subsequent calculations. See the text for other designations.

multiple transitions from forest moisture reserves $(W\text{–}50)_{forest}$ to meadow ones and back – both for a given (*i*–th) year and for a long-term norm. These transitions were carried out according to empirical dependencies (data of the water balance station):

$$(W-50)_{meadow} = 0.978\cdot(W-50)_{forest} - 10.23; \qquad (16.1)$$

$$(W-50)_{forest} = 1.4 + 0.996\cdot(W-50)_{meadow\Gamma}. \qquad (16.2)$$

Calculations of soil humidity at the end of the third decade of May based on the annual humidity factor were made separately for sands and sandy loams (a), light and medium loams (b):

$$\text{a) } (W-50) = xp[2.425 + 1.11\cdot F_{hum}];\ R = 0.905;\ R^2 = 0.828; \qquad (16.3)$$

$$\text{b) } (W-50) = \exp[2.099 + 1.355\cdot F_{hum}];\ R = 0.921;\ R^2 = 0.848; \qquad (16.4)$$

An empirical relationship was used to transfer the hydro-edaphic parameter from the late May values to the middle July ones:

$$(W-50)_{July} = 0.8107\cdot(W-50)_{May} - 17.64;\ R = 0.836;\ R^2 = 0.699. \qquad (16.5)$$

The difference in $t_{50}$ between 1989 and 1990 for the same soil humidity (*W*–50) is 3.0°–3.2° (Figure 16.2). This figure is close to the difference between January average air temperatures in these years according to the Vyksa station – 3.5° (see Table 16.1).

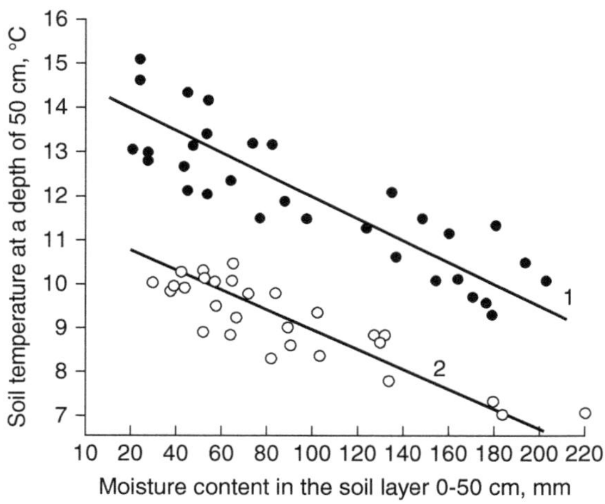


**FIGURE 16.2** Correlation relationships between soil moisture content and temperature during landscape–ecological surveys. Years: 1–1989; 2–1990.

The correlation factors obtained by comparing (according to the data of agrometeorological stations) the average long-term spring–summer reserves of soil humidity and the reserves of moisture in the year of the survey at this test site were used to calculate the long-term norm of soil humidity for all test sites and for all groups of biogeocoenoses of the territory under study. For 1989, this factor was equal to 1.18, and for 1990, it was 0.74. Based on the linear empirical relationship $t_{50}$ with ($W$–50) (Figure 16.2), the correlation factor of 1.18 was also used in the calculation of the long-term average soil temperature.

In the calculation procedure, we used three initial empirical parameters: (1) the reserves of productive moisture in the soil during the spring–summer period (the first10 days of June) at the test site, as well as according to the data of agrometeorological stations for this period in the average long-term regime and in the years of the surveys; (2) the directly measured green mass of the grass cover (in the air-dry state) in the forest on the test areas of the site; (3) humidity factors calculated according to the data of the Vyksa station for $t_{\text{July}}$ <19.5° (a) and $t_{\text{July}}$> 19.5° (b):

$$\text{a)}\ F_{\text{hum}} = 0.001726 \cdot r_{\text{year}} - 0.1954 \cdot t_{\text{July}} + 3.72; R = 0.95; R^2 = 0.90; \tag{16.6}$$

$$\text{b)}\ F_{\text{hum}} = 0.001362 \cdot r_{\text{year}} - 0.1069 \cdot t_{\text{July}} + 2,25;\ R = 0.96;\ R^2 = 0.92. \tag{16.7}$$

In the first case, it was the hydrological year of 1989, and in the second case, it was the year of 1990.

The green mass of the grass stand ($P_{\text{grass}}$) is a key empirical parameter that responds almost instantly to changes in hydrothermal conditions of the air (under the forest canopy) and the soil. Using this parameter, it was possible for us to obtain quantitative estimates of total annual productivity of the forest ecosystem (PC) and carbon deposition in the skeletal mass of the tree and shrub layer, along with the corresponding annual growth (PS) of this mass. The PS and PC parameters were calculated based on forest taxation data from the test areas, followed by the use of the known tables of the growth rate of the tree stands (Zagreev et al. 1992).

## 16.3 INDICATION OF THE ANNUAL INCREASE IN SKELETAL TREE-AND-SHRUB PHYTOMASS BY GREEN MASS OF THE HERBAGE

In general, in the facies groups, the production of above-ground mass of the herb stratum is inversely proportional to both the productivity of skeletal (mainly woody) phytomass and to the total annual production of the forest biogeocenosis (Figure 16.3). This means that if the above-ground herbaceous mass increases with an increase in the spring and summer humidity of the soil, then the production of all other phytomasses, both above-ground and underground, decreases respectively. The fairly clear relationships between $P_{grass}$ with PC and PS, as well as empirically established dependencies of the first parameter on the spring–summer reserves of productive humidity in the soil (Figure 16.4), make it possible to use the green mass of the herbage as an indicator of the productivity of the skeletal tree and shrub phytomass (above-ground+root) and calculate changes in carbon sink in the phytocenotic pool of the forest geo(eco)system.

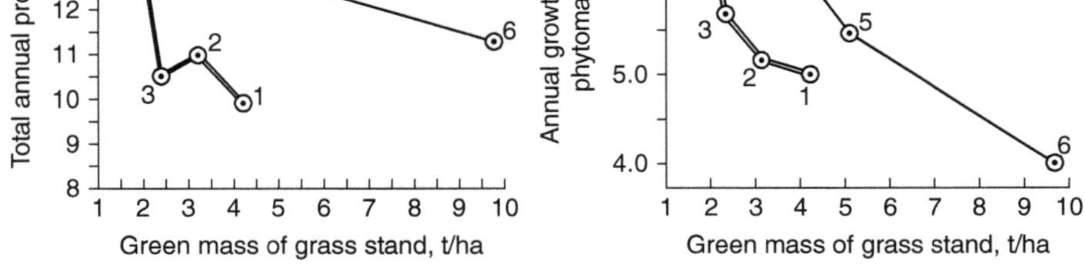


**FIGURE 16.3** Correlation relationships between the green mass of the grass cover and the overall productivity of forest ecosystems (a) and the annual increase in skeletal phytomass (b). Links of local catenas: a – upper and middle (location types E, TE, and T); b – lower (types TA, A, and SA). Explanations in the text.

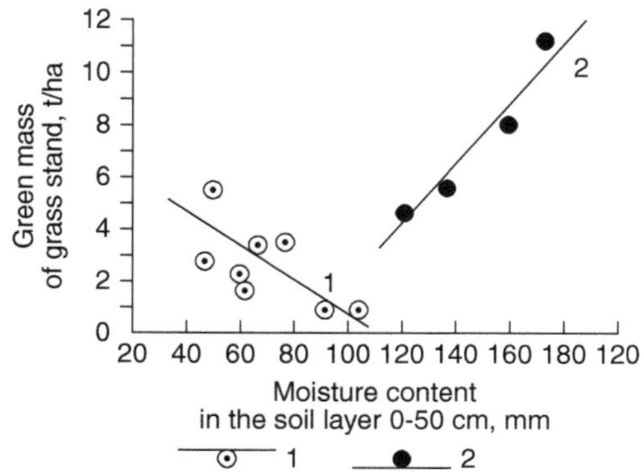


**FIGURE 16.4** Correlation links of green mass of grass stand with spring–summer reserves of productive moisture in the soil. Links of local landscape conjugations: 1 – E, TE, and T (upper "floor" of catena); 2 – TA, A, and SA (lower "floor").

For such an indication, it was necessary to establish the dependence of the parameter $P_{\text{grass}}$ on the hydrothermal conditions of the soil. The relationships were generally found to be too "blurred" in the "frontal" continuum–statistical analysis. Hence, for the Vyksa-89 model, the following was obtained:

$$P_{\text{grass}} = 143.2 - 21.06 \cdot t_{50} + 0.791 \cdot t_{50}^{\ 2};\ R = 0.454;\ R^2 = 0.206; \tag{16.8}$$

$$P_{\text{grass}} = 2.386 + 0.223 \cdot (W - 50);\ R = 0.329;\ R^2 = 0.108. \tag{16.9}$$

The Vyksa-89 model had the following partial relation:

$$P_{\text{grass}} = 5.43 - 0.077 \cdot (W - 50) + 0.791 \cdot (W - 50)^2; R = 0.408; R^2 = 0.166. \tag{16.10}$$

The multiple correlation across both models was generally unreliable. According to Equations (16.9) and (16.10), the green mass of the herbage increases as a whole with an increase in spring–summer moisture reserves in the soil; however, this is just a generalized and very poorly expressed trend, which can be traced mainly when looking at the overall variation in productivity – from minimal to maximum values. Against the background of this trend, the opposite regularity has been revealed in certain parts of the catena – with a negative relationship between these parameters. Let us consider these regularities in more detail. At the same time, we divide each general set of the test areas into samples corresponding to the position of biogeocoenoses in the system of their catenae differentiation, i.e., we use a discrete statistical approach, operating (to obtain a simplified picture of the relationships) with the values of functional parameters averaged for each sample group.

The system of local landscape connections at the test site is clearly differentiated into two "levels", which differ in the sign of the correlation between the mass of the grass canopy $P_{grass}$ and the parameter ($W$–50). The upper level, which covers the largest areas, includes E (placor), TE, and T types of the sites, which include 1–4 groups of biogeocoenoses (GB, see Figure 16.3). There is an inversely proportional relationship between $P_{grass}$ and soil humidity (see Figure 16.4), and this relationship manifests itself, obviously, through tree and shrub layers, the development of which has a direct negative impact on the herb stratum.

On the lower level of the catena, which includes TA, A, and SA types of MP, with excessive edaphic humidification (GB 5 and 6), the development of forest stands and shrubs is weakened, and the forest canopy is sparse, which contributes to the formation of a sufficiently high and dense herbage with a significant green mass. In these local hydro-edaphic conditions, the relationship between the $P_{grass}$ and ($W$–50) parameters turns out to be directly proportional (see Figure 16.4).

It has been previously determined (Kolomyts et al. 1993; Kolomyts 2008) that the outwash marshy woodlands of the Middle Volga Region, on the zonal boundary of the sub-taiga and broadleaf forests, are in conditions of atmospheric humidification from insufficient to slightly excessive ($F_{hum}$=0.94–1.12). The lack of atmospheric moisture affects, first of all, the forest communities of the most well-drained types of sites, belonging to the upper level of the catena. For this reason, any increase in soil hydromorphicity leads to an activation of the production process in the forest stands, undergrowth, and shrubs, which in turn has a depressing effect on the ground cover. As a result, it turns out that with the overall reduced productivity of the grass cover on this catenary level, an increase in atmospheric (and with it, edaphic) humidity leads to an even greater decrease in productivity.

The described regularity manifests in the spatial relationships of $P_{grass}$ with the total annual productivity (PC) of forest communities, as well as with the productivity (PS) of their skeletal mass (see Figure 16.3). We can see here two similar catenary levels. In the system of biogeocoenotic transitions: 1 → 2 → 3 → 4, a decrease in herbage mass is accompanied by an increase in PC and PS parameters, especially sharp in the interval of transitions from TE sites to transit ones on flat slopes of the interfluves. With further transitions to the TA–A and SA locations, already on the lower level of the catena, a significant increase in herbaceous green mass corresponds to an equally sharp decrease in the annual growth of woody and shrubby phytomass and a decrease in the total productivity of the forest phytocoenosis. Based on this, it is assumed that a decrease in green herbaceous mass in summer, compared with the long-term norm, will be adequate (in percentage terms) to an increase in PC and PS parameters, and vice versa.

## 16.4 SIMULATION MODELS OF ECOLOGICAL CONSEQUENCES OF CLIMATE CHANGES

Shifts in the hydrothermal regime and functioning of forest ecosystems under the effect of one or another hydrothermal signal are quite clearly differentiated by the types of the sites and groups of forest types, resulting in the formation of a whole range of responses of forest biogeocoenoses to this signal (Figure 16.5).

The *thermo-arid climate signal* (Vyksa-89 model) caused generally insignificant shifts in the spring–summer moisture content of the soil. Maximal negative anomalies compared with the long-term norm were observed in mesomorphic and hydro-mesomorphic wood sorrel spruce forests TE and E sites, respectively (biogeocoenosis group 3): $(W50) = 68 \rightarrow 61$ mm. In other cases, $\Delta(W50) \leq 5$ mm, which does not exceed the limits of soil humidity measurements themselves. Nevertheless, it is possible to designate some spring–summer desiccation of swampy super-aqual forests, as well as meso-xeromorphic pine forests, with spruce, on gently undulating watersheds (GB 6 and 2).

The soil temperature was 2.1°–2.4° higher than the long-term spring–summer norm, but this increase was partially neutralized by additional precipitation (see Table 16.1), so the temperature component of the thermo-arid signal itself proved to be ineffective. As a result, deviations from the long-term average rate of total annual production of forest communities and, consequently, carbon deposition in their skeletal mass were generally low, although almost universally positive (see Figure 16.5 and Table 16.2). The carbon content of $B_C$ in the living skeletal mass increased most clearly in transit oak–pine–spruce forests (GB 4). Here it could reach 4.45 t/ha, exceeding the norm by 14%. An increase in the carbon sink was much weaker in xeromorphic and

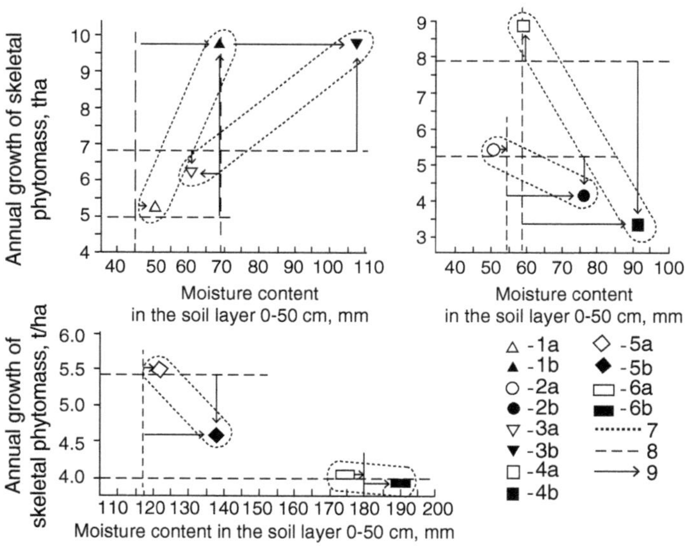


**FIGURE 16.5** Ratios of soil moisture content (*W*–50) and productivity of skeletal phytomass PS in different forest ecosystems with two alternative climatic signals. 1–6 – groups of biogeocoenoses (see Table 5.3). Hydrothermal signals: a – thermo-arid (Vyksa-89 model); b – humid (Vyksa-90 model). 7 – "areas" describing the amplitude between signals. 8 – long-term norms (*W*–50) and 9 – direction and conditional value of the deviation of the specified parameters from their norm.

**TABLE 16.2** Deviations of soil moisture content from the long-term norm and changes in carbon runoff in forest biogeocenoses groups under different climatic signals

| | | *TYPES OF LOCATIONS AND BIOGEOCOENOSES GROUPS (SEE TABLE 5.3)* | | | | | |
|---|---|---|---|---|---|---|---|
| *MODEL* | *PARAMETERS*[a] | *TE–E* 1 | *E–TE* 2 | *E–T* 3 | *T–TA* 4 | *TA–A* 5 | *EA–SAQ* 6 |
| Vyksa-89 | $\Delta(W\text{–}50)$ | 3 | –4 | –9 | 2 | 1 | –6 |
| | $\Delta(B_C)$ | 0,14 | 0,15 | 0,27 | 0,56 | 0,02 | 0,02 |
| Vyksa-90 | $\Delta(W\text{–}50)$ | 25 | 28 | 36 | 32 | 23 | 10 |
| | $\Delta(B_C)$ | 2.43 | –0.54 | 2.06 | –2.22 | –0.28 | 0 |

[a] $\Delta(W\text{–}50)$ – deviation of soil moisture content in the 0–50 cm layer (in mm) in the year of the landscape–ecological survey from the long-term moisture content norm; $\Delta(BC)$ – change in carbon deposit (t/ha) in skeletal (above-ground and root) tree and shrub phytomass in the year of the landscape–ecological survey compared to the long-term norm of its accumulation.

*Positive values of $\Delta(BC)$ for the flat interfluve indicate a positive carbon balance, while negative values indicate a negative balance.*

meso-xeromorphic watershed (TE and E) pine forests (GB 1). The green mass of the herbage had the greatest positive deviations from the norm here: 4.4 → 5.5 t/ha. The carbon sink remained unchanged at the lower level of the catenae, in waterlogged small-leaf spruce–pine forests. The thermal component of the climate signal worked poorly in this case.

Against this background, the exception was made by the eluvial-transit mountain sorrel spruce forests, where there was a decrease in $B_C$ by an average of 9%–10% (down to 6.23 t/ha), which was caused by more intensive desiccation of the soil in the late spring period (see above). At the same time, the green mass of the herbage decreased (on average from 2.3 to 1.75 t/ha).

Thus, winter warming itself, even if very significant, affects insignificantly the functioning of forest ecosystems, including their productivity and carbon balance, if it is not accompanied by a sufficient decrease or increase in atmospheric humidification. This empirical conclusion confirms our statement (Kolomyts 2003) that the "mitigation" of winter cannot serve as a decisive factor in shifts in the functioning and structure of geo(eco)systems.

The *humid climate signal* (Vyksa-90 model) had a much more significant impact on the functional state of forest communities compared to the thermo-aridization described above. In this case, the local contrasts in the dynamics of these states were especially pronounced. With maximal positive efficiency, a pluvial signal manifested itself in the most well-drained water-separated biogeocoenoses, where moisture deficiency is most pronounced. At the same time, a significant increase in the spring–summer moisture content of the soil played a decisive role (see Table 16.1). Thus, in TE, E, and T sites, in xeromorphic lingonberry–mossy pine forests and mesomorphic oxalis spruce forests (GB 1 and 3), parameter shifts ($W50$) were 43 → 68 and 59 → 91 mm, respectively. Consequently, the production process in forest communities has significantly intensified, and the carbon sink has increased in the skeletal phytomass. The sink dynamics were as follows:

$$B_C(1) = 2.50 \rightarrow 4.92\ T/\Gamma a;\ B_C(3) = 2.85 \rightarrow 4.91\ T/\Gamma a.$$

It is important to stress that in the upper and middle links of the catena, an increase in carbon deposit occurred despite the lower soil temperatures, which turned out to be on average 2.70 degrees colder than under last year's thermo-arid conditions. This points to an insignificant role of the thermal component of this climatic signal in the local conditions of initially increased humidity deficiency.

However, with moderate and weakened drainage, we have a completely different picture. A sharp decrease in carbon sink from the atmosphere into the skeletal biomass occurred in poorly drained grassy oak–pine–spruce forests of TA and A sites, i.e., on gentle slopes and at the foots (GB 4). Here, the average changes ($W50$) were 71 mm → 107 mm, which was accompanied by a decrease in soil temperature (see Figure 16.2). This led to a significant weakening of autotrophic biogenesis and, consequently, to a decrease in $B_C$ from 3.90 to 1.68 t/ha.

Eluvial meso-xeromorphic beadruby–lingonberry pine forests with spruce on gently wavy interfluves (GB 2), as well as accumulative long-mossy blueberry pine–spruce forests of gently concave slopes (GB 5), experienced a significantly lower loss of carbon deposit. In these forest communities, the shifts of the parameter ($W$–50) were respectively: 49 to 77 mm and 114 to 137 mm, and the soil temperature decreased by 2.8 to –3.10 compared to the previous year. As a result, the following decrease in carbon sink occurred here: $B_C\ (2) = 5.18 \approx 4.11$ t/ha; $B_C\ (5) = 5.46 \rightarrow 4.60$ t/ha.

In excessively humidified long-moss–sphagnum pine forests and spruce forests of inter-river depressions, a further increase in soil moisture reserves did not have any significant impact on the overall production process and carbon balance. We can only argue for a slight decrease in the carbon deposit here.

The critical value of the spring–summer soil moisture content, at which an increase in the carbon deposit in the skeletal phytomass of forests is replaced by its loss, significantly depends on the form of the microrelief, which determines the degree of drainage and the temperature regime at this site. Hence, in the more drained oxalis spruce forests of wavy watersheds, an increase in $B_C$ continued up to the values ($W$–50) ≈ 110 mm, while in broadleaf coniferous forests of gentle slopes, with their weaker drainage, –a sharp decrease in carbon deposit occurred already at (W–50) ≤ 90 mm (see GB 3 and 4 in Figure 16.5). At the same time, the soils were almost the same in their mechanical composition – sandy loam and light loam.

In the first case, a sufficiently long activation of autotrophic biogenesis and a corresponding increase in carbon sink was facilitated by a decrease in the deficit of edaphic humidification against the low soil temperature. In the second case, there was excessive flooding of hydro-edaphotopes at similarly relatively low temperatures, which led to a weakening of the production process, with a corresponding decrease in the $B_C$ parameter at the earlier stage of the humid trend.

Hence, a rather complicated picture of the functional response of forest ecosystems at the topological level to the same background pluvial signal emerges. The mechanisms of carbon accumulation in tree–shrub phytomass also differ in corresponding spatial diversity. Positive dynamics of the carbon sink are typical for those forest communities where a humid signal eliminates the initial deficit of soil–groundwater moisture, with an indifferent role of the temperature regime. Under conditions of normal and even more excessive initial humidification, a further increase in soil moisture content is accompanied by a decrease in its temperature, causing negative dynamics of the carbon sink.

## 16.5 EXTREME SIMULATION MODEL

If we calculate the difference in the parameters $t_{50}$, ($W$–50), and PS between 1990 and 1989, we obtain a picture of shifts in the hydrothermal regime of forest biogeocoenoses, in their productivity, as well as in some carbon sink, a much more pronounced variant of the thermo-arid climate trend, which is characterized not only by an increase in winter and summer temperatures but also a significant decrease in precipitation (see Figure 6.6 in Chapter 6). According to the average long-term atmospheric humidification conditions ($F_{hum}$), this hypothetical model refers the Lower Oka woodlands in its initial, i.e., current, state (analog of 1990) to the more western territories of the Middle Oka and Upper Volga basins. The final state of the Lower Oka forests (analog of 1989) in some perspective will be due to their certain ecological shift. Such a shift can be expected around the middle of our century, according to one extreme climate model, HadCM3, version A2 (see Chapter 12).

The Vyksa 1990-1989 model simulates a very significant range of possible ecological consequences of the thermo-arid trend. On the territory of the Vyksa site, $t_{50}$ increases by 2.7°–4.2°, and a decrease in ($W$–50) ranges from 6–18 to 31–46 mm. At the same time, the model demonstrates a pronounced differentiation of forest thermo-hydro-edaphotopes by the same background climatic signal. In general, the largest hydrothermal shifts are typical for forest ecosystems of the middle links of landscape conjugations – T and TA types of sites (GB 3, 4, and 2). Minimum changes in temperature and soil humidity will differ in biogeocoenoses of two opposite "poles" of the catena, confined, on the one hand, to E–TE sites, and on the other, to accumulative and super-aquatic ones. Since the spatial relationship between $t_{50}$ and ($W$–*50*) is inversely proportional, their distribution over the catena vector is mirror-like.

We consider the local mosaic of functional responses of forest ecosystems to a given climate signal. The least sensitive to this signal are hydromorphic coniferous–small-leaf forests of interfluve depressions and poorly drained foothills of the slopes (GB 6 and 5). Accordingly, the productivity of the forest community and the amount of carbon sink into skeletal phytomass will change little here, with insignificant positive values ($\Delta$PS $\leq$ 0.9 t/ha).

A significant negative thermo-arid effect ($c\Delta t_{50}$ = 4.2° and $\Delta$($W$–50) = –46 t/ha) will be manifested in mesomorphic flat interfluve oxalis spruce forests and spruce–pine forests on E–T flat–wavy watersheds (GB 3), where $\Delta$PS = –3.6 t/ha is expected. At the same time, lower along the slopes in more humidified hydro-mesomorphic oak–pine–spruce grassy forests of T and TA locations (GB 4), a slightly less thermo-aridization ($\Delta_{t50}$=3.8°, $\Delta(W{-}50)$=–31 t/ha) will cause, on the contrary, a significant activation of the production processes and an increase in carbon sink by 5.6 t/ha – the maximum for all ensembles of biogeocoenotic groups (see Figure 6.6). In the first case, the thermo-arid climate signal causes the removal of forest communities from their state with an optimal ratio of heat and moisture, and in the second, an approach to this state, which leads to negative and positive changes in carbon sink, respectively.

## 16.6 SOME INFERENCES

Empirical simulation of the ecological consequences of climatic changes made it possible to identify specific shifts in the hydrothermal regime and the primary productivity of forest ecosystems under the effect of two alternative atmospheric hydrothermal signals. In the belt of zonal transition from mixed forests to broadleaf ones, a humid climate signal has a more efficient and diverse effect on the functional state of forest ecosystems than the thermo-arid signal. Accordingly, in the first case, a more pronounced and territorially more contrasting picture of changes in carbon sink into the living skeletal phytomass appears, which, as already shown (see Chapters 10 and 12), plays a decisive role in the carbon balance of forest ecosystems. The most vivid and diverse picture of functional shifts in forest ecosystems with expected global warming is provided by an extreme simulation model built on the same empirical data, accounting for both an increase in winter and summer temperatures and a significant decrease in annual precipitation.

# Brief Account of Predictive Landscape–Ecological Conception Using Empirical Simulation Models

# 17

## 17.1 GENERAL PROVISIONS

The book expounds for the first time the regional paleo-prognostic conception in geographical ecology. Prognostic landscape–ecological scenarios of the near future of the biosphere as a single system of global changes in the environment have been considered for the first time. The mechanisms of local and regional response to the anticipated global warming in the 21st century have been assessed using the example of forest ecosystems of the Volga River basin headwaters, which are located near the southern boundaries of the boreal belt on the Russian Plain.

The quantitative methods (analytic and cartographic) of local and regional landscape–ecological forecasts developed by the authors are described. The landscape–ecological approach considers the discrete empirical–statistical models of natural ecosystems. These models describe the behavior of geo(eco)systems as indivisible


DOI: 10.1201/9781003733119-17

complex formations – in the aspect of V.N. Sukachev's doctrine about biogeocoenoses. They make it possible to operate with a relatively small number of the most informative characteristics and yield results that, while possibly not quite exact in time, have a higher level of spatial resolution than known simulation models. Predictive calculations are based on cartographic and satellite data as well as the authors' materials from large-scale landscape–ecological surveys on experimental sets. The methods of information theory, "fuzzy set" theory, Markov chain theory, and new methods of geomorphometry are also used. The local and regional prognostic maps of plant cover and changes in common carbon content in forest ecosystems across various experimental sets and within one large ecoregion are presented. The proposed and realized methodology represents a fundamentally new means of local and regional ecological prognoses, which is based directly on empirical materials and therefore yields the most reliable results.

Two global climate prediction models, GISS and HadCM3, were used for prognosis. With the help of matrices and vector graphs of landscape–ecological transitions, the picture of multitudinous shifts in the patterns of vegetation, local ecosystems, and landscape–zonal systems is revealed. The mechanisms of shifts in the mosaic structure of vegetation and soils have been uncovered in the model territories under different scenarios of disturbing influences of the climatic system anticipated in the foreseeable future.

The future scenarios of intensive thermo-aridization of soil–plant cover as a general process of "savannization" of the broadleaf sub-zone and mixed forest zone on the Russian Plain are forecasted. A significant initial "jump" in the functional transformation of the forest ecosystems on the forest–steppe ecotone due to anthropogenic warming was suggested. Mesophyte broadleaf forests of the flat interfluves will be subjected to various and irreversible transformations. They will be intensively decomposed and substituted by meadow steppe. The most xerophyte forest–meadow–steppe associations on sunny slopes and pine–broadleaf forests will experience aridity to an even greater degree. It is anticipated that these marginal forests may disappear completely. One should expect sequential northward shifts of zonal and sub-zonal boundaries of the plant formations. Desertification of the steppes themselves will also develop intensively.

The strategy of prognostic landscape–ecological studies on climate-genic changes in the biological cycle in forest ecosystems and their carbon balance as leading factors of the biotic regulation of the environment is presented. This makes it possible to reveal those zonal–regional types of forest formation which will promote the sustainability of terrestrial ecosystems in a changing climate (according to the well-known Le Chatelier's principle).

By means of gradient (ordination) analysis, it has been established that the changes in the two branches of the small biological cycle, production and detritus, predicted in broadleaf, mixed, and coniferous forests of the Middle Volga Region, will be caused by the thermo-arid climatic trend. The general model scheme describes the driving forces of climatogenic successions: the fundamental mechanism of transition from functional changes to structural transformation of ecosystems under the effect of global warming. The prognosis of changes in biological turnover, respective values of carbon balance, and biotic regulation of the carbon cycle in forest ecosystems is presented according to HadCM3 scenarios (version A2, which predicts the most powerful global warming).

Prognostic numerical valuations of biotic regulation of the carbon cycle according to well-known scenarios of global warming are presented. It is predicted that under global warming, different temperate forest associations and formations will carry out both positive and negative roles in carbon cycle biotic regulation in the system of soil–vegetation–atmosphere. Broadleaf–coniferous forests of uplands in the nemoral forest sub-zone and in the southern belt of the sub-taiga zone will probably be the only group of boreal forest formations that will essentially support the sustainability of the continental biosphere under forthcoming global warming. Thus, it is predicted that under global warming, different temperate forest associations and formations will carry out both positive and negative roles in carbon cycle biotic regulation in the system of soil–vegetation–atmosphere.

## 17.2 THE SCIENTIFIC AND METHODOLOGICAL BASIS OF PROGNOSTIC ANALYSIS

Ecological safety of large territorial subunits of the continental biosphere significantly depends on the state of the zonal–regional types of natural ecosystems, first of all, forest cover. Therefore, the problem of maintenance of forest ecosystems and reproduction of forest resources on the southern boundary of the temperate forest zone, where forest communities are present in states close to critical, is among the fundamental ecological problems.

The idea about structural levels of landscape organization previously proposed by the author, which is based on a conceptual cybernetic model of the natural complex as a hierarchic system of regulation, was used as a methodical basis for landscape–ecological analysis. An experiment was performed in order to analyze the state of zonal–regional geo-spaces by three units of the model: landscape frame, processor, and landscape pattern.

The solution of multipurpose tasks of ecological forecasting may depend to a great extent on the results of system analysis of natural–territorial structures, which are most sensitive to external effects, including anthropogenic ones. The scientific search in this direction focuses more and more attention on the natural boundaries – both individual and complex – where the most significant natural or anthropogenic shifts in the structure and function of geo(eco)systems are observed. Considering one or another natural boundary as a vector (connection, cascade, para-dynamical, etc.) landscape system with a clearly defined spatial polarization of its different properties, we obtain a "quickly running" model of state response and resistance of geo(eco)systems to the action of certain ecological factors.

The systematic approach to the study of natural environment organization would be expediently realized by the example of marked and large enough biogeographical and landscape borders – ecotones. Such a border on the territory of Eurasia is a broad boundary frontier between the boreal (mainly taiga–forest) and sub-boreal (forest–steppe and steppe) belts of plant formations. The basic watershed of the Volga River basin is included in the boreal ecotone of the Russian plain. As far as it is known, ecotones are the most sensitive and fragile parts of the biosphere in reference to external influence.

A detailed analysis of ecological principles of landscape–zonal organization of the boreal ecotone revealed quite a number of new regularities in the mechanisms of this organization. Zonal boundaries are considered as modern spatial analogs of future landscape changes over time. Besides, the first-priority and most noticeable processes of transformation of natural–territorial structures on zonal boundaries could be anticipated under the action of background climatic perturbations.

The work dwells on the basic "trigger" mechanisms of zonal boundary formation at the interaction of background climatic signals and their refraction by local (mainly lithogenic) factors. The enhancement of the role of local factors to the level of zonal–regional ones is a characteristic feature of the natural zone periphery, where the climatic background reaches its critical values. The revealed functional scheme of zonal boundary formation may be considered as a possible mechanism of differentiated shifts of zonal borders at anticipated global climatic changes.

The most important immanent feature of zonal boundaries is the stronger territorial interrelationship of landscape types on either side of each boundary, as well as the higher spatial orderliness of landscape areas on the boundaries compared to the cores of typical natural zones themselves. This is one of the main reasons for the higher sensitivity of zonal ecotones to background climatic signals.

Two different scales of landscape–ecological connections are considered. Zonal–regional phytocoenological and landscape units show a fairly strong dependence on the annual atmospheric humidity factor. These correlations were used to create predictive maps of natural zones (sub-zones) in the region based on the corresponding predictive maps of atmospheric humidification. The groups of primary plant formations and landscape types delineated in the geobotanical and landscape maps of European Russia were taken as operational units for regional ecological prognoses.

A quite different approach to prediction has been used for the level of local (topological) ecosystems. Landscape facies (biogeocoenosis) corresponding to the categories "Site", "Ecoelement", and "Landtype Phase" represent the smallest natural territorial complex and indicate the structure–functional "core" of the biosphere. It has been established empirically that the main channel of connections between landscape facies and the background climatic system passes through the summer moisture capacity of soil, which is the most powerful direct ecological factor predetermining their state and territorial distribution.

The local forecast analysis was based on the materials from large-scale landscape–ecological surveys carried out by the authors at seven experimental test sites (ranges) in the Middle Volga Region. These test sites embrace the main diversity of zonal transitions on the boreal ecotone of the eastern sector of the Russian Plain. Six facial groups distinguished for each regional ecosystem cover the major topological diversity of forest types and are located in the vector system of landscape couplings (catenas) – in the direction of increasing edaphic moistening. On each test site, the empirical information (statistical sample) collected in 40–50 sample areas included about 80 characteristics of the structural–functional state of forest topo-ecosystems.

The empirical basis of paleo-prognostic constructions is formed by three groups of landscape–ecological models:

a. landscape–geophysical – a system of chain reactions in component interrelations in response to external impacts, which opens the main channels for the transmission of hydrothermal signals and is closed at vegetation, soils, and primary biological productivity;
b. bioclimatic – an ordered ensemble of climatic niches of soil–plant cover in the space of landscape–geophysical factors, which are the starting objects of climatic prognoses;
c. functional – a group of parameters describing the small biological cycle, which experiences primary impacts from the climatic system and therefore must be a top-priority object of landscape–ecological prognosis.

The initial climatic parameters of global and regional prognoses (average January and July temperatures, the annual sum of precipitation, sometimes with a division of the latter into precipitation of cold and warm periods) have been taken from the data of hydro-meteo-net and from known literature sources. The derivative geophysical parameters for the territory of the Volga River Basin (annual radiative balance, annual potential evaporation, annual humidity factor, stored soil summer moisture, annual surface and underground flow, etc.) have been calculated using empirical formulas (with high coefficients of correlation).

The geo-ecological analysis has been performed using a series of empirical statistical models: (1) informational, describing the geo-component interrelations, i.e., monosystem organization of nature complexes, serving as a basis for the regional bank of their ecological niches that characterize their parametric space; (2) "fuzzy" set theoretical models, describing the polysystem units of landscape organization through operations with the ecological niches as descriptive vectors; (3) cluster models, characterizing the hierarchical structure of landscape interconnections. Complex interconnections were revealed, and the territorial ecological space of natural landscapes was described based on "fuzzy" set theory methods and cluster models.

The informational structure of relations in the binary "factor–phenomenon" permits the revelation of a certain system of ecological niches (in this case, hydrothermal niches) of either object. Type schemes of binary ordination of geo-component attributes have been created by the systems of ecological niches.

The ordination of kind landscape unities by the gradient of the "exoticism–trivialism" factor is carried out with the help of "fuzzy" set models. This served as a basis for distinguishing the ecological groups of natural–territorial units with a different degree of reflection of the zonal geophysical background, which determines their corresponding role in the natural–territorial organization of either ecoregion. On the matrices of inclusion relations, the reserve of "homeostaticity" as a measure of the potential stability of geo(eco)systems to a given factor was calculated. The structural schemes of temporal states of regional and local geo(eco)systems, as well as schemes of their climatogenic transformations, have been created based on this measure.

The suggested kind of empirical simulation modeling of global–regional changes by local geo(eco)systems, based on the schemes of binary ordinations, is intended to promote a certain advance toward elucidating the local mechanisms of these changes. With this purpose, the regional systems of local zonality, including vector series of flat-interfluve (placor) and extra-zonal topo-ecosystems, are considered. Consequently, we obtain an ergodic operation system, which (thanks to the properties of its self-similarity) becomes the basis for revealing the real spectrum of possible directions of functional–structural transitions between ecosystems.

## 17.3 EMPIRICAL SIMULATION MODELING OF FOREST ECOSYSTEMS IN PREDICTIVE ANALYSIS

The concepts of discrete empirical simulation forecast modeling of natural ecosystems are presented. A strategy of empirical–statistical simulation modeling of forest biogeocoenoses organization is accounted for as initial objects of terrestrial geosystem monitoring. Spatial functional monitoring of forests is based on empirically established local and regional landscape–ecological connections, which are considered mechanisms of metabolic reactions of forest geo(eco)systems to certain climatic trends. This is a brief formulation of the concept of empirical simulation predictive modeling of natural ecosystems. This represents the novelty of the prognostic empirical simulation concept of local (and regional) geo-ecological forecasting developed by the author. This modeling served as the main tool for pre-forecast landscape–ecological analysis and subsequently entered the algorithm of the forecast calculations themselves.

In landscape ecology, the creation of models of the structural–functional organization and dynamics of forests, where a holistic natural elementary unit – a biogeocoenosis – would be used as an object of modeling, still remains a problem. The predictive dynamics of forest ecosystems under changing climate have been developed quite poorly thus far. By now, methods of *continuous (dynamic) simulation modeling* of ecosystems have been developed, including for forecasting purposes. These models of forest responses to climatic impacts meet the narrow preset framework of habitat conditions; therefore, the results of such modeling are insufficient for predicting the state of the forest community as a whole and do not cover the spatial diversity of succession changes under the same background influence.

Dynamic simulation models, despite their high efficiency, do not actually go beyond the limits of aut- and synecology. They do not cover the territorial diversity of successional phytocenotic changes under the same background impact. Rather, these are "beacon" models that reveal certain invariant features of the organization of the category of natural ecosystems under consideration. However, landscape ecology cannot be limited to such a modeling method, as it requires spatial expansion of basic and predicted situations.

In our studies, monitoring of geosystems was carried out on the basis of the *discrete empirical-simulation models*, which can be attributed to discrete statistical modeling. These are self-organizing a posteriori models. They differ significantly from the a priori simulation models described above. These models make it possible to operate with a relatively small number of the most informative characteristics and to obtain results that are probably not as precise in time as in simulation modeling but with a higher degree of spatial resolution and a more solid (geographical) level of territorial generalizations.

In empirical simulation modeling, the research results are used not to verify certain calculated data but as a factual basis for the modeling itself. This yields results with a much higher spatial resolution. Unlike dynamic simulation models, the method of local discrete modeling of natural ecosystems proposed and implemented by the author can be considered a biogeocoenotic method with much greater justification, in the spirit of the teachings of V.N. Sukachev. Such models have made it possible to conduct a predictive geo-ecological analysis based on the mass material of large-scale landscape–ecological surveys conducted by the author in the Middle Volga region, using various cartographic materials. The ecotone spectrum of natural zones covering the main catchment area of the Volga River basin turned out to be a favorable model territory.

In accordance with known developments, the biosphere is considered a statistical ensemble of biogeocoenoses – weakly interacting with each other but internally highly ordered (based on stabilizing selection) elementary biochorological units. Each geosystem of the topological level is assigned certain functional properties. This allows us to study the behavior of topo-geosystems in various geomorphological and edaphic conditions based on the mass empirical information collected in the field while maintaining statistical methods of analysis. Thus, it is possible to present the mechanisms of biological circulation and regulation of the carbon cycle in empirical–statistical forecast models of local geographic space, which is a fundamentally new aspect of solving this problem.

The landscape–ecological analysis of the spatial organization of natural complexes and their climate-genic dynamics developed by the author is based on the construction of discrete empirical–statistical models of their structure and functioning. As part of these models, a method of empirical imitation of predicted geo-ecological situations is proposed and implemented. In this method, the materials of natural studies are used not as a check of certain calculated data (which is practiced when creating dynamic simulation models) but as a factual basis for the modeling itself. The possibilities of widely using the results of field landscape studies in predictive analysis, as well as various cartographic materials, make it possible to obtain results with a high degree of spatial resolution and to empirically substantiate fairly broad geographic generalizations.

Spatial functional monitoring of forests is based on empirically established local and regional landscape–ecological connections, which are considered mechanisms of metabolic reactions of forest geo(eco)systems to certain climatic trends. The system of these connections used in the forecasting procedure is presented. The local display of the regional bioclimatic trend by the hydrothermo-edaphic system of biogeocoenoses is established. This is the novelty of the prognostic empirical simulation concept of local (and regional) geo-ecological forecasting developed by the author. The empirical simulation landscape–ecological forecast, as the central link in monitoring, has an experimental nature. It was developed as a system of operations with ecological (hydrothermal) niches of the objects under study. Empirical imitation of landscape–ecological forecasts is implemented in the form of reproducing forecast scenarios of biogeocoenotic systems according to the laws of their basic spatial organization, with the replacement of spatial coordinates with time coordinates in the forecasting procedure, in accordance with the fundamental properties of ergodicity of the natural environment. One of the promising areas of research has become the setting up of a series of experiments to study the mechanisms of local response to background hydro-climatic signals, with the identification of a system of transfer and transforming functions in landscape connections.

The following were used in the forecasting procedures: (1) the properties of polyzonality of local (and regional) ecosystems as a way of their response to global climate change; (2) regional structures of localized natural zonality capable of modeling the main directions and scales of geo(eco)systemic restructurings; (3) local empirical imitation of a regional bioclimatic trend based on binary hydro- and thermo-edaphic ordination of forest topo-geosystems; (4) empirical imitation of climatically conditioned changes in discrete parameters of the biological cycle based on their hydrothermal ordination analysis; (5) an algorithm for landscape–ecological forecasting as a reproduction of scenarios of biogeocoenotic systems according to the laws of their basic spatial organization; and (6) functional isomorphism of natural ecosystems, meaning their dynamic similarity, due to which the adjustment (sparseness) of the network of calculated landscape–ecological transitions is carried out.

The property of polyzonality of local geo(eco)systems as a way of their reaction to global climate changes is revealed. An ordination analysis of landscape connections is carried out based on the main discrete parameters of the biological cycle. A local empirical imitation of a regional bioclimatic trend by a hydrothermo-edaphic system of biogeocoenoses is conducted, and patterns of transitions of forest communities to critical states are revealed. Catenary landscape–ecological structures are described that form regional systems of localized natural zonality capable of simulating the main directions and scales of geosystemic reorganizations. The landscape–ecological forecast itself is of an experimental nature. It is presented as a system of operations with ecological niches of the objects under study. In the calculation models, the course of the predicted processes is reproduced using their empirical imitation by spatially distributed parameters of the basic ecological niches.

The landscape–ecological forecast is presented as a system of operations with ecological niches of the objects under study. In the calculation models, the course of the predicted processes is reproduced using empirical imitation by spatially distributed parameters of the basic ecological niches. The basic principle of landscape–ecological forecasting is formulated, corresponding to one of the provisions of general ecology – the Gause law of competitive exclusion.

Models of critical states of forest communities in the south of the forest zone are presented, in which the most abrupt functional–structural changes in the geo(eco)systems of the flat interfluve series, reflecting the zonal–regional background of a given territory, should occur.

The property of polyzonality of local geo(eco)systems, as a way of their reaction to global climate changes, is revealed. An ordination analysis of landscape connections is carried out based on the main discrete parameters of the biological cycle. A local empirical imitation of a regional bioclimatic trend by a hydrothermo-edaphic system of biogeocenoses is conducted, and patterns of transitions of forest communities to critical states are revealed. The article describes catenary landscape–ecological structures that form regional systems of localized natural zonality capable of simulating the main directions and scales of geosystemic reorganizations. The landscape–ecological forecast itself is of an experimental nature. It is presented as a system of operations with ecological niches of the objects under study. In the calculation models, the course of the predicted processes is reproduced using empirical imitation by spatially distributed parameters of the basic ecological niches. The basic principle of landscape–ecological forecasting is formulated, corresponding to one of the provisions of general ecology – the Gause law of competitive exclusion. The description of the future states of biogeocoenotic systems is conducted according to their current spatial organization in accordance with the fundamental properties of ergodicity of the natural environment. Certain time limitations of the empirical–simulation method of landscape–ecological forecasting are noted, and to avoid these limitations, it is recommended to use a more meaningful interpretation of the results of forecast calculations.

As a comparative analysis of models of plant responses to climate change has shown, dynamic simulation modeling provides a detailed "time sweep" of the studied structural or functional parameter of the ecosystem but does not have the ability to sufficiently extrapolate the results spatially. At the same time, discrete empirical–statistical models can describe the territorial diversity of forecast situations but suffer from uncertainty in time.

The essence of this uncertainty is that the forecast can be realistic only for the period of maintaining equilibrium or, at least, stationary processes. However, during the forecast period, the ecosystem, initially linear, can move into the category of nonlinear systems, which will reduce the reliability of the forecast calculations themselves. If we accept the trajectory of the expected functional changes in the ecosystem at each discrete step of the forecast as a stationary non-equilibrium process, then each new step of the predicted trajectory will represent a jump in the functional changes in the ecosystem, expressed to one degree or another.

## 17.4 REGIONAL AND LOCAL MECHANISMS OF STRUCTURAL–FUNCTIONAL ORGANIZATION OF NATURE ECOSYSTEMS

The conception offers quite new ideas of a regional bioclimatic system, which characterizes climate-genic exo-dynamic properties of the soil–phytocenotic "core" of natural complexes and serves as a scientific–methodical basis for paleogeographical reconstructions and landscape–ecological prognoses. The Volga River basin was used as an example for modeling the climatic niches of phytocoenological and soil units and their interpretation as a form of hydrothermal stability of soil–plant cover under changing climatic conditions. The models were verified by modern and retrospective bioclimatic connections in the region. The formulated notions of the bioclimatic system, as well as the methods of its modeling, form the basis of the regional prognostic conception.

The known concept of the bio-productive process, being rather independent of the structural characteristics of ecosystems, has been supported. It is almost directly connected with hydrothermal conditions, of both warm and cold seasons equally, as well as with the annual and summer coefficients of atmospheric humidification. This appreciably simplifies the prognostic calculations of the primary productivity of natural ecosystems based on climatic prognosis data.

The entire prognostic trend under consideration is characterized by an inversely proportional statistical relation between summer temperatures and annual precipitation: the warmer regions have always been and will be drier, and vice versa. Consequently, beginning from the upper Pleistocene, the region has had the same climate-forming mechanism of atmospheric circulation as at present, and this mechanism will obviously persist in the future. The conclusion obtained proves the validity of using the method of actualism for constructing prognostic scenarios of landscape–ecological conditions.

Analytic and cartographic materials for the Volga River basin and its surroundings were created to establish the regularities of functional and structural responses of zonal–regional geo(eco)systems to global climatic signals, producing a development of landscape–geophysical connections that provide wider and more efficient use of the method of actualism in retrospective and numerical geo-ecological prognosis. The models of thermal and pluvial niches showed, for both vegetation and soils, that the zonal range of the most stable climatic niches is very broad, evidencing the presence of "cores" (centers) of high potential resistance to changes in the regional climatic system in all natural zones (sub-zones) of the boreal ecotone. Using the accepted measures of unconformity of niches by temperature and precipitation, it was revealed that most plant and soil systems must have a coordinated reaction to the changes in temperature and precipitation. This makes the predicted (or retrospective) picture of future (or past) conditions of natural ecosystems much more definite.

The models of unconformity of niches also revealed a response of phytocoenological and soil geo-components to changes in summer temperature and annual precipitation. The thermal signal is predominant in the taiga zone, where it causes primary shifts in the most dynamic properties of soils (soil momentum), whereas the reconstruction of forest phytocenoses is slowed. At the same time, in the zones of mixed forests and forest–steppe, where the pluvial signal dominates, the phytobiota of most ecosystems becomes, on the contrary, less stable than the soil and therefore undergoes the quickest structural changes.

In the course of verification of the models of the regional bioclimatic system, it was established that modern middle- and south-taiga plant formations, as a whole, are adequate to their soil differences. Evidently, both natural components in the southern half of the boreal belt have now come into a state of relative stability. The most variegated picture of the relations of phytobiotic and soil areas is found in the typical and southern forest–steppe and northern steppe, which indicates a rather complex transgression dynamics of vegetation in the forest-to-steppe transition zone. This resulted in the wide spreading of southern soil relics (typical and southern chernozems) under broadleaf forests and northern soil relics (ordinary and podzolized chernozems) in the northern steppe.

The model of chain reactions in geo-component interrelations has revealed a system of regional hydrothermal parameters, which was further used for paleogeographical reconstructions and prognostic modeling in light of global climatic changes. These constructions are based on modern landscape–geophysical connections in the region converging at plant cover and soils, as well as primary bio-productivity. The regional spatial connections of temperature and precipitation for modern conditions and for prognostic scenarios of 2010, 2030, and 2050 (by the GISS model) have been comparatively analyzed for the first time.

For the first time, the results of a quantitative assessment of the mechanisms of monosystem (geo-componental) organization of landscape–zonal systems are presented using the example of forest and forest–steppe landscapes of the Volga River basin. The geo-synergistic (synthetical) unity of monosystem landscape structures is expressed by the general partial measure of closeness of connections between different geo-componental attributes. It has been established that natural landscapes and all geosystems of the higher taxonomic orders (landscape zones, provinces, and domains) are characterized by generally low mean statistical values of closeness of component interrelations and a narrow range of their variation. These are the principal conditions for the manifestation of landscape polymorphism.

A differentiated evaluation of the binary interrelations of geo-componental block equivalence allowed us to reveal the main features of landscape monosystem polymorphism and zonal geo-spaces formed by landscapes on the boreal ecotone of the Russian Plain. A weak common componental relationship is the basis of the monosystem polymorphism of the landscape–zonal systems; as a consequence, the partial landscape structures are formed with comparatively high equivalence of the componental blocks.

Two types of landscape polymorphism are described: parametric and territorial. Landscape "projections" (partial sub-systems) with the highest equivalence of geo-componental blocks are considered by the author as geosynergetic "cores" of the general monosystem structure of zonal–regional complexes. In the main watershed of the Volga River basin, only seven partial landscape structures with core properties are formed out of 21 possible paired combinations of geo-componental blocks: thermo-edaphic, oro-edaphic, thermo-productive, edapho-productive, thermo-geochemical, oro-geochemical, and edapho-geochemical.

Using the data on the integration of natural components into different partial landscape structures, as well as the landscape map of the Volga River basin, the author has prepared a series of maps illustrating the degree of manifestation of these structures in different kinds of landscapes and in zonal geo-spaces of the boreal ecotone. A map of averaged equivalence of all considered geo-componental units has been plotted for the same territory. In fact, it is a map of the integrity of the monosystem structures of natural landscapes.

Large-scale empirical–statistical modeling of broadleaf and coniferous forest ecosystems is developed using the example of the southern outpost of the boreal belt on the Russian Plain. Information methods of calculating component interconnections, ecological niches distribution, and measures of forest ecosystem stability have been used. Common information models are considered as a system of chain reactions in landscape connections under external hydrothermal impact. These models demonstrate a rather orderly system of local connections, where biogeocoenoses, forest types, and soil entities are quite closely connected with lithogenic factors and summer soil–ground moistening.

Type schemes of space distribution of ecological niches uncover the connection canals in the system "factor-phenomenon" and permit the revelation of the dynamics of different factors' influence over the state of forest communities. In the ecotone between forest and steppe under conditions of plain relief, the direct action of soil–ground moistening on the zonal groups of landscape facies is very effective. Here, the soil–hydrological factor has quite independent edificatory properties, which not only enhance the role of geomorphological conditions but also possess a certain autonomy. The ecological role of relief – types of local sites, character of meso- and microforms (dispersing, transit, or collecting), solar exposure, and steepness of slopes – drastically increases in low-mountain conditions. At the same time, the direct action of relief forms, solar exposition, and slope steepness on the distribution of biogeocoenoses and soil units becomes stronger.

In lowland outwash forested conditions, the territorial distribution of structural and functional characteristics of coniferous forest ecosystems is strictly determined, first of all, by the depth of the loamy substrate, the level of the groundwater table, and the type of location sites. The revealed local ecological optimum of lowland coniferous forest phytocoenoses characterizes the zonal–regional norm of forest ecosystems' formation and can serve as a "zero" for estimating their different local responses to the background effect.

The study of the mechanisms of auto-regulation in the coniferous forest ecosystem has revealed some principal features of the functional organization of zonal ecotones and allowed for the outlining of methods for the diagnostics of their boundaries. Mechanisms of auto-regulation are conditioned by desuccessional drainage of the territory and determine the age dynamics and population structure of forest phytocoenoses. It was revealed that the southern boundary of the given transitional zone will pass where the local "representatives" of sub-taiga appear to be the extreme members of the hydromorphic series of topo-ecosystems. Accordingly, the northern boundary of the ecotone should be at the northern limit of distribution of extremely xeromorphic topo-geosystems of the adjacent southern zone (sub-zone), in this case, the broadleaf sub-zone.

Using the method described, it is possible to determine on-site or on a large-scale map the width of the zonal ecotone strip, i.e., the size of the "hysteresis loop" in the system of bioclimatic zonal transitions. Such a loop is formed at the mutual overlapping of two adjacent natural zones (sub-zones), and the stronger the local deviations of the background zonal–regional conditions, the greater such overlapping. The landscape–topological method proposed is evidently applicable for establishing the boundaries of any zonal ecotone on the plain.

The analysis of different categories of phytomasses and complex parameters of the biological cycle has shown that the destruction branch of the biological cycle is the leading factor in exchange processes, and the zonal-climax ecosystems (with primary plants) have the highest capacity for letting pass. Lithomorphization and hydromorphization are two competing abiotic factors. The shifts of the dominant effect in this competing pair directly affect the ratio of productivities of the skeletal (mainly arboreous) and green phytomasses. The increase in soil-ground moistening results in a decrease in tree and bush growth and, at the same time, in the intensification of green mass production. The maximum of net primary production is reached under conditions of enhanced hydromorphicity of the soil–grounds.

Elucidation of the mechanisms of formation, functioning, and stability of forest ecosystems at the zonal boundaries, where the sensitivity of biota to external impacts drastically increases, results in a deeper foresight of the regional and local ecological consequences of the forthcoming and obviously already beginning anthropogenic disturbances of the global climatic system. The change in the warmth-to-moisture ratio under global warming in the south of the boreal belt, determined by background climatic shifts, may result in a significant ecological turning point: a change in the sign of biological productivity connection with the main landscape–geophysical parameters and the corresponding drastic reorganization of phytocoenotic and landscape structures.

The involvement of local ecosystems in this reorganization will be quite different not only in the starting time and initial rates but also in the character of transformation. The following factors must play a key role here: (1) the state of natural components as receptors of external effects; (2) the system of internal channels of landscape connections; (3) the initial position of the given biogeocoenosis in relation to the zonal–regional ecological optimum; (4) the relative role of abiotic factors (first of all, lithomorphicity–hydromorphicity) in their effect on primary bio-productivity and other links of the short-term biological cycle; and (5) finally, the degree of stable equilibrium of substance turnover, which determines the levels of resistant and elastic stability of ecosystems. These aspects of landscape–ecological analysis may be research prerequisites for the elaboration of the prognostic topo-ecological concept "Global Changes on the Local Level" as a basis for local ecological monitoring of forests under global anthropogenic climate changes.

## 17.5 METHODOLOGY OF LOCAL AND REGIONAL LANDSCAPE–ECOLOGICAL PROGNOSES

The questions of forecast analysis regarding the state of forest ecosystems in relation to their possible response to global climatic changes have been quite insufficiently covered in the world literature. The known mechanistic simulation models of forest response to climatic impacts satisfy a narrowly specified framework of habitat conditions; therefore, the results of modeling cannot be used to predict the state of forest communities as a whole and do not cover the spatial diversity of succession changes under the same background influence. We have developed a local landscape–ecological forecast based on discrete empirical–statistical models of natural ecosystems, which operate with a relatively small number of the most informative characteristics and produce results that, while probably not entirely exact, have a higher (geographical) level of generalization than simulation modeling.

We will set forth the ideological bases for local and regional landscape–ecological predictions as an abstract.

1. The stable function of geo(eco)systems is provided, first of all, by the soil–biotic mechanisms of their monosystem organization, i.e., *intercomponent interactions* – vertical landscape connections. At the same time, plant cover plays a key stabilizing role.
2. The impulse for the transformation of geo(eco)systems in response to a new climatic signal is given by the aggravation of *intercomplex interactions*, i.e., lateral connections and relations in the local and regional landscape couplings – catenas. This aggravation is determined by the multi-valued refraction of the background signal in the elements of micro- and meso-catenas under the influence of local geomorphological and hydro-edaphic factors.
3. *Spatial diversity of local* functional and then structural *rearrangements* of geo(eco)systems arises under the same background signal. This diversity becomes a subject of independent predictive investigation, where the landscape–ecological approach is most efficient.
4. In conditions of disturbed catenary organization of ecosystems, vegetation *plays the key transforming role.* The main driving force of geo(eco)system rearrangements is the *competitive relations between the ecological niches of phytobiota* in the space of new hydrothermal fields. Thus, under external influence, vegetation is transformed from a stabilizing factor into a geo(eco) system transforming factor.

The strategy of the landscape–ecological prognosis was as follows: first, to carry out an identification of selected ecosystem objects (either zonal types and regional kinds of landscape or local nature complexes – biogeocoenoses (facies) groups) to certain values of contemporary climatic conditions, and then to make an estimate of the most probable transformation of revealed ecological niches of the given objects according to expected climatic changes for a specified prognostic date. It is suggested that there are multiple transformations of ecosystems at a fixed value of the hydrothermal trend, where a new state can encompass several pathways, rather than just one, which are known at present instants in time.

Local hydrothermal prediction includes the following operations:

1. Detection of zonal–regional connections between the July reserves of productive soil moisture of open agricultural fields and the main climatic parameters: the mean January and July temperatures and annual amount of precipitation;
2. Establishment of transitional dependencies from agricultural fields of agro-meteorological stations to forest-covered areas in accordance with the zonal–provincial conditions of the regions;
3. Prognostic calculations of changes in the summer reserves of productive soil moisture in different local ecosystems for all periods of prediction.

Critical values of summer soil moisture content, when there are transitions between the zonal types of plant cover, are determined with the help of the local edaphic humidity factor.

Predictions of ecosystem transformations include two stages of analysis: (1) evaluation of probabilities of changes in the functional states of ecosystems and (2) calculations of the rates of ecosystem transformations. At the first stage, based on the bank of hydrothermal niches, probabilities of either landscape–zonal or biogeocoenosis transition for the given prognostic data are estimated. For this purpose, the measures of inclusion for contemporary and forecasting partial climatic niches of the landscape–zonal types or biogeocoenosis groups are calculated. Each niche is represented in the form of a row vector – a "fuzzy" set of ecosystem states due to the given hydrothermal factor. The row vector consists of normalized partial association coefficients.

The general principle of the prognosis is expressed as follows: the value of the climatic-caused transformation of one ecosystem into another is greater the lesser the degree of intersection of their climatic niches in the initial states; that is, the greater the contemporary contrasts of their functional states, the larger the range of the overlap of niches after the rapprochement of ecosystems by the given climatic factor. By the maximum values of transformation, the most probable directions and the degree of conversion of the ecosystem are estimated. It is supposed that any object *A* must be absorbed by another object *B* if their niches intersect in the final state more than in the initial state.

At the second stage of the analysis, the following procedure is performed: the estimates of potential rates of transitions and average times of the full or partial absorption of the hydrothermal niche of one examined object (facies group, landscape type, etc.) by the niches of other objects. For this purpose, the mathematical apparatus of Markov chain theory is used. The algorithms of predictive calculations are described in detail for both stages of analysis.

The principle of functional isomorphism of ecosystems proposed by us is used for the correction of the network of functional–structural transitions.

The results of the local prognostic analysis are presented in both graphic–analytical models and large-scale maps. The mapping of the basic and predicted structure and metabolic parameters of ecosystems is based on new geomorphometric methods, which describe the statistically significant relationships between these attributes of the ecosystems and the relief.

The transition from the local to the regional level of prognostic modeling is carried out using inductive hierarchical extrapolation, a method that we developed based on the empirically established phenomenon of the polyzonal nature of local ecosystems as a response to global climate changes.

## 17.6 LANDSCAPE–ECOLOGICAL PROGNOSES

Ideologically, predictive analysis and mapping on the scale of landscape–zonal ecosystems (regional level) were based on the fundamental concept of global ecology concerning the close relationship between natural zonality and the heat-to-moisture ratio. The strict and statistically substantiated regularities of the distribution of the annual humidity factor over natural zones (sub-zones) have been obtained for the territory of the Volga River basin, with two longitude-sector versions. Summer moisture content in soil also correlates well with the annual atmospheric humidity factor, determining the latitude-zonal character of its distribution both at present, in the past, and in the future.

The hydrothermal trends for the period up to 2150 have been taken from two global climate models: GISS and HadCM3 (Version A2). The first model predicts relatively temperate global warming. The second model is one of the most extreme models. Ecological prognosis by this model permits evaluation of the forest ecosystem's reaction to such global climatic changes that may exceed the level of tree endurance and cause the disintegration of both nemoral and boreal forests over large areas.

According to the HadCM3 model, by the year 2150, the rise in winter and summer temperatures will be about 6°–7° in the forest–steppe zone (maximum in south-east regions) and 5.5°–6.5° in the zone of mixed forests (minimum in the western sector of the basin). The warming will be accompanied by a very slow increase in precipitation; therefore, the annual and summer humidity factors will start to decrease almost everywhere and quite significantly. The common progressive aridization of the regional bioclimatic system will begin, and this process will become deeper and more total over time. The most significant aridization is expected in the first decades of the 21st century.

Changes will occur in the ratio of expenditure components of the water balance – runoff and evaporation. These changes will consist of a disproportionately great increase in the share of surface river runoff due to a relative decrease in evaporation and subsurface flow. This will indicate a decrease in the extent of closure of the water cycle, which, in turn, should be considered a significant cause of the decrease in tolerance of geo(eco)systems to external impacts.

Across the entire territory of the basin, a significant decrease in soil moisture reserves during the vegetation period is also predicted. By the middle of the 21st century, the July moisture reserves in a 1-m soil layer in the sub-zones of middle and southern taiga will be 5–7 times lower than their minimal moisture capacity (at present, this ratio does not exceed 1.5–1.7). The soil moisture reserves in the zone of mixed forests will fall to a similar level. By 2050, natural reproduction of forest species will have become difficult not only in the northern steppe but also in typical and southern forest–steppe areas. By 2075, such a situation will also have occurred in the sub-zone of broadleaf forests.

In general, by the year 2150, the result of the bioclimatic system rearrangements will be as follows.

1. The most active development will be characteristic of two zonal vegetation types: mixed forests, on the one hand, and typical and southern forest–steppes, on the other. By the year 2075, almost the entire Volga basin will be in the sub-boreal belt; hence, the forest–steppe will become dominant, absorbing other types of plant formations.

2. The nemoral forest sub-zone will sharply narrow and change its location, moving beyond the well-known oro-climatic barrier represented by a latitudinal segment of the Oka-Volga-Kama river valley system. Its present-day territory will be fully covered by typical forest–steppe formations, which will also expand into the sub-taiga zone of the By-Ural Region. Oak–linden communities themselves will transgress the boundaries of the southern and even middle taiga sub-zones, producing new mixed phytocoenotic structures and thereby broadening the total area of the sub-taiga zone.
3. Bioclimatic contrasts will be sharply enhanced in the southeast of the region, which will entail the convergence of zonal and sub-zonal boundaries and even the appearance of the semi-desert sub-zone at the latitude of Orenburg.

In accordance with that, one should expect a sequential northward shift of zonal and sub-zonal boundaries of the functioning of natural complexes. This process must be most efficient in soil–plant cover in the territories of southern forest–steppe, including also northern steppe and middle and southern taiga sub-zones. Motley grass–cereal oak forests of typical forest–steppe will be most unstable and ready to be completely absorbed by higher-continental formations of southern forest–steppe. Desertification of the steppes themselves will develop intensively as well.

Typical forest–steppe conditions will begin to expand northward, so that by the year 2050, the soil–plant cover will have acquired a forest–steppe appearance not only in the sub-taiga but even in the south- and middle-taiga sub-zones. At the same time, the territory of modern broadleaf forests will begin transforming completely into typical forest–steppe. Meanwhile, nemoral communities will show active transgression first to the south and then to the middle taiga, creating mixed phytocoenotic structures and thus extending the total area of the sub-taiga zone in the Russian Plain.

The monograph expounds the main statements of the original topo-ecological concept of prediction: "Global Changes on the Regional and Local Levels", as a basis for regional–local bioecological and geosystem monitoring under global anthropogenic climatic changes. The objects of research are forest biogeocoenoses (landscape facies) – elementary units of geographical ecology.

Regional and local responses to global climatic changes inevitably take the form of multiple reactions of vegetation, soils, and landscapes as a whole to background climatic signals. So far, there is no distinct notion of this multiplicity because the measure of sensitivity of soil-biotic components to climatic changes in different zonal-climatic and geomorphological conditions has not yet been estimated.

The first step towards understanding the local mechanisms of global changes included methodical construction with the working title: "Empirical simulation of the regional bioclimatic trend by local ecosystems". The zonal–regional bioclimatic background is refracted by local geomorphological and hydro-edaphic factors. This phenomenon results in the formation of regional systems of local zonality. They include vector series of flat interfluve (placor) biogeocoenoses reflecting the zonal–regional background of the territory and extra-zonal topo-ecosystems as representatives of other zonal types of geographical environment.

Forest topo-ecosystems are distinctly differentiated by their diverse zonal affiliations. Therefore, the function–structure transitions of local ecosystems must adequately reflect the shifts of the zonal–regional bioclimatic system. Such spatially ordered systems of local zonality are aligned with the vector of predicted climatic changes and are consequently able to simulate the main directions and scales of ecosystem reconstructions, thus creating an empirical basis for environmental forecast constructions.

The predicted thermal-arid trend will initiate an unequivocal and very significant reduction of productive soil moisture resources in the middle of the vegetation period across the entire territory of the Volga River basin. This will be caused by several factors: the reduction of fore-spring snow resources, the earlier disappearance of snow cover, and, finally, which is most important, increased evapotranspiration amid a decrease in atmospheric humidity. The predicted dynamics of summer soil moisture content points to the inevitability of drastic worsening of forest growth conditions in the zone of forest-to-steppe transition on the Russian Plain during the forthcoming global warming.

The main results of local empirical–statistical prediction modeling are as follows:

1. A substantial starting “leap” in the functional transformation of forest ecosystems is anticipated (particularly by HadCM3). The functional stability of boreal and nemoral forest communities may be disturbed as early as the year 2050. The growing intensity of landscape–ecological transitions will be accompanied by an intensification of selectivity in this process.
2. In the marginal forests of zonal forest–steppe ecotones, the widest range of transformation will be typical of the shadowy broad–leaf forests of loamy flat uplands. GISS, with low probability, offers several ways of their transition into xero-mesophyte forest communities of slopes. The more extreme HadCM3 envisages nothing less than a catastrophic transformation of oak forests into xerophyte forest–meadow–forest complexes now typical only of steep sunny slopes. The thermo-arid signal will most significantly affect the state of forest–meadow–steppe complexes and watershed pine/broadleaf forests. By the end of the 21st century, light pine forests will have transformed almost entirely into desert steppe.
3. Overall, progressive acceleration of the aridization of broadleaf forests under global warming is expected in the following dynamical succession order: pure oak forests and oak forests with spruce → pine forests and spruce–pine suborns → typical coniferous and small-leaf forest–steppe → northern and middle steppe → southern (dry) steppe.
4. The mixed forest zone differs from the forest–steppe zone in a generally higher resistance of forest biogeocoenoses to global warming. Aridization of coniferous forest ecosystems under conditions of low-lying outwash terrains will eventually initiate steppification of these communities through intermediate states of typical and southern coniferous–small-leaf forest–steppe.
5. Finally, the sub-zone of broadleaf forests and the southern band of the sub-taiga zone in the Middle Volga Region will be replaced by developing coniferous (pine) forest–steppe as a new (for this region) zonal ecotone of direct contact between boreal and steppe plant formations, with a mosaic

complex of light-coniferous and oak/small-leaf park forests, meadows, and steppes. This will be the general process of "savannization" of the sub-zone of broadleaf and fixed forests of the Russian Plain under global warming.

## 17.7 FORECASTED CHANGES IN BIOLOGICAL CYCLES IN FOREST ECOSYSTEMS AND THEIR CARBON BALANCE UNDER GLOBAL WARMING

The solution to the complex of ecological problems induced by the forthcoming climatic changes is associated, in many respects, with the study of the causative mechanisms of the carbon cycle, which has a regulating effect on global biosphere processes. However, there are currently no predictions of the climatogenic dynamics of biotic factors affecting the regulation of carbon dioxide content in the atmosphere. Further development of the concept of biotic regulation is seen as a study of local–regional regularities of the minor biological turnover, which provides stability to the natural environment in accordance with Le Chatelier's principle.

As is known, spatial diversity of the biological cycles is created primarily by the structural heterogeneity of biogeocoenoses (landscape facies) itself. Gradient (ordination) analysis was used to establish the parametric and spatial correlations of predicted changes in the two branches of the small biological cycle, production and detritus, in broadleaf, mixed, and light coniferous forests of the Middle Volga Region. On this basis, the empirical simulation of climate-genic changes in the biological cycle was carried out.

In the territory of the southern and typical forest–steppe, there is an unambiguous and quite clear decline in the production potential of forest and forest–steppe communities at higher summer soil temperatures and particularly at lower soil moisture content. The ascending branch of the biological cycle will weaken and cause an additional intensification of the atmospheric greenhouse effect, thus aggravating global warming. The role of wood–skeletal phitomas as a sink for atmospheric carbon will diminish.

On the northern boundary of the forest–steppe zone, under slightly excessive atmospheric humidification, broadleaf and mixed forest biogeocoenoses will significantly increase both their primary productivity and reserves of living organic matter with simultaneous rises in temperature and reductions in soil moisture content. Accordingly, the conservation of atmospheric carbon in wood–skeletal phytomass of these topo-ecosystems will increase, and, as a consequence, the significance of this phytocoenotic pool as a carbon sink will increase as well.

The moisture deficit on the zonal forest–steppe ecotone will increase under global warming and impede the decomposition and mineralization of dead organic matter. The growth of forest litter mass results in a higher litter-fall index and a greater share of the dead fraction in the total above-ground phytomass. In the marginal forest communities, a general reduction in the flow capacity of the detritus branch of metabolism will occur, which will induce residual accumulation of dead mass and, as a consequence, an inevitable slowdown of the entire biological cycle.

With the transition from southern marginal forests to the northern band of the forest-steppe zone and further to the zone of mixed forests, the thermo-arid trend increasingly intensifies the processes of decomposition of dead organic matter and correspondingly reduces all detritus fractions (including forest litter) and humus. As a consequence, the degree of detritus utilization will increase. Thus, the "work" of the detritus branch of functioning of the above forest communities will be directed toward the additional release of $CO_2$ from soil–plant cover to the atmosphere, with an inevitable enhancement of the greenhouse effect.

In the outwash forested lowlands of the sub-taiga zone, the limiting role of soil moisture content for autotrophic biogenesis will increase under global warming. The thermo-arid trend is expected to induce a slight growth in timber stand productivity but simultaneously a decrease in total annual forest production, which must result in a lower amount of $CO_2$ bound during autotrophic biogenesis. Accordingly, the efficiency of the production process will drop. Much more unambiguous and intensive changes must occur in the detritus branch of the biological cycle.

A general model scheme is proposed, which describes the driving forces of climate-genic successions – the fundamental mechanism of transition from functional changes to structural transformation of geosystems under the effect of global warming. The main stages of transformation of the forest ecosystem into steppe under the stable development of the thermo-arid climatic signal are established.

Four states of stability of forest ecosystems have been revealed based on the features of synthesis and destruction processes – the coefficient of annual phytomass turnover and the litter-fall index. The first parameter regarding forest stability is much more effective compared to the second one. A method for calculating the indexes of elastic and resistant stability was proposed by the authors, and local and regional maps of these indexes were created. Stability is differentiated depending on both individual geo-components and forest type groups. The amount of young growth, density of undergrowth, and the mass of grass cover play a significant role in increasing the stability of the forest ecosystem. The interaction of abiotic and phytocoenotic factors creates a specific order of the index of elastic stability of forest biogeocoenoses and more complicated landscape units – stows, or land type associations (urotshistshes, in Russian terminology).

Maximal loss of stability under global warming should be expected in forest communities on the zonal ecotone between forest and steppe, where the thermo-arid trend will inhibit both metabolic branches. The tempo of stability decline will be essentially lower northwards, in the direction of the mixed forest zone. Stability may even increase in the most humidified links of catenas, where it will be aided by the higher rate of detritus genesis caused by soil desertification. Eventually, activation of the detritus branch of metabolism on low outwash plains will induce an insignificant general increase in the stability of coniferous forests.

The boreal ecotone of the Russian Plain is a rather favorable model territory for the development of the topological and regional aspects of the new global ecological concept of stability of the continental biosphere: the concept of biotic regulation of terrestrial ecosystems and the determinative (landscape–geophysical, hydrological, geochemical, etc.) natural processes. Further development of this concept is associated, in our opinion, with the study of local–regional regularities of the small biological cycle in natural ecosystems providing environmental stability in accordance with Le Chatelier's principle.

To assess the changes in carbon content in different biotic components, we have used the so-called forestry method based on the dynamics of live and dead phytomasses. This method gives the best results in balance assessments of the carbon cycle components over long periods of time.

The change of $CO_2$ flow above the plant cover, i.e., the carbon balance of the soil–plant–atmosphere system, is the algebraic sum of different living and dead phytomass changes: skeletal tree–shrub mass, root mass, total green mass, forest litter mass, dead skeletal mass (brushwood and deadwood), and finally, labile humus mass. This balance must show whether this group of forest biogeocoenoses absorbs an additional amount of $CO_2$ from the atmosphere due to shifts in the biological cycle induced by global warming (realization of Le Chatelier's principle) or, on the contrary, becomes a source of additional emissions.

Detritus genesis is a metabolic branch most sensitive to climatic impacts; nevertheless, during sufficiently long periods of time, the maximal (by absolute values) changes occur not in the detritus branch but in autotrophic biogenesis: general net production of forest communities. At the same time, skeletal tree–shrub phytomass, mainly wood gain, constitutes a large part of the shifts in productivity. In 2100, the changes in carbon content in skeletal mass in its extreme values will range from $-25 \div 40$ t/ha at the zonal ecotone of forest and steppe to $+85 \div 105$ t/ha in coniferous–broadleaf forests near the northern boundaries of the nemoral forest sub-zone. In the coming century, the content of carbon conserved in skeletal phytomass of forest ecosystems of the Middle and Upper Volga River may change by ±20%–40% or more of the base value on average. This fact points to a significant regulating role that the productivity of mixed and nemoral forests of the region must play in their carbon exchange with the atmosphere.

The prognosis of changes in the biological cycle and biotic regulation of the carbon cycle according to the scenarios of global warming is presented. Two mutually antithetical phenomena of carbon cycle biotic regulation will develop on the Russian Plain. The extreme south of the forest zone will be the site of significant additional accumulation of carbon in the atmosphere. The main positive biotic regulation of carbon exchange between the above-ground ecosystems and the atmosphere, aimed at reducing its greenhouse effect (according to Le Chatelier's principle), will be performed by broadleaf–coniferous forests of high plains in the nemoral forest sub-zone and in the southern belt of the sub-taiga zone.

Thus, it is obvious that not all formations and associations of boreal forests may be considered "lungs of the planet", as has been postulated in the literature for a long time. Coniferous, mixed, and broadleaf forests of temperate latitudes may cause such biotic regulation of the carbon cycle under global climatic changes that will exert both positive and negative influences on the state of the continental biosphere.

Predictive scenarios of carbon balance in cartographic expression are considered as well. As examples, the cartographic models of carbon balance of forests for the period up to 2050, with the expected doubling of atmospheric $CO_2$, are reviewed. Morphometric parameters were derived from NASA data. At the local level, the large-scale mapping was based on predetermined, statistically significant linkages of the structural and functional characteristics of local ecosystems with the morphometry of the relief. Maps of basic carbon contents in forest ecosystems and their carbon balance to 2050 (according to the HadCM3 model) were constructed for two experimental sites – Zhyguli and the By-Oka-Terrace reserve.

The regional level is represented by the Oka River basin. The modern forest cover of this territory, under global warming, will perform positive (as a whole) carbon cycle regulation; however, the value of such regulation here will not be significant. The contribution of separate groups of forest formations to the total biotic carbon cycle regulation was assessed.

Carbon deposits in dark coniferous forests will exceed $CO_2$ emissions by nearly 70%; however, because of the insignificant area of these formations, their contribution to the regional carbon balance will be minor. Pine forests, occupying the greatest areas in the Oka River basin, will make a considerable positive contribution to carbon cycle regulation.

Secondary small-leaf forests will experience changes in carbon content that differ by sign. The substitution of secondary birch and aspen forests for spruce and broadleaf–spruce forests, i.e., the permanent presence of initial stages of restorative successions, must contribute to excess accumulation of carbon mass in the growing timber. This seems to highlight the leading role of the boreal forest cover of dark coniferous taiga and sub-taiga types in positive carbon cycle regulation by Le Chatelier's principle.

# Conclusion

The single key feature of this book (which we might term as its unique work): *In the basic and predictive models presented in the book, geographical ecology acquires an effective formalized tool for analysis and prediction using the methods of discrete mathematics for the processing and generalization of bulk empirical data obtained from field and laboratory landscape studies.*

A strategy for predictive empirical–statistical modeling of forest ecosystems as objects of terrestrial geosystem monitoring is presented in the given monograph. Local mechanisms of global changes are revealed through an empirically established display of the background bioclimatic trend by a catenary system of biogeocoenoses. Regional structures of localized natural zonality capable of simulating the main directions and scales of geo(eco)system restructurings are considered. Empirical simulation of forecasts is implemented as a reproduction of predicted scenarios of biogeocoenotic systems according to the laws of their basic spatial organization, with the spatial coordinates replaced by time coordinates in the forecast procedure, in accordance with the fundamental properties of ergodicity (spatio-temporal relationships) of the natural environment.

The book presents the mechanisms of functional and structural organization of forest ecosystems as objects of terrestrial geosystem monitoring. The spatial–functional monitoring of forests is based on empirically established local and regional landscape–ecological connections, which are considered mechanisms of metabolic reactions of forest ecosystems to certain climatic trends. This is the novelty of the basic and predictive empirical simulation concept of regional and local geo-ecological monitoring developed by the author. Ordination analysis of landscape connections is aimed at identifying transitions of forest communities to critical states according to the main discrete parameters of biological circulation. Catenary landscape–ecological structures that form regional systems of localized natural zonality are capable of imitating the main directions and scales of geosystemic restructurings.

Landscape–ecological forecasting is experimental in nature. It is presented as a system of operations with ecological (hydrothermal) niches of the objects under study. In calculation models, the course of predicted processes is reproduced using their empirical imitation by spatially distributed parameters of basic ecological niches. The description of future states of biogeocoenotic systems is carried out according to their current spatial organization in accordance with the fundamental properties of ergodicity of the natural environment. The ambiguous nature of the transformation of forest topogeosystems is revealed at a fixed value of the regional geophysical trend, when their new state can have features of not one, but several basic states. A working algorithm for predictive landscape–ecological calculations is presented. Based on geoecological forecasts, the problem of regulating the quality of the natural environment under global climate change through carbon balances and the functional stability of forest ecosystems is covered.

The results of geoecological forecasts allow us to approach the solution of issues of regulating the *quality* of the natural environment, i.e., certain management of it, with the establishment of the effects of permissible impacts, as well as the probabilities of the risk of certain environmental consequences. One of the effective areas of this final stage of geosystem monitoring can be the concept of carbon balances and functional sustainability of forest ecosystems under global climate change presented in the work (Kolomyts 2020). This concept describes the absorption and adaptive capacity of forest, mainly boreal, biomes of European Russia in the context of modern anthropogenic warming.

The conducted experience in landscape ecological analysis of the biotic regulation of the carbon cycle has demonstrated the efficiency of the modern method of biogeocoenology in solving systemic problems of ecological prediction related, in particular, to the forthcoming global climate warming. First, various mechanisms of the reflection of global biosphere processes in the functions and structure of ecosystems on the topological scale have been identified. Second, the directions and intensities of the feedback effect of the forest phytobiota on the chemical composition of the atmosphere and the greenhouse effect, as well as its ecological consequences, have been estimated.

Prognostic models serve as a basis for determining the local series and zonal–regional types of regulation of the atmospheric $CO_2$ content by the forest cover of each ecoregion in the course of climatic thermo-aridization. The distribution of these types with respect to the area is used to estimate the total regional contribution of the forest phytobiota to the stability of the continental biosphere. It is predicted that under global warming, different temperate forest associations and formations will carry out both positive and negative biotic regulation of the carbon cycle in the system of soil–vegetation–atmosphere.

The results presented in the given monograph are not only of scientific and/or methodical interest. They obviously will have certain practical significance also for the economic regions of forest and forest–steppe zones of the Volga River basin. The data on the condition of forest and forest–steppe ecosystems and prognostic assessments of their forthcoming changes may be used to develop the bases for preservation, reproduction, and rational use of forest (and accordingly water) resources, to form a network of protected and recreational territories (nature reserves, wildlife refuges, national parks), which is particularly important for the regions with critical existence conditions for vegetation, including the zone of forest-to-steppe contact.

As a whole, the results of scientific search presented in the monograph prove that the regional and local approach to the assessment of global changes may become one of the promising methodological aspects of geographical ecology called to solve global ecological problems. The outcome of geoecological projections allows us to approach the issues of regulating the quality of the natural environment, i.e., managing it by establishing the effects of permissible impacts and the probabilities of risk of particular environmental consequences. One of the efficient directions of this final stage of geosystem monitoring can be served by the concept of carbon budgets and functional stability of forest ecosystems under global climate change. This operational concept predicts the absorptive and adaptive capacity of the forest, mainly boreal, biomes of European Russia under current warming trends, along with an ecological justification for the transition to adaptive forest management.

# References


Aber, J., Nelson, R.P., Mcnulty, St., et al. 2001. "Forest processes and global environmental change: predicting the effects of Individual and multiple stressors." *BioScience*, 51 (9): 735–751.

Agakhanyants, O.E. 1981. *Arid Mountains of the USSR: Nature and Geographical Models of Florogenesis*. Moscow: Mysl (in Russian).

Albritton, D.L., Barker, T., Bashmakov, I., et al. 2003.*Changing of the Climate 2001: Synthesis Report MGEIK*, edited by R.T. Whotson. Geneva: World Meteorological Organization (in Russian).

Alcamo, J., Van den Born, D.J., Bouwman, A.F., et al. 1994. "Modeling the global society-boisphere-climate system. Part 2: computed scenarios." *Water Air & Soil Pollution*, 76: 37–78.

Alekhin, V.V. 1951. *Vegetation of the USSR in the Main Zones: A Textbook for Universities and Pedagogical Universities*, 2nd edition. Moscow: "Soviet Science" (in Russian).

Alferov, A.M., Blinov, V.G., Gitarsky, M.L., et al. 2017. *Monitoring of Greenhouse Gas Flows in Natural Ecosystems*. Saratov: Amirit (in Russian).

Alisov, B.P. 1956. *Climate of the USSR*. Moscow: Publishing House of MGU (in Russian).

Andreev, V.L. 1980. *Classification Constructs in Ecology and Systematics*. Moscow: Nauka (in Russian).

Antomonov, Yu. G. 1977. *Modelbrovanie biologicheskih System*. Kiev: Naukova dumka (in Russian).

Anuchin, R.P. 1982. *Forest Inventory*. Moscow: Timber Industry (in Russian).

Armand, A.D. 1975. *Information Models of Natural Complexes*. Moscow: Nauka (in Russian).

Armand, A.D. 1983. "Stability (homeostaticity) of geographic systems of various types of external influences." *Stability of Geosystems*, edited by A.D. Armand and I.Yu. Dolgushin, 14–32. Moscow: Nauka (in Russian).

Armand, A.D. 1988a. "Landscape as a construction." *Izvestiya VGO,* 120 (2): 120–125 (in Russian).

Armand, A.D. 1988b. *Self-Organization and Self-Regulation of Geographic Systems*. Moscow: Nauka (in Russian).

Armand, A.D. 1989. "General views." *Ecosystems in Critical States*, edited by Yu.G. Puzachenko, 10–41. Moscow: Nauka (in Russian).

Armand, A.D. 1992. "Sharp and gradual timberlines as a result of srecies interaction." *Landscape Boundaries: Consequences for Biotic Diversity and Ecological Flows,* edited by A.J. Hansen and F. di Castri, 360–378. New York, Berlin, Heidelberg: Springer.

Armand, A.D. and Dolgushin, I. Yu., eds. 1983. *Stability of Geosystems*. Moscow: Nauka (in Russian).

Armand, A.D. and Kaydanova, O.V. 1976. "Landscape triggers." *Izvestiya Rossiiskoi Akademii Nauk Seriya Geograficheskaya,* 3: 22–28 (in Russian).

Armand, A.D. and Kupriyanova, T.P. 1976. "Types of natural systems and physical-geographical zoning." *Bulletin of the USSR Academy of Sciences. Geography Series*, (5): 15–31 (in Russian).

Armand, A.D. and Vedyushkin, M.A. 1989. *Trigger Geosystems (Preprint)*. Moscow: Institute geographii AN SSSR (in Russian).

Armand, D.L. 1975. *Science About Landscape*. Moscow: Mysl' (in Russian).

Armand, D.L., Preobrazhensky, V.S. and Armand, A.D. 1969. "Natural complexes and modern methods of their study." *Izvestiya Rossiiskoi Akademii Nauk Seriya Geograficheskaya,* 5: 5–16 (in Russian).

Arsky, Yu. M., Danilov-Danilyan, V.I., Zalihanov, M.Ch., et al. 1997. *Environmental Problems: What Is Happening, Who Is to Blame and What to Do*. Moscow: MNEPU (in Russian).

Ashby, W.R. 1956. *An Introduction to Cybernetics*. London: Oxford University Press.

Bailey, R.B. 1995. *Ecosystem Geography*. New York, Berlin, Heidelberg: Springer Verlag. Inc.

Bartalev, S.A., Ershov, D.V., Isaev, A.S., Lupyan, E.A. and Uvarov, I.A. 2010. *Vegetation Map of Russia*. Moscow: IKI RAN and CEPL RAN (in Russian).

Bartalev, S.A. and Isaev, A.S. 2004. "Modern possibilities of satellite monitoring of dynamics of forest boreal ecosystems of northern Eurasia". *Anthropogenic Transformation of Taiga Ecosystems of Europe: Ecological, Resource and Economic Aspects*, 11–19. Petrozavodsk: Publishing House of the Karelian Research Centre of the Russian Academy of Sciences.

Bartholome, E. and Belward, A.S. 2005. "GLC2000: a new approach to global land cover mapping from Earth observation data" *International Journal of Remote Sensing*, 26: 1959–1977.

Bazhenova, O.I. and Martyanova, G.N. 2000. "Response of steppe and forest-steppe morphodynamic systems to modern climate changes." *Geography and Natural Resources,*4: 23–32 (in Russian).

Bazilevich, N.I. 1993. *Biological Productivity of Ecosystem of the Northern Eurasia*. Moscow: Nauka (in Russian).

Bazilevich, N.I., Grebenshchikov, O.S. and Tishkov, A.A. 1986. *Geographic Patterns of the Structure and Functioning of Ecosystems*. Moscow: Nauka (in Russian).

Bazilevich, N.I. and Rodin, L.E. 1971. "Productivity and circulation of elements in natural and cultural phytoceonoses (based on materials from the USSR)." *Biological Productivity and Circulation of Chemical Elements in Plant Communities,* 5–32. Moscow: Nauka (in Russian).

Bazilevich, N.I., Titlyanova, A.A., Smirnov V.V., et al. 1978. *Methods for Studying Biological Cycle in Different Natural Zones*. Moscow: Nauka (in Russian).

Bee, J.N., Kunstler, G. and Coomes, D.A. 2007. "Resistance and resilience of New Zealand tree species to browsing." *Journal of Ecology,* (95): 1014–1026.

Beets, P.N., Kimbereley M.O., Paul T.S., et al. 2011. "Planted forest carbon monitoring system – forest carbon model validation study for *Pinus radiate*." *New Zealand Journal of Forest Science,* (41): 177–189.

Begon, M., Harper, J.L. and Townsend, C.R. 1986. *Ecology: Individuals, Populations and Communities*. Ozford, Boston, Melbourn: Blackwell Scientific Publications.

Berg, L.S. 1947. *Geographic Zones of the Soviet Union*. Moscow: Geographgiz (in Russian).

Beruchashvili, N.L 1990. *Geophysics of the Landscape*. Moscow: Nauka (in Russian).

Beruchashvili N.L. and Zhuchkova V.K. 1977. *Methods of Complex Physical and Geographical Research*. Moscow: Publishing House of Moscow State University.

Blüthgen, I. 1973. *Geography of Climates, Vol.* 2. Transl. from Germ. Moscow: Progress (in Russian).

Bogdanov A.P. and Aleshko Z.A. 2011. "Development of a methodology for monitoring the state of forests based on the use of space imagery multispectrum data." *Problems of Ecological Monitoring and Modeling of Ecosystems,* XXVII (1): 98–109.

Bogdanovich, A. Yu. and Lipka, O.N. "Synergy of the climate global goal of sustainable development and the national adaptation plan in Russia." *Problems of Ecological Monitoring and Modeling of Ecosystems,* XXII (3–4): 7–32.

Bobkova, K.S. 1987. *Biological Productivity of Coniferous Forests of the European North-East*. Leningrad: Science (in Russian).

Bonan, G.B., Polland, D. and Thompson, S.L. 1992. "Effect of boreal forest vegetation on global climate warming." *Nature,* 359: 716–718.

Borzenkova, I.I. 1992. *Climate Change in the Cenozoic*. St. Petersburg: Gidrometeoizdat (in Russian).

Brillouin, L. 1956. *Science and Information Theory*. New York: Academic Press INC Publishers.

Brown, Ch.J., Schoeman, D.S., Sydeman, W.J., et al. 2011. "Quantitative approaches in climate change ecology." *Global Change Biology,* 17: 3697–3713.

Buccolini, M., et al. 2012. "Relationships between a new slope morphometric index and Calanchi erosion in northern Sicily, Italy." *Geomorphology,* 149–150: 41–48.

Budnik, S.V. 2017. "Basin nature management and environmental monitoring." *Problems of Ecological Monitoring and Modeling of Ecosystems*, XXVII (4): 94–104.

Budyko, M.I. 1971. *Climate and Life*. Leningrad: Gidrometeoizdat (in Russian).

Budyko, M.I. 1977. *Global Ekology*. Moscow: Mysl' (in Russian).

Budyko, M.I. 1980. *Climate in the Past and in the Future*. Leningrad: Gidrometeoizdat (in Russian).

Budyko, M.I. 2002. "Global warming." *Climate Change and Its Consequences*. St. Petersburg: Nauka, 7–12 (in Russian).

Budyko, M.I., Borzenkova, I.I., Menzhulin, G.V. and Selyakov, K.I. 1992. "Forthcoming changes in the region al climate." *Izvestiya Rossiiskoi Akademii Nauk, Seriya Geograficheskaya,* 4: 36–53 (in Russian).

Budyko, M.I. and Hecht, A.D., eds. 1991. *Upcoming Climate Changes: Joint Soviet-American Report on Climate and Its Changes*. Leningrad: Hydrometeoizdat (in Russian).

Buks, I.I. 1977. "Landscape-ecological characteristics of the Baikal_Amur Mainline zone and resistance of environment." *Baikal-Amur Mainline. Problems of Geography,* 105: 81–97. Moscow: Mysl' (in Russian).

Buyvolov, Yu. A., Paramonov, S.G. and Gromov, S.A. 2021. "Comprehensive background monitoring in Russia's biosphere reserves: triumph or fiasco?" *Issues of Geography,* 152: Man and Biosphere 101–134 (in Russian).

Byallovich, Yu. P. 1960. "Biogeocenotic horizons." *Proceedings of the Moscow Society of Natural Sciences,* III: 43–60 (in Russian).

Bykov, B.A. 1973. *Geobotanical Dictionary*. Alma-Ata: "Science" of the Kazakh-SSR (in Russian).

Cairns, J. Jr. and van der Schaile, W.Y. 1980. "Biological monitoring: Part I – Early warning systems." *Water Research*, 14: 1179–1190.

Calogero Schillaci, et al. 2017. "Modelling the topsoil carbon stock of agricultural lands with the Stochastic Gradient Treeboost in a semi-arid Mediterranean region." *Geoderma,* 286: 35–45.

Chapin, F.S., Walker, B.H., Hobbs, R.J., et al. 1997. "Biotic control over the functioning of Ecosystems". *Science*, 277: 500–504.

Cherkashin, A.K. 1997. *Polysystem Analysis and Synthesis. Application in Geography.* Novosibirsk: Nauka (in Russian).

Chertov, O., Komarov, A., Kolström, M., et al., 2003. "Modeling the long-term dynamics of populations and communities of trees in boreal forests based on competition on light and nitrogen." *Forest Ecology and Management,* 76: 355–369.

Chi, H., Sun, G., Huang, J., Guo, Z., Ni, W. and Fu, A. 2015. "National forest above ground biomass mapping from ICESat/GLAS data and MODIS imagery in China." *Remote Sensing*, 7: 5534–5564.

Chorley, B.J. and Kennedy, B.A. 1971. *Physical Geography: A System Approach.* London: Prentice-Hall.

Dahl, E. 1993. "Impact of climatic change on terrestrial ecosystems." *Impacts of Climatic Change on Natural Ecosystems (With Emphasis on Boreal and Arctic/Alpine Areas).* 81–83. Trondheim: NINA and DN.

Danilov-Danilyan, V.I. and Losev, K.S. 2000. *Ecological Challenge and Sustainable Development*. Moscow: Progress – Tradition (in Russian).

Dedkov, A.P. 1972. "On denudation cuts and ancient planation surfaces in the Middle Volga region". *Exogenous Processes in the Middle Volga Region,* 3–20. Kazan: Publishing House of Kazan University.

Diakonov, K.N. 2002. "Interaction of structural, functional and evolutionary directions in landscape studies." *Vestnik MGU*, 1: 13–21 (in Russian).

di Castri, F. and Hansen, A.J. 1992. "The environment and development crises as determinants of landscape dynamics." *Landscape Boundaries: Consequences for Biotic Diversity and Ecological Flows,* 3–18. New York, Berlin, Heidelberg: Springer–Verlag.

Diaz, S. and Cabido, M. 2001. "Vive la difference: plant functional diversity matters to ecosystem Processes." *Trends in Ecology & Evolution,* 16: 646–655.

Dinesman, L.G. 1976. "Holocene history of biogeocenoses of the Russian plain." *History of Biogeocenoses of the USSR in the Holocene*. Moscow: Nauka, 122–132 (in Russian).

Dobrovolsky, G.V. and Nikitin, E.D. 2006. *Ecology of Soils*. Moscow: Publishing House of MGU (in Russian).

Dokuchaev, V.V. 1899. *To the Doctrine of Natural Zones. Horizontal and Vertical Soil Zones*. St.-Petersburg: Printing House of St.-Petersburg City Administration (in Russian).

Dokuchaev, V.V. 1948. *Doctrine about Natural Zones*. Moscow: Geographgiz (in Russian).

Dokuchaev, V.V. 1949. *Selected Works. Vol.* III. *Cartography: Genesis and Classification of Soils*. Moscow: State Publishing House of Agricultural Literature (in Russian).

Dorman, F.C. 2007. "Promising the future & global change projections of species distributions." *Basic and Applied Ecology,* 8: 387–397.

Duchaufour, Ph. 1965. *Precis de Pedologie: L'evolution des Soils* (Precis de Pedologie: L'evolution of Soils). Paris: Deuxieme (in French).

Dumitrashko, N.V., ed. 1966. *Caucasus: Natural Conditions and Natural Resources of the USSR*. Moscow: Nauka (in Russian).

Dylis, N.V. 1978. *Basics of Biogeocoenology*. Moscow: Publishing House of Moscow State University (in Russian).

Efimova, N.A. 1977. *Radiation factors in productivity of vegetation cover.* Leningrad: Gidrometeoizdat (in Russian).

Efimova, N.A., Strokina, L.A., Baykova, I.M. and Malkova, I.V. 1990. "Changes in the main climate elements on the territory of the USSR in 1967–1990." *Meteorology and Hydrology*, 4: 34–41 (in Russian).

Emanuel, W.R., Shugart, H.H. and Stevenson, M.R. 1985. 'Climatic changes and the boreal-scale distribution of terrestrial ecosystem complexes." *Climatic Change,* 7: 29–43.

Fatichi, S., Papas, Gh., Zscheischler, J. and Leuzinger S. 2019. "Modeling carbon sources and sinks in terrestrial vegetation." *New Phytologist,* 221(2): 652–668.

Fedorov, V.D. 1974. "Stability of ecological systems and its measurement." *Izvestiya Academy of Sciences of the USSR, Seriya Biologist,* 3: 402–415 (in Russian).

Fedorov, V.D. 1975. "The concept of stability of biological systems." *Comprehensive Analysis of the Environment: Proceedings of the Council-American Symposium*, 207–217. Leningrad: Gidrometeoizdat (in Russian).

Forman, R.T.T. 1995. *Land Mosaics: The Ecology of Landscapes and Regions.* Cambridge: Cambridge University Press.

Forman, R.T.T. and Gordon, M. 1986. *Landscape Ecology*. New York: John Wiley & Sons.

Franklin, J., Serra-Diaz, J.M., Syphard, A.D. and Regan, H.M. 2017. "Big data for forecasting global change impact on plant communities." *Global Ecology and Biogeography,* 26 (1): 6–17.

Gamon, J.A., Huemmrich, K.E., Peddle D.R., et al. 2004. "Remote sensing BOREAS: Lessons learned." *Remote Sensing of Environment,* 89: 139–162.

Ganzei, K.S. 2024. "Island geosystems of the Kuril Islands and the islands of Peter the Great Bay. Features of functioning and landscape sustainability assessment." *Problems of Geography,* 158: 233–254. Moscow: Media PRESS (in Russian).

Gartsman, I.N. 1977. "Systemic aspects of modeling in hydrology." *Proceedings of the Far Eastern Scientific Research Institute of Hydrological Research,* 63: 3–84. Leningrad: Gidrometeoizdat (in Russian).

Gauthier, S., Bernier, P., Kuuluvainen, T., Shvidenko, A.Z. and Schepaschenko, D.G. 2015. "Boreal forest health and global change." *Science,* 349 (6259): 819–821.

Gerasimov, I.P. 1975. "Scientific foundations of modern environmental monitoring." *Izvestiya Akademii Nauk SSSR, Seriya Geograficheskaya,* (3): 13–26 (in Russian).

Gerasimov, I.P. 1985. *Environmental Problems in the Past, Present and Future Geography of the World.* Moscow: Nauka (in Russian).

Gerasimov, I.P. 1986. *The Doctrine of Dokuchaev and the Present.* Moscow: Mysly' (in Russian).

Gerasimov, I.P. and Velichko, A.A., eds. 1982. *Paleogeography of Europe Over the Last Hundred Thousand Years (Atlas-Monograph).* Moscow: Nauka (in Russian).

Gerrard, A.J. 1981. *Soils and Landforms: An Integration of Geomorphology and Pedology.* London: George Allen & Unwin.

Gettner, A. 1930. *Geography: Its History, Essence and Methods.* Transl. from German. Leningrad, Moscow: "Gosizdat" (in Russian).

Gil'manov, T.G. and Bazilevich, N.I. 1983. "Construction and analysis of ecosystem models." *Questions of Geography,* 127: 55–96. Moscow: Mysl ' (in Russian).

Girs, A.A. 1971. *Long-term variations in atmospheric circulation and long-term hydrometeorological forecasts.* Leningrad: Gidrometeoizdat (in Russian).

Glazovskaya, M.A. 1964. *Geochemical Principles of Typology and Methods of Studies on Natural Landscapes.* Moscow: Publishing House of MGU (in Russian).

Glazovskaya, M.A. 1981. *General Soil Science and Soil Geography.* Moscow: Vysshaya Shkola (in Russian).

Glazovskaya, M.A. 1988. *Geochemistry of Natural and Anthropogenic Landscapes of the USSR.* Moscow: Vysshaya Shkola (in Russian).

Glazovskaya, M.A. 1992. "Biogeochemical organization of ecological space in natural and anthropogenic landscapes as a criterion of their stability." *Izvestiya Rossiiskoi Akademii Nauk Seriya Geograficheskaya,* (5): 5–12 (in Russian).

Glazovskaya, M.A., Gavrilova, I.P. and Solntseva, N.P. 1993. "Soil-geochemical maps and their use for environmental purposes." *Geographiya i kartographiya pochv*, edited by V. V. Dokuchaeva, 81–92. Moscow: Pochvenny inst-t im (in Russian).

Glazovsky, N.F. 2009. *Selected Works, Volume 2: Sustainable Development of the Biosphere.* Moscow: KMK (in Russian).

Glazovsky, N.F. and Armand, A.D., eds, 1992. *Mechanisms of Geosystem Stability.* Moscow: Nauka (in Russian).

Global Change. 1990. *The International Geosphere – Biosphere Programme: A Study of Global Change (IGBP).* Paris: International Council of Scientific Union (ISCU), (2).

Golubyatnikov, L.L., Mokhov, I.I., Denisenko E.A. and Tikhonov V.A. 2005. "Model estimates of the impact of climate change on vegetation cover and carbon sink from the atmosphere." *Izvestiya RAN, Physics of the Atmosphere and Ocean,* 41 (1): 25–35 (in Russian).

Gorshkov, V.G. 1994. *Physical and Biological Bases of Life Stability. Man, Biota, Environment.* New York: Springer Verlag.

Grace, J. 2004. "Understanding and managing the global carbon cycle." *Journal of Ecology,* 92 (2): 189–202.

Grodzinsky, M.D. 1987. "Resistance of geosystems: theoretical approach to quantitative analysis." *Izvestiya Rossiiskoi Akademii Nauk Seriya Geograficheskaya,* 6: 5–15 (in Russian).

Global Change. 1990. *The International Geosphere – Biosphere Programme: A Study of Global Change (IGBP)* (2). International Council of Scientific Union (ISCU).

Gokhman, V.M., Gurevich, B.L. and Saushkin, Yu. G. 1968. "Problems of metageography." *Questions of Geography,* 78: 3–14 (in Russian).

Golitsyn, G.O. and McBain, G.A. 1992 "Atmosphere and climate changes." *Izvestiya RAS, Geograficheskaya Series*, 2: 33–43 (in Russian).

Gordon, C., Cooper, C., Senior, C.A. et al. 2000. "The simulation of SST sea ice extents and ocean heat transport in a version of the Hadley Centre coupled model without flux adjustments" *Climate Dynamics,* 16: 147–168.

Gorshkov, V.V., Gorshkov, V.G., Danilov-Danilyan, V.I., Losev, K.S. and Makarieva, A.M. 1999. "Biotic regulation of the environment." *Russian Journal of Ecology,* 2: 105–113 (in Russian).

Green, A.M., Ananyeva, L.M., Kulikov, Yu. N., and Utekhin, V.D. 1980. "Comparative studies of the structure and functioning of natural and natural-technical ecosystems." *Structure Dynamics and Development of Landscapes,* 82–100. Moscow: Nauka (in Russian).

Gribova, S.F., Isachenko, T.I. and Lavrenko, E.M. eds. 1980. *Vegetation of the European Part of the USSR.* Leningrad: Nauka (in Russian).

Grigor'ev, A.A. 1966. *Regularities of the Structure and Development of the Geographic Environment.* Moscow: Mysl' (in Russian).

Grigoryev, A.A. 1965. *Rasvitie teoreticheskih problem sovetskoy phizicheskoy geographii.* Moscow: Nauka (in Russian).

Grigor'ev, A.A. 1970. *Types of the Geographical Environment.* Moscow: Mysl' (in Russian).

Gulisashvili, V.Z. 1980. *Savannahs Are Woodlands.* Tbilisi: "Metsniereba" (in Russian).

Gunderson, L. 2000. "Ecological resilience: in theory and application." *Annual Review of Ecology and Systematics,* 31: 425–439.

Haggett, P. 1975. *Geography: A Modern Synthesis,* 2nd edition. New York, London: Harper International Edition.

Hansen, J., Lacia, A., Rind, D., et al. 1984. "Climate sensitivity; analysis of feedback mechanism." *Climate Processes and Climate Sensitivity.* Geophysical Monograph series, Vol. 29, 130–163. Washington, DC: American Geophysical Union.

Hansen, A.J. and di Castry, F., eds. 1992. *Landscape Boundaries. Consiquence for Biotic Diversity and Ecological Flows.* New York, Berlin, Heidelberg: Springer-Verlag.

Hansen, J.G., Hung, I., Lasis, A., et al. 1988. "Global climate changes as forecasted by goddard institute for space studies three-dimentional models." *Journal of Geophysical Researchearch,* 92: 9341–9354.

Hansen, J., Sato, M., Ruedy, R., et al. 2007. "Dangerous human-made interference with climate: a GISS model E study." *Climate Dynamics,* 7: 2287–2312.

Harbaugh, J., Bonham-Carter, G. 1974. *Computer Modeling in Geology.* Transl. in Engl. Moscow: Mir (in Russian).

Helbert, S.H. 2008a. "Imaginary replicates and planning of ecological field experiments." *Problems of an Ecological Experiment (planning and analysis of observations),* edited by G.S. Rosenberg and D.B. Gelashvili, 9–45. Togliatti: Institute of Ecology of the Volga River Basin RAS. Lobachevsky University of Nizhny Novgorod (in Russian).

Helbert, S.H. 2008b. "An incredible interpretation of imaginary repetitions and related problems. Answer to L. Oksanen." *Problems of an ecological experiment (planning and analysis of observations),* edited by G.S. Rosenberg and D.B. Gelashvili, 61–65. Togliatti: Institute of Ecology of the Volga River basin RAS. Lobachevsky University of Nizhny Novgorod (in Russian).

Holling, C.S. 1973. "Resilience and stability of ecosystems." *Annual Review of Ecology,* 4: 1–23.

Holten, J.I., Paulsen, G. and Oechel, W.C., eds. 1993. *Impacts of Climatic Change on Natural Ecosystems (With Emphasis on Boreal and Arctic/Alpine Areas).* Trondheim: NINA and DN.

Houghton, L.G., Meira Filho, L.G., Callander, B.A., et al., eds. 1996. *Climate Change 1995: The Science of Climatic Change.* Cambridge: Cambridge University Press.

Humboldt, A. von. 1807. *Ideen zu einer Geographie der Planzen nebst einen Naturge malde der Tropenllănder (Ideas for a Geography of the Plants Along with a Natural Painting of the Tropical Countries).* Tűbingen: F.G. Cotta. Paris F. Schnoell.

Isachenko, A.G., 1965. *Fundamentals of Landscape Science and Physiographic Zoning.* Moscow: Vysshaya shkola (in Russian).

Isachenko, A.G. 1985. *Landscapes of the USSR.* Leningrad: Publishing House of Leningrad State University (in Russian).

Isachenko, A.G. 1988. "The system of landscapes and the content of the landscape map of the World." *Izvestia Vsesoyuznoye Geograficheskoye Obshchestvo,* 120 (6): 489–501 (in Russian).

Isachenko, A.G. 1991. *Landscape Science and Physical-Geographical Zoning: Textbook for Universities.* Moscow: Higher School (in Russian).

Isachenko, T.I. and Lavrenko, E.M., eds. 1974. *Vegetation Map of the European Part of the USSR.* M-b 1: 2 500 000. Moscow: GUGK (in Russian).

Isaev, A.S., ed. 2012. *Diversity and Dynamics of Forest Ecosystems in Russia.* Book 1. Moscow: KMK (in Russian).

Isaev, A.S. and Korovin, G.N. 1999. "Carbon in the forests of Northern Eurasia." *Carbon Cycle in Russia,* 63–95. Moscow: Ministry of Science and Technology of the Russian Federation (in Russian).

Isakov, Yu. A., Kazanskaya, N.S. and Tishkov, A.A. 1986. *Zonal Patterns of Ecosystem Dynamics.* Moscow: Nauka (in Russian).

Ivanova, Yu. R. and Skok, N.V. 2019. "Seasonal development of plant communities in the context of changing weather conditions in the low mountains of the Middle Urals." *Problems of Ecological Monitoring and Modeling of Ecosystems,* XXX (1–2): 70–85 (in Russian).

Ivashutina L.I. and Nikolaev V.A. 1972. "Study of the contrast of landscape conjugations for the purpose of substantiating physical-geographical zoning." *Landscape Science.* Moscow: Moscow State University Press, 59–73 (in Russian).

Izrael, Yu. A. 1984. *Ecology and Control of the State of the Natural Environment.* Leningrad: Gidrometeoizdat (in Russian).

Izrael, Yu. A., ed. 1992. *Estimates of the Environmental and Socio-Economic Impacts of Climate Change. IPCC Working Group II Report.* Sankt Petersburg: Gidrometeoizdat (in Russian).

Izrael, Yu. A., ed. 2001. *Condition and Comprehensive Monitoring of the Natural Environment and Climate: Limits of Change.* Moscow: Nauka (in Russian).

Izrael, Yu. A., Filippova, L.M., Insarov, G.E., Semevsky, F.N. and Semenov, S.M. 1986. "Influence of background pollution of the natural environment on biota: problems of assessment and forecast." *Problems of Ecological Monitoring and Modeling of Ecosystems,* V: 6–18. Leningrad: Gidrometeoizdat (in Russian).

Izrael, Yu. A. and Munn, R.T. 1985. "Environmental and renewable resource monitoring." *In Sustainable Development of the Biosphere,* edited by W.C. Clark and R.E. Munn, Dordrecht: D. Reidel Pub. Co.

Izrael, Yu. A. and Ryaboshapko, A.G. 2011. "Climate geoengineering: implementation opportunities." *Problems of Ecological Monitoring and Modeling of Ecosystems,* XXIV: 11–22 (in Russian).

Izrael, Yu. A., Semenov, S.M. and Khachaturov, M.A. 1993. "Bio-climatological aspects of integrated global monitoring." *Problems of Ecological Monitoring and Modeling of Ecosystems,* XV: 8–20. Leningrad: Gidrometeoizdat (in Russian).

Jeffers, J. 1981. *Introduction to Systems Analysis: Application in Ecology.* Transl. in Engl. Moscow: Mir (in Russian).

Kellomäki, S. 1995. "Computation on the influence of changing climate on the soil moisture and productivity in scots pine stands in southern and northern Finland." *Climate Change* (29): 35–51.

Kerzhentsev, A.S. 1999. "The mechanism of spatio-temporal variability of soils and ecosystems." *Ecology and Soil. Favored Lectures of VIII-IX All-Russian Schools (1998–1999)*, III: 31–58. Moscow: Institute of Fundamental Problems of Biology RAS (in Russian).

Kerzhentsev, A.S. 2006. *Functional Ecology*. Moscow: Nauka (in Russian).

Kerzhentsev, A.S. 2010. "Soil functioning mechanism and ecosystem sustainability." *RAS Bulletin*, 80 (8): 704–709 (in Russian).

Khamedov, V.A. and Mazurov, B.T. 2015. "Development of methodological issues of creating a satellite monitoring system for the state of forest ecosystems under the influence of the oil and gas complex of the territory of Western Siberia." *Vestnik GUG and T*, 3 (31): 16–31 (in Russian).

Kharitonychev, A.T. 1968. *Nature of the Nizhny Novgorod Volga Region*. Gor'kiy: Volgo-Vyatka Book Publishing House (in Russian).

Khotinsky, N.A. and Savina, S.S. 1985. "Paleoclimatic schemes of the territory of the USSR in the boreal, atlantic and subboreal periods of the Holocene." *Izvestiya RAN, Seriya geographich*, (4): 18–34 (in Russian).

King, D., Bourennane, H., Isampert, M. and Macaire, J.J. 1999. "Relationship of the presence of a n,n-calcareous clay-loam horizon to DEM attributes in a gently sloping area." *Geoderma*, 89 (1–2): 95–111.

Kislov, A.V. 2006. "Changes and variability of the global climate." *Modern Global Changes in the Natural Environment*, 1: 118–129. Moscow: Nauchny Mir (in Russian).

Klige, R.K. 2000. "Global hydro-climatic changes." *Global and Regional Climate Changes and Their Natural and Socio-economic Consequences*, 6–23. Moscow: GEOS.

Klijn, F. and Udo de Haes, H. 1994. "A hierarchical approach to ecosystems and its implycation for ecological land classification." *Landscape Ecology*, 9 (2): 89–104.

Klimanov, V.A., Khotinsky, N.A., and Blagoveshchenskaya, N.V. 1995. "Climate fluctuations over the historical period in the center of the Russian Plain." *Izvestiya. RAN. Ser. Geographical* (1): 89–96 (in Russian).

Kljun, N., Black, T.A., Griffis, T.J., et al. 2007. "Response of net ecosystem productivity of three boreal forest stands to drought." *Ecosystems*, 10: 1039–1055.

Kobak, K.I. 1988. *Biotic Components of Carbon Cycle*. Leningrad: Gidrometeoizdat (in Russian).

Kobak, K.I. and Kondrasheva, N. Yu. 1992. "Global warming and natural zones." *Meteorology and* Hydrology, 8: 91–98 (in Russian).

Kobak, K.I., Kondrasheva, N. Yu and Turchinovich, I.E. 2002. "Impact of climate change on natural zonality and Russian ecosystems." *Climate Changes and Their Consequences*, 205–210. St-Petersburg: Nauka (in Russian).

Kobeleva N.V. 1975. "An Experience of Mathematical Analysis of a Geobotanical Map." *Modeling of Elementary Geosystems*, 129–143. Irkutsk: Institute of Geography of Siberia and the Far East, Siberian Branch of the USSR Academy of Sciences (in Russian).

Kolesnikov, B.P. 1956. "Cedar forests of the Far East." *Transactions of Far East Filial Akademii of Sciences of USSR, Seriya Botanicheskaya*, 2 (4). Moscow–Leningrad (in Russian).

Kolomyts, E.G. 1987. *Landscape Research in Transition Zones*. Moscow: Nauka (in Russian).

Kolomyts, E.G. 1995. "Organization and sustainability of coniferous forest ecosystems in the boreal ecotone of the Russian Plain." *Izvestiya RAN. Ser. geogr*. 3: 37–51 (in Russian).

Kolomyts, E.G. 1998. *Polymorphism of Landscale-Zonal Systems*. Pushchino: OPTI PSC RAS (in Russian. Abstract in English, 266–277).

Kolomyts, E.G. 2003. *Regional Model of Global Environmental Changes*. Moscow: Nauka (in Russian. Abstract in English, 306–318).

Kolomyts, E.G. 2005. *Boreal Ecotone and Geographic Zonality: Atlas-Monograph.* Moscow: Nauka (in Russian. Abstract in English, 380–384).

Kolomyts, E.G. 2008. *Local Mechanisms of Global Changes in Natural Ecosystems.* Moscow: Nauka (in Russian. Abstract in English, 374–380).

Kolomyts, E.G. 2009. "Forest Ecosystems of the Volga River basin under global climatic Changes (Experience of Landscape-Ecological Prediction)." *International. Journal of Ecology & Development* 13 (S09): 3–24.

Kolomyts, E.G. 2018. *Experimental Geographic Ecology. The Record of Geographer-Naturalist.* Moscow: KMK (in Russian).

Kolomyts, E.G. 2020. *Carbon balance and sustainability of forest ecosystems under global climate changes. Ecological Resources of Boreal Forests.* Moscow: Nauka (in Russian).

Kolomyts, E.G. 2022. *Experimental Geographical Ecology: Problems and Methods.* Newcastle upon Tyne: Cambridge Scholars Publishing.

Kolomyts, E.G. 2024. *Mechanisms of Forfest Ecosystems Sustainability in a Chamging Climate: Complete Triad of Ecological Monitoring.* New York City: Springer.

Kolomyts, E.G., Rozenberg, G.S., Saksonov, S.V. and Sharaya, L.S. 2012. *Forests of Volga River Basin under Global Warming (Landscape-Ecological Analysis and Prognosis).* New York: Nova Science Publishers Inc.

Kolomyts, E.G., Yunina, V.P., Sidorenko, M.V. and Vorotnikov, V.P. 1993. *Ecosystems of Coniferous Forest on the Zonal Boundary: Organization, Stability, Anthropogenic Dynamics.* Nizhny Novgorod: Institute of Ecology of the Volga River basin of the Russian Academy of Sciences.

Komarov, A., Chertov, O., Zudin, S., et al. 2003. "EFIMOD 2 – system of simulation models of forest growth and elements cycles in forest ecosystems." *Ecological Modeling,* 170: 373–392.

Kondrat'ev, K.Ya. and Demirchyan, K.S. 2000. "Global climate change and the carbon cycle." *Izvestiya Russkogo geograficheskogo obshchestva,* 4: 1–20 (in Russian).

Kondrat'ev, K.Ya., Losev, K.S., Ananicheva, M.D. and Chesnokova, I.V. 2003. *Natural Scientific Foundations of the Sustainability of Life.* Moscow: VINITI. Institute of Geography RAS (in Russian).

Kononova, M.M. 1963. *Soil Organic Matter: Its Nature, Properties and Methods of Study.* Moscow: Publishing house of the Academy of Sciences of the USSR (in Russian).

Köppen, V. 1938. *Fundamentals of Climatology (Climates of the Globe).* Moscow: Uchpedgiz (in Russian).

Koptyug, V.A. 1994. *United Nations Conference on Environment and Development (Rio de Janeiro, June 1992): Briefing Note.* Novosibirsk: SO RAN (in Russian).

Korotkov, I.A. 1978. *Forests of the Mongolian People's Republic (Geography and Typology).* Tr. joint Soviet-Mongolian expedition. Moscow: Nauka (in Russian).

Korotkov, S.A., Stonozhenko, L.V., Kiseleva, V.V. and Glazunov, Yu. B. 2020. "Influence of environmental and socio-ecological factors on the formation of forests near Moscow." *Problems of Ecological Monitoring and Modeling of Ecosystems,* XXXI (1–2): 91–110 (in Russian).

Korzukhin, M.D. and Korotkov, V.N. 2018. "Modification of the ROBUL model for calculating the carbon balance of forests in Russia." *Fundamental and Applied Climatology,* (3): 30–53 (in Russian).

Kotlyakov, V.M. 1992. "Geography and survival of mankind." *Izvestiya Rossiiskoi Akademii Nauk, Seriya Geograficheskay,* 2: 8–14 (in Russian).

Kotlyakov, V.M. 1997. *Science, Society, Environment.* Moscow: Nauka (in Russian).

Kotlyakov, V.M. 2001. *Selected Works. Book 3. Geography in a Changing World.* Moscow: Nauka (in Russian).

Kotova, T.V. ed. 1987. *Vegetation of the European Part of the USSR and the Caucasus.* M-b 1: 2000 000. Moscow: GUGK (in Russian).

Kovda, V.A. 1971. "Biogeochemical concept of the biosphere." In *Biosphere Resources on the Territory of the USSR* 16–26. Moscow: Nauka (in Russian).

Krapivin, V.F., Svirezhev, Yu. M. and Tarko, A.M. 1982. *Mathematical Modeling of Global Biospheric Processes*. Moscow: Nauka (in Russian).

Krasheninnikov, I.M. 1951. *Geographical Works.* Moscow: Geografgiz (in Russian).

Krauklis, A.A. 1969. "Structural-dynamic facies analysis of the southern taiga landscape of the Lower Angara region." *Southern Taiga of the Angara Region,* 32–119. Leningrad: Nauka (in Russian).

Krauklis, A.A. 1974. "Features of geographical gradations of the topical order." *Topological Aspects of the Doctrine of Geosystems*, edited by V.B. Sochava, 87–137. Novosibirsk: Nauka (in Russian).

Krauklis, A.A. 1979. *Experimental Landscape Science Problems.* Novosibirsk: Nauka (in Russian).

Krauklis, A.A., ed. 1991. *Geosystems of the Contact of Taiga and Steppe: South of Central Siberia.* Novosibirsk: Nauka (in Russian).

Krenke, A.N., Chernavskaya, M.M., Bradzil, R., et al. 1995. *Climate Variability of Europe in the Historical Past.* Moscow: Nauka (in Russian).

Krištofovič, A.N. 1946. "Evolution of plant cover in the geological past and its main Factors". *Materialy po istorii flory i rastitelynosty SSSR, Vyp. II,* edited by V.L. Komarov, 21–87. Moskva, Leningrad: Izd-vo AN SSSR (in Russian).

Krut'ko, V.N., Pegov, S.A., Khomyakov, D.M. and Khomyakov, P.M. 1982. *Formalization of the Assessment of the Quality of Environmental Components*. Moscow: VNII sistemnyh issledovany [All-Russian Research Institute of System Research] (in Russian).

Kubota, Y., Murata, H. and Kikuzawa, K. 2004. "Effects of topographic heterogeneity on tree species richness and stand dynamics in a sub-tropical forest in Okinawa Island, southern Japan" *Journal of Ecology,* 92: 230–240.

Kudeyarov, V.N. ed. 2007. *Modeling the Dynamics of Organic Matter in Forest Ecosystems.* Moscow: Nauka (in Russian).

Kukhta, A.E. and Popova, U.N. 2020. "Climatic signal in the linear growth of Scots pine in boreal phytocenoses of the White Sea coast." *Problems of Ecological Monitoring and Modeling of Ecosystems,* XXXII (3–4): 33–42 (in Russian).

Kurganova, I.N., Lopez de Gorenyu, V.O., Rozanova, L.N., et al. 2007. "Long-term monitoring of $CO_2$ emission from soddy-podzolic soil." *Problems of Ecological Monitoring and Modeling of Ecosystems,* XXI: 23–47 (in Russian).

Kupriyanova, T.P., ed. 1989. *Factors and Mechanisms of Stability of Geosystems.* Moscow: Institute of Geography of the Academy of Sciences of the USSR (in Russian).

Kurnaev, S.F. 1980. *Shady Broad-leaved Forests of the Russian Plain and the Urals.* Moxcow: Nauka (in Russian).

Kurz, W.A., Apps, M., Banfield, E. and Stinson, G. 2002. "Forest carbon accounting at the operation scale." *The Forestry Chronicle,* 78: 672–679.

Kustler, G. 1957. "ABC of information theory." *Information Theory in Biology*, edited by Hubert P. Yockey, 5–48. London, New York, Los Angeles: Pergamon Press.

Lavrenko, E.M. and Isachenko, T.I. 1976. "Zonal and provincial botanical-geographical division of the European part of the USSR." *Izvestiya Vsesoyuznogo Geograficheskogo Obshchestva*, 108 (6): 469–483.

Leemans, R. 1992. Modelling ecological and agricultural impacts of global change on a global scale. *Journal of Science & Industrial Research*, 51: 709–724.

Leemans, R. 1999. "Modeling for species and habitats: New opportunities for problem solving." *The Science of the Total Environment,* 240: 51–73.

Le Quere, C., Moriary, R., Andrew, R.M., et al. 2015. "Global carbon budjet 2014." *Earth System Science Data*, 7 (I): 47–85.

Leskinen, P., Lindner, M., Verkerk, P., et al., eds. 2020. *Forests of Russia and Climate Change: What Can Science Tell Us?* Bonn: European Forest Institute (in Russian).

Lischke, H., Guisan, A., Fischlin, A. and Bugmann, H. 1998. "Vegetation response to climate change in the alps: modeling studies." *Views from the Alps: Regional Perspectives on Climate Change,* 309–350. Cambridge, MA: MIT Press.

Liverovsky, Yu. A. 1974. *Pochvy SSSR. Geogreficheskaya kharakteristika.* Moscow: Mysly (in Russian).

Logofet, D.O. and Svirezhev, Yu. M. 1983. "The concept of stability of biological systems." *Problems of Ecological Monitoring and Modeling of Ecosystems,* VI: 159–171. Leningrad: Gidrometeoizdat (in Russian).

Loreau, M., Downing, A., Emmerson, M., et al. 2002. "A new look at the relationship between diversity and stability." *Biodiversity and Ecosystem Functioning,* 79–91. Oxford: Oxford University Press.

Losev, K.S. 2001. *Environmental Problems and Prospects for Sustainable Development of Russia in the XXI Century.* Moskva: Publishing House "Kosmosinform" (in Russian).

Lovelock, J.T. 1989. *The Ages of Gaia: A Biography of Our Living Earth.* Oxford, New York, Toronto: Oxford University Press.

Lukina, E.V. 1974. "Forests: Steppes." *Nature of the Gorky Region,* 193–226. Gor'kiy: Volga-Vyatka Book Publishing House

Lukina, N.V. 2020. "Global challenges and forest ecosystems." *Bulletin of the Russian Academy of Sciences,* 90 (6): 528–532 (in Russian).

Lukina, N.V., Isaev, A.S., Kryshen, A.M., et al. 2015. "Priority directions for the development of forest science as a basis for sustainable forest management." *Forestry,* 4: 243–254 (in Russian).

MacArtur, R.U. 1955. "Fluctuations of animal populations and a measure of community stability." *Ecology,* 36: 533–536.

MacMillan, R.A., Torregrosa, A., Moon, D., Coupé, R. and Philips, N. 2009. "Chapter 24. Automated predictive mapping of ecological entities." *Geomorphometry Concepts Software Applications,* edited by T. Hengl and H.T. Reuter, 551–578. Amsterdam: Elsevier.

Maksimova, O.V., Minin, A.A. and Kukhta, A.E. 2021. "On the connection between seasonal phenomena and processes in trees in the conditions of the Central Forest Biosphere Reserve." *Problems of Ecological Monitoring and Modeling of Ecosystems,* XXXII (1–2): 56–66 (in Russian).

Mamay, I.I. 1992. *Landscape Dynamics.* Moscow: Moscow University Press (in Russian).

Margalef, R. 1958. "Information theory in ecology." *General Systems,* 3: 36–71.

Martin-Benito, D. and Pederson, N. 2015. "Convergence in drought stress, but a divergence of climatic drivers across a latitudinal gradient in a temperate broadleaf forest." *Journal of Biogeography,* 42: 925–937.

Maslov, A.A. and Peterson Yu. V. 1993. "Succession trend in forest communities of protected areas: 10 years of observations in the Moscow region." *Forests of the Russian Plain,* 122–125. Moscow: Scientific Council for Forest Problems (in Russian).

Matveev, N.M., Novikova. E.V. 1988. "Dynamics of seed renewal of natural and artificial forests in the steppe Trans-Volga Region." *Issues of Forest Biogeocoenology Ecology and Nature Protection in the Steppe Zone,* 37–53. Kuybyshev: Kuybyshev. State University (in Russian).

Menzhulin, G.V. and Savvateev, S.P. 2002. "World food problem and modern global warming". *Climate Changes and Their Consequences,* 122–151. St. Petersburg: Nauka.

Mezentsev, V.S., Karnatsevich, I.V., Belonenko, G.V., et al. 1974. *Modes of Moisture Supply and Conditions for Irrigation and Drainage of the Steppe Region.* Moscow: Kolos (in Russian).

Mikheev, V.S. 1974. *Verkhnecharskaya Depression. Experience in Topological Study of the Landscape.* Novosibirsk: Nauka (in Russian).

Mil'kov, F.N. 1967. *Natural Zones of the USSR.* Moscow: Mysl (in Russian).

Mil'kov, Ph.N. 1981. *Physical Geography: Current State, Patterns, Problems*. Voronezh: Voronezh Publishing House University (in Russian).

Mil'kov, Ph.N. 1986. *Physical Geography: The Doctrine of the Landscape and the Geographic Zonality*. Voronezh: Voronezh Publishing House University (in Russian).

Mil'kov, Ph.N. and Gvozdetsky, N.A. 1986. *Physical Geography of the USSR*. Moscow: Vysshaya Shkola (in Russian).

Mirkin, B.M., Rosenberg, G.S. and Naumova, L.G. 1989. *Dictionary of Concepts and Terms of Modern Phytocoenology*. Moscow: Nauka (in Russian).

Mitropol'sky, A.K. 1971. *Technique of Statistical Calculations*. Moscow: Nauka (in Russian).

Moiseev, N.N., Alexandrov, V.V. and Tarko, A.N. 1985. *Man and the Biosphere: Experience in Systems Analysis and Experiments with Models*. Moscow: Nauka (in Russian).

Mokhov, I.I. 2006. "Model assessments of possible climatic changes in the XXI century in comparison with climatic changes in the past and present." *Possibilities of Preventing Climate Change and its Consequences. The Problem of the Kyoto Protocol,* 75–93. Moscow: Nauka (in Russian).

Mokhov, I.I., Dufresne, J.-L. Le Tret, E., et al. 2005. "Changes in drought regimes and bio-productivity of terrestrial ecosystems in the regions of Northern Eurasia according to calculations with a global climate model with a carbon cycle." *Reports of RAS,* 405 (6): 1–5 (in Russian).

Molchanov, A.A. 1961. *Forest and Climate.* Moscow: Publishing House of the USSR Academy of Sciences (in Russian).

Montgomery D.C. and Peck E.A. 1982. *Introduction to Linear Regression Analysis,* 504 p. New York: John Wiley & Sons.

Mordkovich, V.G. 1982. *Steppe Ecosystems*. Novosibirsk: Nauka (in Russian).

Morozov, G.F. 1949. *The Doctrine of the Forest*, 7th edition. Moscow, Leningrad: Goslesbumizdat (in Russian).

Murzaev, E.M. 1984. *Dictionary of Popular Geographical Terms*. Moscow: Mysl (in Russian).

Naeem, S. 1998. "Species redundancy and ecosystem reliability." *Conservation Biology,* 12: 39–45.

Neef, E. 1968. "On some questions of comparative landscape ecology." *Reports of Institute of Geography of Siberia and the Far East,* 19: 44–53 (in Russian).

Nefedyeva, E.A. and Yashina, A.V. 1985. *The Role of Snow Cover in the Differentiation of the Landscape Sphere.* Moscow: Nauka (in Russian).

Neigh, C.S.R., Nelson, R.F., Ranson, K.J., et al. 2013. "Taking stock of circumboreal fores carbon with ground measurements. airborne and spaceborne LiDAR." *Remote Sensing of Environment*, 137: 274–287.

Nelso, R., Ranson, K.J., Sun, G., et al. 2009. "Estimating Siberian timber volume using MODIS and ICE Sat/GLAS." *Remote Sensing of Environment,* 113: 691–701.

Nikolaev, V.A. 1978. *Classification and Small-Scale Mapping of Landscapes*. Moscow: Moscow State University Publishing House (in Russian).

Nikolaevsky, V.S. and Nikolaevskaya, T.V. 1993. "Reasons for the drying up of oak forests in the Volga region." *Meeting "Forests of the Russian Plain". Abstracts of Reports*, 143–146. Moscow: Nauch. Forest Council.

Nitsenko., A.A. 1971. *Plant Association and Plant Community as Primary Objects of Geobotanical Research.* Moscow: Nauka (in Russian).

Odeh Inakwu, O.A., McBratney, A.B. and Chittleborough, D.J. 1992. "Fuzzy-c-means and kriging for mapping soil as a continuous system." *Soil Science Society of American Journal,* 56: 1848–1854.

Odum, Eu.P. 1971. *Fundamentals of Ecology*, 3rd edition. Philadelphia, London, Toronto: W.B. Saunders Company.

Odum, Eu.P. 1983. *Basic Ecology*. Philadelphia: CBS College Publishing.

Ontl, T.A., Janowiak, M.K., Swanston, C.W., et al. 2020. "Forest management for carbon sequestration and climate adaptation." *Journal of Forestry,* 118 (1): 86–101.

Orlov, D.S., Biryukova, O.N. and Sukhanova, N.I. 1996. *Organic Matter of Soils of the Russian Federation*. Moscow: Nauka (in Russian).

Osipov, V.V., ed. 2004. *Ecosystems of the Tellerman Forest*. Moscow: Nauka (in Russian).

Paffengolts, K.N., ed. 1973. *Geological Dictionary*. Moscow: Nedra (in Russian).

Paris Agreement. 2015. *Conference of the Parties Twenty-First Session*. Paris. 30 Nov. to 11 Dec. 2015. 19 p.

Perel'man, A.I. 1975. *Landscape Geochemistry*. Moscow: Vysshaya Shkola (in Russian).

Perevedentsev, Yu. P., Kolobov, N.V. and Vereshchagin, M.A. 1985. "Some features of the statistical structure of the surface air temperature field over the territory of the European part of the USSR, South Western Siberia, Kazakhstan." *Problems of Complex Geography,* 89–99. Kazan: Publishing house of Kazan University (in Russian).

Perevedentsev, Yu. P., Vereshchagin, M.A., Naumov, E.P. and Shanta linsky, K.M. 2001. "Long-term fluctuations in the main indicators of the hydrometeorological regime of the Volga basin." *Meteorology and Hydrology,* 10: 16–22 (in Russian).

Peterson, G., Allen, C.R., and Holling, C.S. 1998. "Ecological resilience biodiversity and scale." *Ecosystems,* 1: 6–18.

Pianka, E.R. 1978. *Evolutionary Ecology*. New York, Hagerstown, San Francisco, London: Harper and Row Publishers.

Pike, R.A. 2002. *Bibliography of Terrain Modelling (Geomorphometry) the Quantitative Representation of Topography – Supplement 4.0. Open-file report 02–465*. U.S. Geological Survey.

Polozova, L.G. 1954. On the Characteristics of Climate Continentality. *News of the All-Union Geographical Society*, 86 (5): 412–422 (in Russian).

Polynov, B.B. 1956. *Selected Works*. Moscow: Publishing House of the USSR Academy of Sciences (in Russian).

Ponomareva, V.V., and Plotnikova, T.A. 1980. *Humus and Soil Formation: Methods and Results of the Study*. Leningrad: Nauka (in Russian).

Pope, V.D., Gallani, M.L., Rowntree, P.R. and Stratton, R.A. 2000. "The impact of new physical parametrizations in Hadley Centre climate model – HadCM3." *Climate Dynamics*, 16: 123–146.

Popova, V.V. 2001. "Precipitation variations on the Russian plain over the last millennium." *Izvestiya Rossiiskoi Akademii Nauk Geograficheskaya Seriya,* 1: 42–49 (in Russian).

Preobrazhensky, V.S. 1969. "O sisteme metodov obshchey fizicheskoy geografii." *Metody landshaftnyh issledovany,* edited by V.S. Preobrazhensky, 7–34. Moscow: Nauka (in Russian).

Preobrazhensky, V.S., ed. 1978. *Nature Technology Geotechnical Systems*. Moscow: Nauka (in Russian).

Preobrazhensky, V.S. 1986. *Organization of Landscapes (Preprint)*. Moscow: Institute of Geography RAS (in Russian).

Printice, I.C., Cramer, W., Harrison, S.P., et al. 1992. "A global bio model based on plant physiology and dominance soil properties and climate." *Journal of Biogeography,* 19: 117–134.

Pushin, A.V. 1997. "Study of the relationship between the variability of oak growth and climatic conditions at the northern boundary of its range on the Russian Plain." *Materialy Meteorological Studies. No.* 16. *Problems of Climate*, 165–170. Moscow: Institute of Geography of the Russian Academy of Sciences.

Puzachenko, Yu. G. 1983. "Invariance of geosystems and their components." *Stability of Geosystems*, edited by A.D. Armand and I. Yu. Dolgushin, 32–41. Moscow: Nauka (in Russian).

Puzachenko, Yu. G. 1985. "Prostranstvennaya ierarhiya geosystem s pozitsii teorii kolebany." *Voprosy geographii*. 127. *Modelirovanie geosystem*. 96–111. Moscow: Mysly. (in Russian).

Puzachenko, Yu. G. 1990. "Problems of stability and regulation." *Structural-Functional Organization and Stability of Biological Systems,* 122–147. Dnepropetrovsk: Scientific. Advice on Problems of Ecology and Anthropology, and Dynamics Biology System (in Russian).

Puzachenko, Yu. G. 1998. *Methodological Foundations of Geographical Forecasting and Environmental Protection*. Moscow: Publishing House of the Ural Regional Educational Institution (in Russian).

Puzachenko, Yu. G. and Skulkin, V.S. 1981. *Structure of Forest Zone Vegetation for USSR: System Analysis.* Moscow: Nauka (in Russian).

Puzachenko, Yu. G. and Sankovsky, A.G. 2005. "Climatic conditioning of pure biosphere production." *Izvestiya Rossiiskoi Akademii Nauk Seriya Geograficheskaya,* 5: 5–19 (in Russian).

Quegan, S., Beer, C., Shvidenko, A., et al. 2011. "Estimating the carbon balance of central Siberia using a landscape-ecosystem approach. Atmospheric inversion and dynamic global vegetation models." *Global Change Biology*, 17: 351–365.

Rabotnov, T.A. 1978. *Phytocoenology*. Moscow: Moscow University Press (in Russian).

Ramenskiy, L.G. 1971. *Selected Works: Problems and Methods of Studying the Vegetation Cover.* Leningrad: Nauka (in Russian).

Ran'kova, E.Ya., Samokhina, O.F., Antipova, U.I. and Smirnov, V.D. 2024. "Features of the temperature regime at the surface of the Globe in 2023." *Fundamental and Applied Climatology,* 10 (2): 288 –323 (in Russian).

Rauner, Yu. L. 1981. *Climate and Grain Yield.* Moscow: Nauka (in Russian).

Ricklefs, R.E. 1976. *A Textbook in Basic Ecology.* Portland, OR: University Pennsylvania, Chiron Press Inc.

Rode, A.A. 1965. *Fundamentals of the Theory of Soil Moisture. Vol. I. Water Properties of Soils and the Movement of Soil Moisture*. Leningrad: Gidrometeoizdat (in Russian).

Rodoman, B.B. 1999. *Territorial Areas and Networks*. Smolensk: Oikumena (in Russian).

Rodriguez-Veiga, P., Saatchi, S., Tansey, K. and Baltzer, H. 2016. "Magnitude, special distribution and uncertainty of forest biomass stoks in Mexico." *Remote Sensing of Environment,* 183: 265–281.

Rosen, R. 1969. *Optimality Principle in Biology*. Moscow: Mir (in Russian).

Romanovskaya, A.A. 2018. "Needs and ways of development of adaptation monitoring." *Problems of Ecological Monitoring and Modeling of Ecosystems,* XXIX (1): 107–125.

Romanovskaya, A.A. 2019. "Towards the concept of public administration and monitoring in the field of climate change in Russia." *Problems of Ecological Monitoring and Modeling of Ecosystems,* XXX (3–4): 61–83.

Rozenberg, G.S. 1984. *Models in Phytocoenology*. Moscow: Nauka (in Russian).

Rozenberg, G.S. 1991. "Towards the construction of a system of concepts of modern ecology." *Journal of General Biology* 52 (3): 422–440 (in Russian).

Rozenberg, G.S. 2004. *Ecology Faces*. Togliatti: Volzh. Ecology Institute. Basin RAS (in Russian).

Rozenberg, G.S., Mozgovoy, D.P. and Gelashvili, D.B. 1999. *Ecology: Elements of Theoretical Constructs in Modern Ecology*. Samara: Samarsk. Nauchn. Tsentr RAN (in Russian).

Rosenberg, G.S., Shitikov, V.K. and Zeitlin, N.A. 2008. "Theoretical foundations of optimal planning of an experiment and processing of multifactorial experiments." *Problems of an Ecological Experiment (Planning and Analysis of Observations),* 170–211. Tolyatti: Institute of Ecology of the Volga Basin of the Russian Academy of Sciences, Nizhny Novgorod. State University named after N.I. Lobachevsky (in Russian).

Rozhkov, A.A. and Kozak, V.T. 1989. *Ustoichivost' lesov* (Resistance of Forests). Moscow: Agropromizdat (in Russian).

Rubinstein, E.S. and Polozova, L.G. 1966. *Modern Climate Changes*. Leningrad: Gidrometeoizdat (in Russian).

Santer, B. 1985. "The use general circulation models in climate impact analysis – a preliminary stay of the impacts of a $CO_2$ – indicated climatic change on west European agriculture." *Climatic Ghanges,* 7 (1): 71–93.

Santoro, M., Cartus O., Carvalhais N., et al. 2021. "The global forest above-ground biomass pool for 2010 estimated from high-resolution satellite observations." *Earth System Science Data,* 13 (8): 3927–3950.

Savitsky, P.N. 1927. *Geographical Features of Russia. Part I: Vegetation and Soil.* Prague: Eurasian Book Publishing (in Russian).

Schimal, D.S., House, J.L., Hibbard, K.A., et al. 2001. "Recent patterns and mechanisms of carbon exchange by terrestrial ecosystems." *Nature*, 414 (8) 169–178.

Schmidt, G.A., Ruedy, R., Hansen, J.E. et al. 2006. "Present day atmospheric simulations using GISS Model E: Comparison to in-situ, satellite and reanalysis data." *Journal of Climate,* 19: 153–192.

Semenov-Tyan-Shansky, V.P. 1928. *Region and Country*. Moscow, Leningrad: State Publishing House (in Russian).

Shanin, V. and Chertov, O. 2020. "Simulating the effect of forest fires, cuttings, and increased nitrogen deposition on dynamics of key forest ecosystem properties and processes in Russian North-West." *European Journal of Forest Research*, 139: 1–19.

Sharaya, L.S. 2009. "Predictive mapping of forest ecosystems in geoecology." *Povolzhskiy Ecological Journal,* 3: 249–257 (in Russian).

Shary, P.A. 2001. *Analytical GIS Eco*. Web site: www.giseco.info.

Shary, P.A. 2006. *Personal research website on geomorphometry and applications*. Web site: https:// www. giseco.info/.

Shary, P.A., Sharaya, L.S. and Mitusov, A.V. 2002. "Fundamental quantitative methods of land surface analysis." *Geoderma,* 107 (1–2): 1–32.

Shcherbakov, Yu. A. 1970. "Experience in studying the role of exposure in landscape formation." *Influence of Exposure on Landscapes: Uchenye Zapiski Permskogo Gosudarstvennogo Universiteta (Perm State University)*, 240: 3–99 (in Russian).

Shnitnikov, A.V. 1957. "Variability of general moisture content of the continents of the northern hemisphere." *Zap. VGO. T.* 16. *New Series.* Moscow, Leningrad: Publishing House of the USSR Academy of Sciences (in Russian).

Shugart, G.G., Antonovsky, M.Ya., Yarvis, P.J. and Sanford, A.P. 1989. "Carbon dioxide, climate change and ecosystems." In: *Greenhouse effect, climate change and ecosystems*, 513–547. Leningrad: Gidrometeoizdat (in Russian).

Shvidenko, A.Z. 2012. "Global changes and Russian forest inventory." *Forest Inventory and Forest Management,* 47 (1): 52–75 (in Russian).

Shvidenko, A.Z., Kraxner, F., Obersteiner, M. and Schepaschenko, D.G. 2014. "The problem of transition to sustainable forest management in Russia; potential and risks." *Materials of the All-Russian Conference with International Participation "Forest Biogeocoenoses of the Boreal Zone: Geography Structure Functions Dynamics"*. Novosibirsk: Publishing House of SB RAS (in Russian).

Shvidenko, A.Z., Schepaschenko, D.G., Nilsson, S. and Boluy, Yu. I. 2008. *Tables and Models of the Course of Growth and Productivity of Plantations of the main Forest-Forming Species of Northern Eurasia (regulatory and reference materials).* Moscow: International Institute of Applied System Analysis (in Russian).

Shvidenko, A.Z., Schepaschenko, D.G., Kraksner, F. and Onuchin, A.A. 2017. "Transition to sustainable forest management in Russia: theoretical and methodological prerequisites." *Siberian Forest Journal,* 6: 3–25 (in Russian).

Serebryannaya, T.A. 1982. "On the dynamics of the forest-steppe zone in the Center of the Russian Plain in the Holocene." Natural Development of the USSR in the Late Pleistocene and Holocene, 179–186. Moscow: Nauka (in Russian).

Simonov, Yu. G. 1966. "Ways of using the ergodic theorem for paleogeographical analysis of Continents." *Vestnik Moskovskogo Universiteta Series* 5. Geography, 5: 3–18 (in Russian).
Simonov, Yu. G. 1982. "Basic properties of objects of geographical forecasting and methods of their formalized description." *Problemy regionalynogo geograficheskogo prognoza,* edited by A.P. Kapitsa and Y.G. Simonov, 112–193. Moscow: Nauka. (in Russian).
Sirotenko, O.D. and Abashina, E.V. 1998. Agroclimatic resources and physical-geographic zoning of the territory of Russia under global warming. *Meteorology and Hydrology*, 3: 92–103 (in Russian).
Smirnova, O.V., ed. 2004. *Eastern European Forests: History and Modernity*. Books 1 and 2. Moscow: Nauka (in Russian).
Smith, J.M. 1974. *Models in Ecology*. Cambridge: At the University of Sussex Press.
Smith, T.M., Leemance, R. and Shugart, H.H. 1992. "Sensitivity of terrestrial carbon storage to $CO_2$-induced climate change: comparison of four scenarios based on general circulation models." *Climatic Change*, 21: 367–384.
Smith, T.M. and Shugart, H.H. 1993. "The transient response of terrestrial carbon to perturbed climate." *Nature,* 361 (6412): 523–526.
Solntsev, N.A. 1973. "In defense of Dokuchaev's law". *Landscape Collection*, 5–28. Moscow: Moscow State University Publishing House.
Sochava, V.B. 1962. "Initial provisions of typification of taiga lands on a land scape-geographical basis." *Reports of Institute of Geography of Siberia and the Far East,* 2: 14–23. Irkutsk (in Russian).
Sochava, V.B. 1969. "Experimental geographical research and taiga development." *Information Bulletin on the Complex Development of Taiga Territories*, 4; 5–14. Irkutsk (in Russian).
Sochava, V.B. 1974. "Geotopology as a section of the theory of geosystems." *Topological Aspects of the Study of Geosystems,* 3–86. Novosibirsk: Nauka (in Russian).
Sochava, V.B. 1978. *Introduction into the Theory of Geosystems*. Novosibirsk: Nauka (in Russian).
Sochava, V.B. 1979. *Vegetation Cover on the Subject Maps*. Novosibirsk: Nauka (in Russian).
Sochava, V.B., Bachurin, G.V., Krauklis, A.A. and Nechaeva, E.G. 1970. "Topological features of heat and moisture in taiga geosystems." *Reports of Institute of Geography of Siberia and the Far East,* 26: 39–53 (in Russian).
Sokolov, V.E. and Bazilevich, N.I., eds. 1983. *Theoretical Foundations and Experience of Environmental Monitoring*. Moscow: Nauka (in Russian).
Solntsev, N.A. 1973. "In defense of Dokuchaev's law." *Landscape Collection,* edited by N.A. Solntsev, 5–28. Moscow: Publishing House of Moscow State University (in Russian).
Solntsev, N.A. 2001. *The Doctrine of the Landscape: Selected Works*. Moscow: Publishing House of Moscow State University (in Russian).
Solntsev, V.N. 1981. *Systemic Organization of Landscapes*. Moscow: Mysl' (in Russian).
Solntsev, V.N. 1997. *Strukturnoe landshaftovedenie. Osnovy kontceptcii, nekotorye argumenty*. Moscow: MGU (in Russian).
Solovieva, A.N. and Shikhova, T.G. 2011. "Phenological responses of the biota of the east of the Russian Plain to weather anomalies." *Problems of Ecological Monitoring and Modeling of Ecosystems,* XXXII (1–2): 37–49 (in Russian).
Sprygin, I.I. 1986. "Materials for the knowledge of the vegetation of the Middle Volga Region." *Scientific Heritage*, 11. Moscow: Nauka (in Russian).
Stankov, S.S. 1951. *Essays on the Physical Geography of the Gorky Region.* Gor'kiy: Regional State Publishing House (in Russian).
Steffen, W. and Tyson, P., eds. 2001. *Global Change and the Earth System: A Planet under Pressure*. Stockholm: IGBP Science 4.
Stinson, G., Kurz, W.A., Smyth, C.E., et al. 2011. "An inventory-based of Canada's managed forest carbon dynamics 1900 to 2008". *Global Change Biology*, 17: 2227–2244.

Stoll, R.R. 1968. *Sets Logics Axiomatic Theories*. Transl. from English Moscow: Prosveshchenie (in Russian).

Stone, L., Gabric, A. and Berman, T. 1996. "Ecosystem resilience. Stability. and productivity: Seeking a Relationship". *The American Naturlist,* 148: 892–903.

Stupishin, A.V., ed. 1964. *Physical and Geographical Zoning of the Middle Volga Region*. Kazan': Kazan Publishing House. University (in Russian).

Sukachev, V.N. 1972. *Selected Works. Vol.* 1. *Basics of Forest Typology and Biogeocoenology.* Leningrad: Nauka (in Russian).

Sukachev, V.N. 1975. *Selected Works. Vol.* III. *Phytocoenology Problems.* Leningrad: Nauka (in Russian).

Sukachev, V.N., Dylis, N.V. 1964. *Fundamentals of Forest Biogeocenology*. Leningrad: Nauka (in Russian).

Svetlosanov, V.A. 1990. *Sustainability and Stability of Natural Ecosystems (Model Aspect)*. Vol. 8. Moscow: VINITI, Results of Science and Technology (in Russian).

Svetlosanov, V.A. 2009. *Stability of Natural Systems to Natural and Anthropogenic Impacts (Textbook)*. Moscow: Publishing House of Moscow State University (in Russian).

Svirezhev, Yu. M. 1977. "Matematicheskoe modelirovanie biology icheskih soobshchestv". *Chislo i Mysly,* 63–117. Moccow: Znanie (in Russian).

Svirezhev, Yu. M. 1982. "Mathematical models in ecology." Number and Thought, (5): 16–55. Moscow: Znanie (in Russian).

Svirezhev, Yu. M. 1983. "Stability and complexity in mathematical ecology." *Stability of Geosystems*, 41–50. Moscow: Nauka (in Russian).

Svirezhev, Yu. M. and Elizarov, E.Ya. 1972. "Mathematical modeling of biological systems." *Problems of Space Biology,* XX. Moscow: Nauka (in Russian).

Svirezhev, Yu. M. and Logofet, L.O. 1978. *Sustainability of Biological Communities*. Moscow: Nauka (in Russian).

Sykes, V.T. and Prentice, C. 1993. "Modeling the response of Nordic forest to climatic change." *Impacts of Climatic Change on Natural Ecosystems (with Emphasis on Boreal and Arctic/ Alpine Areas),* edited by J.I. Holten, G. Paulsen, and W.C. Oechel, 75–79. Trondheim: NINA and DN Norway.

Syomkin, B.I. 1973. "Descriptive sets and their applications." *Research of Systems. Vol. 1: Analysis of Complex Systems,* 83–94. Vladivostok: Pacific Institute of Geography Far East Scientific Center of the Academy of Sciences of the USSR (in Russian).

Targulyan, V.O. and Sokolov, I.A. 1978. "Strukturny i funktcionalny podhod k pochve: pochva-pamyaty i pochva-moment." *Matematičeskoe modelirovanie v ecologii*, edited by A.M. Molchanov, 17–38. Moscow: Nauka (in Russian).

Tarko, A.M. 2005. *Anthropogenic Changes in Global Biospheric Processes: Mathematical Modeling*. Moscow: Phizmatlit (in Russian).

Tchebakova, N.M., Parphenova, E.I. and Soja, A.J. 2009. "The effects of climate, permafrost and fire on vegetation change in Sibiria in a changing climate." *Environmental Research Letters,* 4 (4): 45–63.

Thompson, I., Mackey, B., McNulty, S. and Mosseler, A. 2009. "Forest resilience biodiversity and climate change: a synthesis of the biodiversity/resilience/stability relationship in forest ecosystems." *Technical Series*, 43: 1–67. Montreal: Secret. Conv. Biol. Div.

Timofeev-Resovsky, N.V. 1970. "Structural levels of biological systems." *System Research Yearbook,* 80–113. Moscow: Nauka (in Russian).

Timofeev-Resovsky, N.V. and Tyuryukanov, A.N. 1966. "On elementary bichorologic subdivisions of the biosphere." *Byulleten' MOIP (Moscow Society of Nature Investigators). Biological Section,* LXXI (1): 123–132 (in Russian).

Tishkov, A.A. 1988. "Approaches to the study of the dynamics of biota as an object of geographic forecasting." *Geographic Forecasting and Environmental Problems,* 49–60. Moscow: Institute of Geography of the USSR Academy of Sciences (in Russian).

Tishkov, A.A. 1991. "Roly suktsssiy vo vremennoy organizatsii geosistem." *Geosistemy vo vremeny*, 118–132. Moscow: Institute Geografii AN SSSR (in Russian).

Tishkov, A.A. 2005. *Biosphere Functions of Natural Ecosystems in Russia.* Moscow: Nauka (in Russian).

Tishkov, A.A. 2012. "Actual biogeography as a methodological basis for conserving biodiversity." *Voprosy Georgii,* 134: Actual Biogeography, 15–57. Moscow: "Kodeks" (in Russian).

Tishkov, A.A. 2021. "The concept of biosphere reserves of the MAB Program and the tasks of biodiversity conservation: achievements and problems after 50 years." *Questions of Geography,* (152) Man and Biosphere 62–100 (in Russian).

Tselniker, Yu. L. 2006. "$CO_2$ gas exchange in forest biogeocenoses." *Ideas of Biogeocoenology in Forestry and Afforestation*, 213–229. Moscow: Nauka (in Russian).

Tsyganov, M.S. 1950. "Soil formation process and displacement of landscape zones of Western Siberia." *Soil Science,* 5: 262–272 (in Russian).

Tyuryukanov, A.N. 2001. *Selected Works.* Moscow: REFIA Publishing House (in Russian).

Tyuryukanov, A.N. and Fedorov, V.M. 1996. *Biospheric Reflections.* Moscow: RAEN (in Russian).

Udra, I.F. 1988. *Plant Distribution and Issues of Paleo- and Biogeography.* Kyiv: Naukova Dumka (in Russian).

Usoltsev, V.A. 2003. *Phytomass of Forests in Northern Eurasia. Marginal Productivity and Geography.* Ekaterinburg: Bot. Garden of the Ural Branch of the Russian Academy of Sciences (in Russian).

Utkin, A.I. 1995. "Carbon cycle and forestry". *Lesovedenie*, (5): 3–20 (in Russian).

Vasil'ev, V.N. 1946. "Conformity to natural laws of process of plain changes)." In *Materialy po istorii flory i rastitelynosty SSSR. Vyp. II,* by edited V.L. Komarov: 365–403. Moscow–Leningrad: Publishing House AS USSR (in Russian).

Velichko, A.A., ed. 2010. *Climates and landscapes of Northern Eurasia in the context of global warming. Retrospective analysis and scenarios. Atlas-Monograph.* Moscow: GEOS (in Russian).

Velichko, A.A., Grichuk, V.P., Gurtovaya, E.E., and Zelikson, E.M. 1983. "Paleoclimate of the USSR Territory during the Optimum of the Last (Mikulino) Interglacial." *Izvestiya AN SSSR. Geograf. Ser.*, (6): 30–45 (in Russian).

Velichko, A.A. and Klimanov, V.A. 1990. "Climatic conditions of the Northern Hemisphere 5-6 thousand years ago." *Izvestiya AN SSSR. Geograf. Ser.* (5): 38–52.

Vereshchagin, M.A., Perevedentsev, Yu. P. and Shantalinsky, K.M. 1997. "On probable changes in air temperature and atmospheric circulation in the east of the Russian Plain at the end of the current and at the beginning of the XXI century." *Izvestiya RAS Geograficheskaya Series,* 1: 103–109 (in Russian).

Verigo, S.A., and Razumova, L.A. 1973. *Soil moisture (in relation to the needs of agriculture).* Leningrad: Gidrometeoizdat (in Russian).

Vernadsky, V.I. 1994. *Living Matter and Biosphere.* Moscow: Nauka (in Russian).

Viktorov, A.S. 1986. Landscape drawing. Moscow: Mysl (in Russian).

Viktorov, A.S. 2006. *Basic Problems of Mathematical Morphology of Landscape.* Moscow: Science (in Russian).

Vlasov, A.A., Ryzhkov, O.V. and Puzachenko, A. Yu. 2021. "Monitoring of ecosystems in the Central Chernozem Biosphere Reserve: scientific and prognostic significance." *Questions of Geography,* 152*:* Man and Biosphere 333–356 (in Russian).

Volobuev, V.R. 1963. *Soil Ecology (Essays).* Baku: Publishing House of AN Azerb. SSR (in Russian).

Volobuev, V.R. 1974. *Introduction to the Energetics of Soil Formation.* Moscow: Nauka (in Russian).

Vygodskaya, N.N. 1981. *Radiation Regime and Structure of Mountain Forests.* Leningrad: Gidrometeoizdat (in Russian).

Vysotskiy, G.N. 1909. "On phyto-typological maps methods of their compilation and their practical significance." *Pochvovedenie,* 2: 98–119 (in Russian).

Vysotskiy, G.N. 1960. *Selected Works.* Moscow: Sel'khozgiz (in Russian).

Walter, H., 1968. *The Vegetation of the Earth: From an Eco-physiological Perspective* (Die vegetation der Erde: in öko-physiologischer Betrachtung). Bd. II. Die gemäßigten und arctischen Zonen. Jena: Veb Gustav Fischer Verlag.

Walter, H. 1976. *Die öcologischen Systeme der kontinente (Biogeosphäre). Prinzipien ihrer Gliederung mit Biespielen.* Stuttgart: Fischer.

Wetherald, R.T. and Manabe, S. 1995. "The mechanisms of summer dryness induced by greenhouse warming." *Journal Climate,* 8: 3096–3108. University Press.

Whittaker, R.H. 1975. *Communities and Ecosystems.* New York: Macmillan Publishing Co., Inc.

Yachi, S. and Loreau, M. 1999. "Biodiversity and ecosystem productivity in a fluctuating environment: the insurance hypothesis." *Proceedings of the National Academy of Sciences of the United States of America*, 96: 1463–1468.

Zagreev, V.V., Sukhikh, V.I., Shvidenko, A.Z. and Moshkalev, A.G. 1992. *All-Union Standards for Forest Taxation. Handbook.* Moscow: Kolos.

Zaletaev, A.S. 1984. "Ecotonnye ecosystemy kak geographicheskoe yavlenie i problemy ekotonizatsii biosphery". *Sovremennye problemy geographii ecosystem*, edited by R.I. Zlotin et al., 53–55. Moscow: Institut geographii AN SSSR (in Russian).

Zaletaev, A.S., ed. 1997. *Ecotones in Biosphere.* Moscow: Water Problems Inst. RAS (in Russian).

Zalihanov, M.Ch., Losev, K.S. and Shelekhov, A.M. 2006. "Natural ecosystems are the most important resource of humanity." *Vestnik RAN,* 76 (7): 612–634 (in Russian).

Zamolodchikov, D.G., Grabovskiy, V.I. and Kraev, G.N. 2011. "Dynamics of the carbon budget in Russian forests over the last two decades." *Lesovedenie,* 6: 16–28 (in Russian).

Zamolodchikov, D.G., Grabovsky, V.I., Korovin, G.N., et al. 2013. "Carbon budget of managed forests in the Russian Federation in 1990–2050: retrospective assessment and forecast." *Meteorology and* Hydrology, (10): 73–78 (in Russian).

Zamolodchikov, D., Grabovsky, V. and Kurts, V. 2014. "Forest carbon balance management in Russia: past, present and future." *Sustainable Forest Management,* 2 (39): 23–31 (in Russian).

Zamyatin, D.N. 1999. "Methodological analysis of the chorological concept in geography." *Izvestia RAN. Series Geographical,* (5): 7–16 (in Russian).

Zaret, T.M. 1982. The stability/diversity controversy: a test of hypotheses. *Ecology,* 63: 721–731.

Zavarzin, G.A., ed. 2007. *Pools and Fluxes of Carbon in Terrestrial Eco Systems in Russia.* Moscow: Nauka (in Russian).

Zeidis, I.M., Kruzhalin, V.I. and Simonov, Yu. G. 2001. "General properties of the dynamics of geosystems." *Vestnik Moskovskogo Universiteta Series* 5. Geography, (4): 3–8 (in Russian).

Zheltukhin, A.S., Kurbatova, Yu. A., Sandlersky, R.B. and Puzachenko, M. Yu. 2021. "Stationary research in the Central Forest State Natural Biosphere Reserve". *Questions of Geography,* 152*:* Man *and the Biosphere* 357–378 (in Russian).

Zlotin, R.I. 1987. "Ecological problems of biota and stability of geosystems." *Izvestiya AN SSSR, Seriya Geograficheskaya,* (6): 45–51 (in Russian).

Zonn, S.V., Chikishev, A.G., eds. 1976. *Natural Resources of the Russian Plain in the Past. Present and Future.* Moscow: Nauka (in Russian).

Zonn, S.V. and Mints, A.A., eds. 1967. *Middle Zone of the European Part of the USSR.* Moscow: Nauka (in Russian).

# Index


## A

Actualism method in landscape ecology 12, 172
Adaptive structures of the nature complexes 44, 266
Alexandr Humboldt's comparative geographical method 121
Algorithm of landscape-ecological prognosis 12, 152, 181–182
Allometric coefficient 95
Allometric low 176
Anabolism 120
Annual atmospheric humidity factor (Vysotsky–Ivanov's index) 27, 104, 133–134, 226
Annual evaporativity 27, 133
Autotrophic biogenesis 120, 129

## B

Biogeocoenotic process 217
Biogeocoenoses (landscape facies) 51, 89
Biologically equivalent habitats 172
Boreal biogeographic ecotone 30, 34, 314–315
Budyko's radiation dryness index 104

## C

Calculating climatic models 69–71
Carbon balances of forest ecosystems under climate change 269–270, 288–290, 297
Catabolism 120
Chain reactions in geo-component correlations 38, 102–103, 169
Changes in the carbon cycle under global warming 97–98
Climate-genic monitoring of forests 7, 49, 229–230, 265–266
Climate-genic successions 132, 170
Concept "Adaptation monitoring" of forest ecosystems 10, 13–14, 23, 44, 264–265
Concept "Global changes on the regional and local levels" 5, 96, 148, 216–217
Concept of biotic regulation of terrestrial ecosystems 40, 41, 97–98, 265–266
*Continuous* (dynamic) simulation modeling of ecosystems 9, 50, 317
Critical ecological zones 40, 269
Critical reserves of productive moisture in the soil 175, 227

## D

Decrease in the ecological resources of forest cover under replacement of primary forests by derivative ones 288–290
Detritus-genesis 130
Development of mixed forests and "savannization" of broadleaf forests under global warming 230–232, 313
Disadvantages of paleogeographic reconstructions of predicted environmental scenarios 147
Discrete empirical–statistical models of natural ecosystems 24, 50–51, 318
Dokuchaev's paradigm about the nature zonality 26, 172
Dominant of the ecological niche 56, 135, 160, 173, 237–238
Dual problem: adsorption and adaptation in the Paris Agreement on Climate Change 293

## E

Ecological efficiency of the soil–phytocoenotic nucleus of an ecosystem 154
Ecological groups (ecotypes) of the ecosystems 62, 64–65, 67–68, 74, 140–141, 218
Ecological niches of the phenomenon in the space of values of the factor 55, 58, 218
Ecological resources for forest cover carbon cycle regulation 266
Edaphic humidity factor 138, 143–145, 174
Efficiency of the off-ecological parameter Pprim and CY in the transformation of forest ecosystems under climatic aridization 159, 161–162
Empirical-simulation basic and forecast modeling of natural ecosystems 10, 13–14, 120–121
Empirical simulation of climate-genic changes in the biological cycle 168
Empirical simulation of regional bioclimatic trend by ecosystems at the local level 91, 121–123, 150, 153, 168, 171
Ergodic properties of the natural environment 7–8, 10, 116, 120–121, 170
Experimental character of the geographical forecasting 12, 149, 217

Experimental test sites 82–84
Extra-zonal topo-ecosystems 121

**F**

Features of the arid phytocoenogenesis 200–201
Final stage of endo-ecogenetic (restorative) successions, according to Sukachev 267
Flat-interfluve (placor) landscape facies 89, 120
Focal or area transformation of the natural complexes under global warming 203, 207–208
Forecast landscape maps as the maps of future landscape-ecological conditions 207
Forest covers as the decisive factors for the stability of the continental biosphere 265
Forest type, by Kolesnikov 74
Forms of atmospheric circulation (by the Vangengeim–Dzerdzeevsky–Girs classification) 242
Forms of micro-relief 89
Functional and structural relaxation of nature ecosystems 153–154
Functional isomorphism as a way to maintain the initial rates of biological circulation 154
Fundamentally probabilistic nature of geoecological forecasts 149
"Fuzzy" links of the ecological niche 56, 135, 160, 173, 237–238

**G**

General increase in boreal forest productivity under expected global warming 195–106
General principle of the landscape-ecological forecast–13, 152
"Generic barrier" for landscape transitions 205, 207
Geographical belts and natural zones 29–30
Geographical ecotone 31–32
Geosystem monitoring 5, 20–21, 49–50
Gerasimov-Izrael conception of ecological monitoring 9, 20–21, 24, 73
Global coupled atmosphere–ocean general circulation models (AOGCMs) 69–70, 224–225

**H**

Hydro-thermal ordination of metabolic parameters of topo-ecosystems 124–126
Homeostasis of geo(eco-)systems 68
Humid climate signal 308

**I**

Ideology of local and regional landscape ecological predictions 150–151
Inapplicability to landscape systems of ecosystem stability models developed in general ecology 247
Induction transformation of landscape types 211, 213–214
Inertial soil–biotic stability 251–252
International Geosphere-Biosphere Program 17, 146

**L**

Labile phytocoenotic stability 248–249
"Landscape-generic drift" of the phytocoenological core of geosystems 205
Landscape map 75–76
Landscape kinds 75
Le Chatelier's principle 6, 8, 40, 265, 278
Lithomorphism and hydromorphism as two local environmental factors in the meso-relief 90–91
Local coefficient of common humidification 139
Local landscape connections (catenas) 89
Local landscape–ecological prediction 227–228, 230–231

**M**

Main Landscape Border of the Russian Plain 35–36
Main problem is the transformation of global climatic signals into regional and local ecological ones 147
Main prognostic intervals (steps) 69
Matrix maps that describe the studied properties of an ecosystem by quantitative parameters 81–82, 89–90, 255–257, 271, 273, 278, 280–281
Mean July temperature and annual precipitation as the main climate factors of environmentel changes 235–236
Measure of functional isomorphism of the ecosystems 62, 63, 67, 155
Measure of similarity between plants based on their primary productivity 162–163
Measures of similarity of forest ecosystems by these relationship with bioproductivity 163
Mechanism of probability estimation of landscape–ecological transitions 179–180
Metabolic index of biogeocoenosis 62, 64
Method of inductive–hierarchic extrapolation 258–260, 273

Methods of experimental forest biogeocoenology 94
Method of local discrete modeling of natural ecosystems 11, 51–52
Method of spatio-temporal analogies 123
Models for empirical-imitation of environmental consequences of climate change 301–302, 307
Models of zonal-regional phytocoenological transitions under global warming 189–190
Monosystem landscape paradigm 4, 52–54, 104

## N

Negative and positive effects on the carbon balance in primary boreal and sub-boreal forests of the Volga River basin 283–285
New geomorphometric methods for creating spatiotemporal maps 255–257, 271, 273
Normalized coefficient of interrelation 52, 54
Numerical landscape-ecological forecasting methodology 148–149, 179–180

## O

Ordination analysis in the geographical ecology 123–124
Organismic concept of plant communities 216–217
Organization of landscape systems 50

## P

Parameters of biological turnover 15, 95–96
Partial coefficient of connection 52, 55
Phase space of the system of factor–phenomenon 57
Phytocoenological transformation 203
Phytocoenotic induction 203
Polymorphism properties of plant formation 167
Polysystem landscape paradigm 52–53, 60–61
Potential rates of transformations nature ecosystems 183–184
Predictable regional bioclimatic system 189–190
Primary bioproductivity 105, 110–112
Primary landscape facies (biogeocoenoses) 120
Primary plant formations 80–82
Principle of the functional isomorphism of ecosystems 166
Priority nature of functional changes in ecosystems 11, 153
Probabilistic nature of geoecological forecasts 149--150
Problem of forest stability 244–246
Problem of transformation of global climatic signals into regional and local levels 146–147, 217–218
Prognostic calculations of the changes in summer soil moisture reserves 227

## Q

Quantitative evaluations of forest ecosystem potential sustainability 11, 249, 251

## R

Regional bio-climatic system with zonal–provincial phyto-climatic structure 78, 82–83, 118, 282–283
Regional hydrothermal parameters 101–102
Regional paleo-prognostic conception 18, 101
Regional systems of localized natural zonality 172, 221
Regional types of location 203
Relative reserve of homeostaticity of the biogeocoenoses 65–66
Resistance and resilience stabilities of the ecosystem 249–251

## S

Sharp and multidirectional divergence of the climatic areas of the geosystems in forecast landscape–ecological maps 209
Shifts in landscape-zonal boundaries under expected global warming 193–194
Simulation of local representation of the background climatic trend 121, 171
Simulation models in ecological-geographical forecastings 165–166
Simulation of projected phytocoenological transitions by the connections of plant formations with annual atmospheric humidity factor 134–135
Strategy of sustainable forest management 267
Structural schemes of the dominance of the objects 61
Summer soil moisture content as main channel of connections between local ecosystems and the background climatic system 136–137, 168, 217–218

## T

Theory of geographical zoning 25–26
Thermo-arid climatic trend 189–190, 307
Three-dimensional phytogeographic ecotone 240

"Textural coefficient" 211
Timofeev-Resovsky conception defined biogeocoenosis as an elementary unit of the biological cycle 73, 98, 148, 265
Topological polyzonality 172–173, 222
Total aridization as a general direction for forecasting forest ecosystems under global warming 190–192, 198, 200, 228–229, 235–237
Traditional forest management method of estimate the carbon balance 269–270
Transformation of the zonal structure of the Volga River basin under global warming 193–194
Trigger systems with hysteresis properties 34, 39
Types of location sites (geotopes) 84, 89
Typological landscape groups 77

**U**

Unprecedentedly high global warming in the late 20th century and in the modern period 233–234

**V**

Vector graphs of functional transitions between facies groups 220–221
Violation of the basic principles of statistical in relation to complex ecosystems with biological components 124, 126

**Z**

Zonal–regional biogeographic ecotone 300
Zonal–sectoral types and sub-types of landscapes 75
Zonal-provincial phyto-climatic structure 83

For Product Safety Concerns and Information please contact our EU
representative GPSR@taylorandfrancis.com
Taylor & Francis Verlag GmbH, Kaufingerstraße 24, 80331 München, Germany

www.ingramcontent.com/pod-product-compliance
Ingram Content Group UK Ltd.
Pitfield, Milton Keynes, MK11 3LW, UK
UKHW021513250926
475702UK00043B/353

* 9 7 8 1 0 4 1 2 2 5 9 5 9 *